TECHNOLOGICAL ADVANCEMENT IN CLEAN ENERGY PRODUCTION

Constraints and Solutions for Energy and Electricity Development

TECHNOLOGICAL ADVANCEMENT IN CLEAN ENERGY PRODUCTION

Constraints and Solutions for Energy and Electricity Development

Edited by

Amritanshu Shukla, PhD

Kian Hariri Asli

Neha Kanwar Rawat, PhD

Ann Rose Abraham, PhD

A. K. Haghi, PhD

First edition published 2025

Apple Academic Press Inc.
1265 Goldenrod Circle, NE,
Palm Bay, FL 32905 USA

760 Laurentian Drive, Unit 19,
Burlington, ON L7N 0A4, CANADA

CRC Press
2385 NW Executive Center Drive,
Suite 320, Boca Raton FL 33431

4 Park Square, Milton Park,
Abingdon, Oxon, OX14 4RN UK

Library and Archives Canada Cataloguing in Publication

Title: Technological advancement in clean energy production : constraints and solutions for energy and electricity development / edited by Amritanshu Shukla, PhD, Kian Hariri Asli, Neha Kanwar Rawat, PhD, Ann Rose Abraham, PhD, A.K. Haghi, PhD.
Names: Shukla, Amritanshu, editor. | Asli, Kian Hariri, editor. | Rawat, Neha Kanwar, editor. | Abraham, Ann Rose, editor. | Haghi, A. K., editor.
Description: First edition. | Includes bibliographical references and index.
Identifiers: Canadiana (print) 20240349202 | Canadiana (ebook) 20240349237 | ISBN 9781774915585 (hardcover) | ISBN 9781774915592 (softcover) | ISBN 9781032684420 (ebook)
Subjects: LCSH: Nanostructured materials. | LCSH: Nanoparticles. | LCSH: Clean energy—Technological innovations. | LCSH: Energy storage—Technological innovations.
Classification: LCC TA418.9.N35 T43 2025 | DDC 621.042028/4—dc23

Library of Congress Cataloging-in-Publication Data

CIP data on file with US Library of Congress

ISBN: 978-1-77491-558-5 (hbk)
ISBN: 978-1-77491-559-2 (pbk)
ISBN: 978-1-03268-442-0 (ebk)

About the Editors

Amritanshu Shukla, PhD
Professor in Physics, University of Lucknow, Lucknow, Uttar Pradesh, India

Amritanshu Shukla, PhD, is currently working as a Professor in Physics at the University of Lucknow (NAAC accredited A++ University), India. Prior to this, he worked as an Associate Professor at the Rajiv Gandhi Institute of Petroleum Technology (RGIPT) (set up through an Act of Parliament by the Ministry of Petroleum & Natural Gas, as an Institute of National Importance on the lines of IITs). His research interests include nuclear physics and physics of renewable energy resources. He has published more than 150 research papers in various international and national journals and conferences. He has delivered invited talks at national and international institutes. Currently, he is working on several national and international projects and has active research collaborations with institutions in India and abroad. His recent book publications include *Low Carbon Energy Supply: Trends, Technology, Management* (Springer); *Sustainability Through Energy-Efficient Buildings* (CRC Press/Taylor and Frances); *Energy Security and Sustainability* (CRC Press, Taylor and Francis); and *Latent Heat-Based Thermal Energy Storage Systems* (Apple Academic Press). Dr. Shukla earned his master's in Physics and completed his PhD from IIT Kharagpur, India. He did his postdoctoral research work at some of the premier institutes across the globe, namely at the Institute of Physics, Bhubaneswar (Department of Atomic Energy, Govt. of India); University of North Carolina Chapel Hill, NC, USA; University of Rome/Gran Sasso National Laboratory, Italy; and Physical Research Laboratory, Ahmedabad (Department of Space, Govt. of India).

Kian Hariri Asli
Research Scholar, Department of Electronic Engineering, University of Rome "Tor Vergata," Rome, Italy

Kian Hariri Asli is a Research Scholar in the Department of Electronic Engineering, University of Rome "Tor Vergata," Rome, Italy. He has expertise in Electronic and Control Engineering. He works toward the development of energy saving by networked sensors, fuzzy logic, artificial intelligence (AI) and Internet of Things (IoT) based on Geospatial Information System (GIS).

Neha Kanwar Rawat, PhD
Researcher, Department of Materials Science and Engineering, Indian Institute of Science, Bangalore, Karnataka, India

Neha Kanwar Rawat, PhD, is a recipient of a prestigious DS-Kothari Postdoctoral Fellowship and DST Young Scientist Postdoctoral Fellowship and is presently a Researcher in the Department of Materials Science and Engineering, Indian Institute of Science, Bengaluru, India. She received her PhD in Chemistry from Jamia Millia Islamia (a central university), India. Her main interests include nanotechnology and nanostructured materials synthesis and characterization, with main focus on green chemistry; novel sustainable chemical processing of nano-conducting polymers/nanocomposites; conducting films, ceramics, silicones, and matrices: epoxies, alkyds, polyurethanes, etc. She is also pursuing her interest in fusing new technology in areas that include electrochemistry and organic-inorganic hybrid nanocomposites for biomedical applications, which also includes protective surface coatings for corrosion inhibition and MW shielding materials. She has published numerous peer-reviewed research articles in journals of high repute. Her contributions have led to many books and chapters in international books published with the Royal Society of Chemistry, Wiley, Elsevier, Apple Academic Press, Nova US, and many others in progress. She is a peer reviewer for many international books and is a member of many groups, including the Royal Society of Chemistry and the American Chemical Society (USA), and a life member of the Asian Polymer Association.

Ann Rose Abraham, PhD
Assistant Professor, Department of Physics Sacred Heart College (Autonomous), Thevara, Kochi, Kerala, India

Ann Rose Abraham, PhD, is currently an Assistant Professor at the Department of Physics, Sacred Heart College (Autonomous), Thevara, Kochi, Kerala, India. Her PhD thesis was titled, "Development of Hybrid Mutliferroic Materials for Tailored Applications". She has expertise in the field of condensed matter physics, nanomagnetism, multiferroics, and polymeric nanocomposites, etc. She has research experience at various reputed national institutes including Bose Institute, Kolkata, India; SAHA Institute of Nuclear Physics, Kolkata, India; and UGC-DAE CSR Centre, Kolkata, India and has collaborated with various international laboratories. She is the recipient of a Young Researcher Award in the area of physics and Best Paper Awards–2020, 2021, a prestigious forum for showcasing intellectual capability. She served as assistant professor and examiner at the Department of Basic Sciences, Amal Jyothi College of Engineering, under APJ Abdul Kalam Technological University, Kerala, India. Dr. Abraham is a frequent speaker at national and international conferences. She has a good number of publications to her credit in many peer-reviewed high impact journals of international repute. She has authored many book chapters and edited more than 10 books with Taylor and Francis, Elsevier, etc. Dr. Abraham received her MSc, MPhil, and PhD degrees in Physics from the School of Pure and Applied Physics, Mahatma Gandhi University, Kerala, India.

A. K. Haghi, PhD
Research Associate, Department of Chemistry, University of Coimbra, Portugal

A. K. Haghi, PhD, is a retired professor and has written, co-written, edited or co-edited more than 1000 publications, including books, book chapters, and papers in refereed journals with over 3800 citations and h-index of 32, according to the Google Scholar database. He is currently a research associate at the University of Coimbra, Portugal. Professor Haghi has received several grants, consulted for several major corporations, and is a frequent speaker to national and international audiences. He is Founder and former Editor-in-Chief of the *International Journal of Chemoinformatics and Chemical Engineering* and *Polymers Research Journal*. Professor Haghi has acted as an editorial board member of many international journals. He has served as a member of the Canadian Research and Development Center of Sciences & Cultures. He has supervised several PhD and MSc theses at the University of Guilan (UG) and co-supervised international doctoral projects. Professor Haghi holds a BSc in urban and environmental engineering from the University of North Carolina (USA) and holds two MSc degrees, one in mechanical engineering from North Carolina State University (USA) and another one in applied mechanics, acoustics, and materials from the Université de Technologie de Compiègne (France). He was awarded a PhD in engineering sciences at Université de *Franche-Comté* (France). He is a regular reviewer of leading international journals.

Contents

Contributors

K. B. Akhila
Department of Chemistry, CMS College, Kottayam, Kerala, India

Abhishek Anand
Non-Conventional Energy Laboratory, Rajiv Gandhi Institute of Petroleum Technology, Jais, Amethi, India

Kian Hariri Asli
Department of Electronic Engineering, University of Rome "Tor Vergata," Rome, Italy

Ahumuza Benjamin
Department of Mechanical Engineering, GITAM School of Technology, Gandhi Institute of Technology and Management (Deemed to be University), Bangalore, Karnataka, India

Brahmananda Chakraborty
High Pressure and Synchrotron Radiation Physics Division, Bhabha Atomic Research Centre, Trombay, Mumbai, India
Homi Bhabha National Institute, Mumbai, Maharashtra, India

Chanchal Liz George
Department of Physics, CMS College (Autonomous), Kottayam, Kerala, India
Nanotechnology and Advanced Materials Research Centre, CMS College (Autonomous), Kottayam, Kerala, India

Divya Neravathu Gopi
Bharata Mta College, Thrikkakara, Edappally, Cochin, Kerala, India

K. V. Arun Kumar
Department of Physics, CMS College (Autonomous), Kottayam, Kerala, India
Nanotechnology and Advanced Materials Research Centre, CMS College (Autonomous), Kottayam, Kerala, India

Rohini Manoj
Department of Physics, CMS College (Autonomous), Kottayam, Kerala, India
Nanotechnology and Advanced Materials Research Centre, CMS College (Autonomous), Kottayam, Kerala, India

Saju M. Mohammed
Bharata Mata College, Thrikkakara, Edappally, Cochin, Kerala, India

Brinti Mondal
Department of Energy Science and Engineering (DESE), Indian Institute of Technology (IIT), Bombay, Maharashtra, India

G. P. Nayaka
Physical and Materials Chemistry Division, CSIR-National Chemical Laboratory, Pune, Maharashtra, India
Academy of Scientific and Innovative Research (AcSIR), Ghaziabad, Uttar Pradesh, India

Saurabh Pandey
Non-Conventional Energy Laboratory, Rajiv Gandhi Institute of Petroleum Technology, Jais, Amethi, India

Rony Rajan Paul
Department of Chemistry, CMS College, Kottayam, Kerala, India

G. Santhosh
Department of Mechanical Engineering, NMAM Institute of Technology, NITTE (Deemed to be University), Nitte-off-campus Center, NITTE, India

Abhijith Sharma
Cochin University of Science and Technology, South Kalamassery, Ernakulam, Cochin, India

Atul Sharma
Non-Conventional Energy Laboratory, Rajiv Gandhi Institute of Petroleum Technology, Jais, Amethi, India

B. P. Shivamurthy
Physical and Materials Chemistry Division, CSIR-National Chemical Laboratory, Pune, Maharashtra, India
Academy of Scientific and Innovative Research (AcSIR) Ghaziabad, Uttar Pradesh, India

Amritanshu Shukla
Non-Conventional Energy Laboratory, Rajiv Gandhi Institute of Petroleum Technology, Jais, Amethi, India
Department of Physics, Lucknow University, Lucknow, India

Mamata Singh
CeNSE (Center for Nano Science and Engineering), Indian Institute of Science, Bangalore, Bangalore, Karnataka, India

N. P. Singh
Department of Chemistry, GITAM School of Sciences, Gandhi Institute of Technology and Management (Deemed to be University), Bangalore, Karnataka, India

Riju K. Thomas
Bharata Mata College, Thrikkakara, Edappally, Cochin, Kerala, India

Rinsy Thomas
Department of Physics, CMS College (Autonomous), Kottayam, Kerala, India
Nanotechnology and Advanced Materials Research Centre, CMS College (Autonomous), Kottayam, Kerala, India

Abbreviations

AIC	Akaike's information criterion
BE	binding energy
BHJ	bulk heterojunction
BIC	Bayesian information criterion
CF	carbon foam
CMC	carboxymethyl cellulose
CNFs	carbon nanofibers
CNTs	carbon nanotubes
CNW	carbon nanowalls
CVD	chemical-vapor deposition
DEA	data envelopment analysis
DET	direct electron transfer
DFT	density functional theory
DMUs	decision-making units
DSC	differential scanning calorimetry
DSSCs	dye sensitized solar cells
EC	electrochromic
EG	epoxy, ethylene glycol
EMA	effective mass theory
EUI	energy use intensity
EVs	electric vehicles
FECM	free exciton collision model
FTIR	Fourier transform infrared spectroscopy
GA	graphene aerogel
GIS	geospatial information system
HDPE	high-density polyethylene
HFC	hydrogen fuel cells
HGY	hole graphene
HOMO	highest occupied molecular orbitals
HTM	hole transport materials
HVACR	heating, ventilating, air-conditioning, and refrigeration
IIoT	industrial Internet of Things
IoT	Internet of Things
IPCE	incident photon-to-current efficiency

IPCE	induced photocurrent efficiency
LHSR	latent heat storage tank
LIB	Li-ion batteries
$LiMO_2$	lithium metal oxides
LiPAA	lithium polyacrylate
LMO	$LiMn_2O_4$
LSPR	localized surface plasmon resonance
LTO	$Li4Ti_5O_{12}$
LUMO	lowest unoccupied molecular orbital
MF	melamine formaldehyde
MlCNTs-	multilayer carbon nanotubes
MOFs	metal organic frameworks
MPCM	microencapsulated PCM
MSSCs	mesosuperstructured solar cell
MWCNT	multiwalled carbon nanotube
MWNT	multiwalled nanotubes
NoGNr	non-oxidized graphene nanoribbons
NPs	metallic nanoparticles
NT	nanotube
OGNr	oxidized graphene nanoribbons
OPVs	organic photovoltaic cells
OSSE	octahedral-site stabilization energy
PAA	polyacrylic acid
PAN	polyacrylonitrile
PCE	power conversion efficiency
PCL	polycaprolactone
PCM	phase change material
PDOS	partial density of states
PDSSC	plasmonic dyesensitized solar cells
PEC	photoelectrochemical
PECs	photoelectrochemical cells
PEG	polyethylene glycol
PEO	polyethylene oxide
PGNs	porous graphene nanosheets
PhCh	phthaloylchitosan
PLA	polylactide
PLD	pulsed laser deposition
PMMA	polymethyl methacrylate
PPD	percentage of dissatisfied

PRV	pressure-reducing valve
PSC	perovskite solar cell
PV	photovoltaic
PVD	physical vapor deposition
PVDF	polyvinylidene fluoride
QDSCs	quantum dot solar cells
RS	remote sensing
SDS	sodium dodecyl sulfate
SEI	solid-electrolyte interface
SEM	scanning electron microscopic
SIBs	sodium-ion batteries
SPP	surface plasmon polariton
SPR	surface plasmon resonance
SWCNT	single-walled carbon nanotube
TEM	transmission electron microscope
TES	thermal energy storage
TM	transition metals
UV-NIR	ultraviolet/visible/near infrared
XRD	X-ray diffraction

Preface

This book illustrates the most recent developments and progress on clean energy technology and related materials. It presents recent advances in theoretical and experimental research on devices that can be used in the production of new types of solar cells and hydrogen generation for pollution control. The book also examines the potential applications that can be used to promote green processes and techniques for energy and environmental sustainability. It discusses the pathways to pollution reduction.

The book reviews the new developments and characterization of modern energy materials. It provides new insights into novel forms of energy, also discussing the latest progress in the field of green energies and nanomaterial technology with methodologies designed to solve engineering issues in this important field.

The focus of each chapter is on a different technology as chapters discuss crucial green technology, reflecting on various challenges and possible solutions for them. The topics covered include improving the efficiency of solar cells, battery materials, and technologies, major challenges toward the development of nanomaterial and nanoparticles for energy efficient devices and many more.

This reference volume provides in-depth analysis of various issues related to improvement of technological significance for clean energy production and covers current developments and both basic and advanced concepts in clean energy production.

This volume is a valuable resource for postgraduate researchers, engineers, and policy makers in both energy and environmental field and is intended for individuals pertaining to different disciplines who are interested in the various fields of energy technology and related materials.

CHAPTER 1

Technological Advancement in HVAC&R System Energy Efficiency

KIAN HARIRI ASLI

Department of Electronic Engineering, University of Rome "Tor Vergata," Rome, Italy

ABSTRACT

The World Wide Web is a web-based geographic information system (GIS) that enables the distribution, sharing, and exchange of data at anytime, anywhere, and for any person through the World Wide Web. Therefore, engineers by analyzing online data and advanced technology received signal error through remote reading networked sensor systems and managed thermal and cooling facilities through the World Wide Web. Quick intercommunication of geo-reference data is a new technology that has a positive impact on the efficiency of HVAC&R's facility systems including heating, ventilating, air-conditioning & refrigeration (HVAC&R). The technology of signal detection through remote sensing (RS) provides the management of facility HVAC&R through the geo-reference world area network controlled by networked sensors (RS), and Internet of Things (IoT) based on (GIS). This method as a new concept in the control science and optimized energy consumption can be an effective factor in design, maintenance, energy consumption management, and commissioning of the facilities and building industry. This method provides online control of energy consumption of the facilities and building industry. So, in this work, the model of HVAC&R control in the context with Web-based GIS showed that the regression mathematical analysis is able to predict consumption and evaluate energy loss. This work also investigated hydrogenerator as an energy-saving case named clean energy. The benefit and cost for the

Technological Advancement in Clean Energy Production: Constraints and Solutions for Energy and Electricity Development. Amritanshu Shukla, Kian Hariri Asli, Neha Kanwar Rawat, Ann Rose Abraham, & A. K. Haghi (Eds.)

application of hydro-generator as pressure reducing valve (PRV) or pressure relief valve (PRV) showed the income increasing and energy saving.

1.1 INTRODUCTION

In the 21st century, the international energy-saving attitude of the international community necessitated the use of networked sensors, Internet of Things (IoT), and artificial intelligence (AI) based on the geospatial information system (GIS) for the management of thermal, refrigeration, and air-conditioning industries.

The heating, ventilating, air-conditioning, and refrigerating (HVAC&R) engineering emphasizes on intelligent operation method.

1.1.1 WHY THE ARTIFICIAL INTELLIGENCE?

Artificial intelligence programs are coded in terms of thinking, reasoning, learning, acquiring abilities, and making decisions by simulating humans. Artificial intelligence is used in computer science, engineering science, biological science, medical science, and social science.

Artificial intelligence begins to analyze and make decisions after correctly recognizing the environment and human states. For this reason, it is one of the cases that is used in the field of medicine to diagnose human health during surgery.

One of the most important things in artificial intelligence programming is choosing the right programming language according to our goal. In this way, research is done with the help of intelligence. The next important issue is the intelligentization of goods used by humans. This has caused information to be transferred to neural networks.

On the other hand, some things have become more flexible and interesting by presenting new methods. For example, all kinds of smart devices have developed various features with the help of Python or the most used programming languages in the world.

1.1.2 THE IMPORTANCE OF ARTIFICIAL INTELLIGENCE PROGRAMMING WITH PYTHON

With artificial intelligence, operations are performed automatically. Programming artificial intelligence with Python also helps us to avoid wasting time by providing programming strategies.

Artificial intelligence programming with Python is the science of telling computers to behave like humans. Artificial intelligence imitates humans by studying human behavior and being in different situations, and different patterns in computer science have become problem-solving algorithms.

Artificial intelligence has taken advantage of existing algorithms and placed them in the structure of the Python language. That is why this science is used in today's technologies. Programming artificial intelligence with Python has specific steps, which must first be fully familiar with mathematics. Then he understood the concepts of artificial intelligence and neural network well.

Artificial intelligence takes help from humans with the help of environmental data processing and then it is mixed with programming concepts. Artificial intelligence is actually programmed to examine human states and thinking.

1.1.3 ARTIFICIAL INTELLIGENCE ALGORITHMS

One of the interesting aspects of artificial intelligence is its compatibility with other programming languages. In general, after learning artificial intelligence with Python, it will be possible to learn a set of special conditions, various algorithms, and other things.

1.1.4 ARTIFICIAL INTELLIGENCE PERFORMANCE

Using artificial intelligence with Python, data is programmed with different algorithms. This is one of the stages of building an artificial intelligence and it automatically analyzes the data, and after combining them with each other, the level of artificial intelligence increases. This function leads to the advancement of algorithms. With machine learning, we can build an analytical model.

1.1.5 APPLICATION OF ARTIFICIAL INTELLIGENCE

Sometimes, it becomes similar to human abilities, for example, the ability of vision, which the machine analyzes by learning different patterns and types of algorithms.

Also, the computer recognizes human losses so as to be able to produce human language and use it. Artificial intelligence can be implemented with the help of different languages. Choose according to your needs and goals, for example, artificial intelligence programming with Python is one of the most used ways, which is an efficient and optimal language after Matlab.

1.1.6 DATA MINING IN MACHINE LEARNING

The concept of data mining deals with extracting specific information or patterns in high volume. Now, in machine learning and the process of programming artificial intelligence with Python, we use the concept of data mining in order to discover new patterns among the data. Among other factors, it is used to predict the results. Artificial intelligence should be able to make decisions after analyzing the information.

With data mining, we should be able to find useful data in a pile of surface data so that it can be turned into useful information after processing. The main focus of data mining is on big data to be able to take the required part from each data and turn it into useful data. In this way, the selection level of the system will also increase. Artificial intelligence processes data with the help of computational algorithms.

1.1.7 THE IMPACT OF DEEP LEARNING ON THE PERFORMANCE OF ARTIFICIAL INTELLIGENCE

Deep learning is a hierarchical concept in which it selects high-level data from a set of data. As a result, complex concepts are transformed into simpler concepts. In the continuation of this process, the searched concepts become more basic concepts that can be implemented without human supervision.

1.1.8 NEURAL NETWORKS IN ARTIFICIAL INTELLIGENCE

Neural networks have parallel computing devices. The main purpose of having a neural structure is to create more computational models so that the process of performing computational functions becomes faster. These functions include pattern recognition and organization, estimation, data optimization, and data classification.

1.1.9 THE MAIN STRUCTURE OF ARTIFICIAL INTELLIGENCE

The basic structure of neural networks is an efficient computing system. They are also known as parallel distributed processing systems. The structure of the neural network is a large set of data whose units and parts are interconnected.

To communicate between them in this way, each of these units is called a neuron. Each neuron is a simple processor that works in parallel. Each neuron communicates with other neurons through connections.

Each connection link has a weight that receives information with the input signal. The best way to solve specific problems is to use information from neurons. Each neuron has an internal state and an external state. The internal state of the neuron is called the activation signal. Output signals are generated after combining input signals with rules.

Sometimes this information is sent to other neurons. According to the weight of each neuron, the information changes effectively. In this way, after applying the connection between the neurons, a learning algorithm containing the information of the neurons is formed.

1.1.10 WHY NETWORKED SENSORS, IOT, AND AI BASED ON THE GIS?

The installations of the HVAC&R facilities, proper operation, useful efficiency, and minimum cost depend on the mapping. The purpose of operation is to make proper use of the HVAC&R equipments over their useful life. Scientific utilization of the facilities requires updating the facility's map information in the form of (GIS). Due to the large volume of exploitation information, while updating the facility map information in the GIS format, different data can be extracted in the least amount of time. A robust, efficient database of servers as the primary server can solve many of the problems of social life. If the information about the facility is in the memory of experienced people, it will be out of the system over time. GIS is not just software, but a science that classifies and locates geographic and urban information by different software. This science has recently found its place in urban sciences and is being used by urban sciences and urban planners. The use of GIS in urban development plans by employers is also on their agenda. In many countries, Urban Development Association has used GIS in urban development plans, providing training to urban professionals and urban utilities. Therefore, experts should first prepare scanning, editing, and layered facility maps. It also needs to reflect subsequent changes to these maps so that they do not lose their performance over time. Up-to-date facilities and equipment components are reflected in GIS maps, mainly including:

- Pipes: In the computer maps prepared, the pipes in the grid are carefully updated and reflected on the maps by gender and diameter in GIS format.
- Valves: The location of the valves in GIS format is reflected on the facility maps.
- Meters: The locations of metering equipment such as barometers and flowmeters, etc. are reflected in the GIS in the facility maps.
- Facility maps are updated to suit GIS requirements and are implemented as follows:
- Exchange graphical information from CAD space to GIS space.
- Fixes errors in CAD space.
- Convert graphics data from DWG format to SHP.
- Completing the descriptive and spatial information layers and fixing the errors in the GIS space (descriptive and spatial.(
- Eliminate tolls that are in the wrong place.
- Create primary and external keys for the toll table.
- Creating appropriate tolerances and exchanging effects from spaghetti to topology.
- Preparation of conceptual model for modeling network in GIS space.
- Creating a proper ground database.
- Establishment of traceability and execution of facility analysis.

1.1.11 FUZZY LOGIC

Theory of fuzzy sets and fuzzy logic for the first time was introduced by Professor Lotfali Askar Zadeh. Information and control was introduced in 1965. He describes himself as "an American, mathematically oriented, electrical engineer of Iranian descent, born in Russia." His concepts such as fuzzy sets, fuzzy events, fuzzy numbers, and mathematics and engineering sciences are notable. The most interesting application of fuzzy logic, which is an interpretation of the structure of intelligent decision-making, and, above all, human intelligence, will lose. This logic clearly shows why classical mathematical logic, two values zero and one are not able to explain and describe biological concepts that form the basis of many smart decisions. More than twenty years after 1965 of the fuzzy logic theory to practice, it was not the one who can understand what it means. In the mid-1980s, Japanese craftsmen over the past century have understood the meaning and value of industrial and fuzzy logic were employed. Europeans in the mid-1990s began industrial use. Strategic management scholars, (such as Barney Porter in the 1980s and 1990s) by presenting the five forces model of competitive models (linking internal resources and sustainable competitive advantage), were the pioneers of the SWOT technique. So today as one of the most common techniques of strategic management tools organizations have the capacity to achieve sustainable competitive advantage in the dynamic environments they use.

SWOT is:

Threats (T: Threat)—Opportunities (O: Opportunity)—Weaknesses (W: Weakness)—Strengths (S: Strength).

In the management of heat and cooling facilities based on metadata management, the development and application of new and advanced technologies in all areas of software and hardware in the facility can have a positive impact on system performance and efficiency. The application of state-of-the-art technologies such as IoT can also provide scientific guidance and enhance the technical and hygienic safety factor of the installation systems. IoT is a new concept in the world of technology and communications, but IoT was first used by Kevin Ashton in 2007, describing a world in which everything, including inanimate objects, is used to have a digital identity and allow computers to organize and manage them. The Internet now connects all people, but with the Internet of Things all things are connected. Prior to that, however, Kevin Kelly in his book, "The New Economic Law in the Age of Networks in the Year 4" addressed the issue of small smart nodes (such as open and closed sensors) that are connected to the World Wide Web. The present work showed that the management of HVAC&R air conditioning installations in the GIS field is a new topic in the field of control and optimization of energy consumption internationally, and awareness of this is especially important for facility engineers.[1–15]

1.2 RESEARCH METHOD

In this work, the location-based information in the management, analysis, control, and optimization of energy consumption were applied. Therefore, the types of complications, classes, and subclasses of facilities were on the agenda. Finally, the implementation of these operations led to the introduction of the intelligent facility management model by Artificial Intelligent.[6]

In present work, Artificial Intelligence began its development process with analysis of spatial information through Geo-database. Thus exposure to different environmental conditions was measured. In this learning process, our data were increased and the deepening of the data led to the formation of more layers. This increased the accuracy of artificial intelligence performance.

This work used the GIS as a system consisting of data, hardware, software, methods and algorithms, and networks in order to manage, analyze, and display spatial information (Figures 1.1 and 1.2).

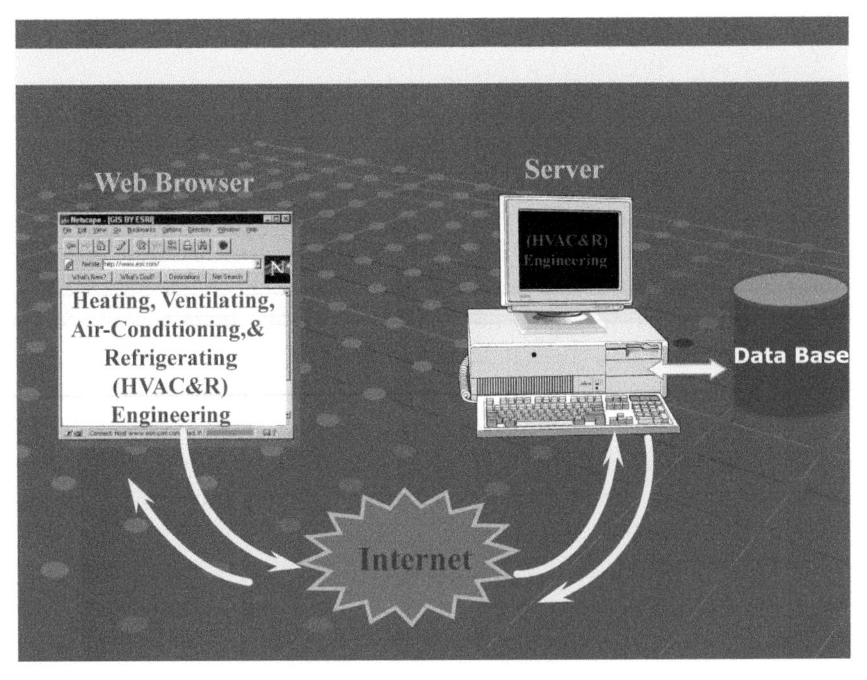

FIGURE 1.1 Managing HVAC&R facilities through the World Wide Web.

1.2.1 FUZZY LOGIC IN THE MANAGEMENT OF COMPLEX SYSTEMS

A quick sightseeing fuzzy logic in the management of complex internal and external factors, more complex mathematical algorithms in the context of Geographic Information Systems GIS research agenda will be used (1.1–1.4):

$$[\mathbf{T} \cdot \mathbf{O} \cdot \mathbf{W} \cdot \mathbf{S}] \quad (1.1)$$

Internal and external factors are variables in the calculation of two types: quantitative values that can be specified with a numeric value of a property are based on quality. The fuzzy set membership functions defined on the basis of each of these attributes and values are between zero and one and the creation of a new region required a new logic. The fuzzy logic is one of them. Thus, the mechanism of dynamic units can be controlled by using fuzzy logic and the best guidance. Fuzzy logic is a new technology for designing and modeling a system as a new way that develops complex systems by using linguistic values and expert knowledge.

In fact, by considering various factors based on deductive thinking values are defined by the patterning of their words and language. The mathematical formulas, if not impossible, would be very complicated, such as multi-valued logic, fuzzy logic and fuzzy sets theory relies. Generalized and extended fuzzy sets of results are conclusive in nature. The definitive collection (Crisp sets) are the same as ordinary sets that are introduced at the beginning of the classical theory of sets. Adding a definite character to make the distinction that is critical to its innovative concepts is one of the so-called fuzzy logic membership functions to bring easily in the mind. In the final set, membership function has only two values in the range (in mathematics, the range of a function is equal to the set of all outputs). Yes, or No (one or zero), which are the two possible values of classical logic, so:

$$\mu_A(x) = \begin{cases} 1 & \text{if } x \in A, \\ 1 & \text{if } x \notin A, \end{cases} \quad (1.2)$$

Board membership function of the fuzzy set is converted into a definitive close range.

$$\tilde{A} = \{(\text{x}, \mu_A(x)) \mid x \in X\} \quad (1.3)$$

For example, linguistic variables such as:

Energy Use Intensity (EUI), Percentage of dissatisfied (PPD), Data Envelopment Analysis (DEA) etc can be considered in which values such as low, high, low, moderate, or strong can take place. To have a mathematical language (EUI = Energy Use Intensity):

$$(\text{EUI} = \text{Energy Use Intensity}) = \{\text{low, high, low, medium, high}\} \quad (1.4)$$

Join the degree of membership of an element to a fuzzy set. If the membership degree of an element is equal to zero, and if the member is totally out of membership degree equal to one is a member, the member is quite complex. If the degree of membership of a member is between zero and one, this number represents the degree of membership is gradual.[16–24]

1.2.2 ENERGY CONSUMPTION

Energy management is the science of controlling and optimizing energy consumption through the integration of Networked Sensors which are used to optimize energy consumption (Figure 1.3). The IoT devices (switches, control valves, actuators, etc.) are integrated and capable of communicating with the utility company to balance power generation effectively and energy consumption. They also give users the chance to remotely control their equipments. These equipments are centrally managed by an equal text interface while enabling advanced programming functions (such as turning on and off the remote heating appliances, controlling the stove, changing lighting conditions, etc). IoT devices can be used for surveillance and the mechanical, electrical, and electronic systems used in a variety of buildings (e.g., public and private, industrial, industrial, and residential) in home and building automation systems. In present work, there were two main fields covered in this area:

- Integrate the Internet with building energy management systems to create energy saving and IoT-based "smart buildings."
- Real-time monitoring to reduce energy consumption and monitor resident behavior.

By Creating Extensions in HVAC&R/ArcGIS-ArcMap software, the work enabled troubleshooting, control, and optimization of power consumption through the online control model in context with the World Wide Web (Figures 1.1 and 1.2).

An online control model optimized energy consumption by providing the information needed to determine Energy Intensity (EUI). For the investigation of energy, this work used intensity based on the Web-based GIS, for better visualization of system components; ease of change, and ability to filter data for the design (Table 1.1). This procedure provided maintenance, energy management, and commissioning of the building and compared the intensity of the building's energy consumption while optimizing its energy consumption (Figure 1.3). The online control model for HVAC&R has heating, cooling, and air conditioning using its and technique information reading techniques and remote control instruments in GIS base for lighting, cooling, heating, security system for buildings. This operation was at its simplest possible by remote control over a very long distance through the telephone line, mobile phone and internet, tablet, and computer. In addition, smart building made the home modern and comfortable. It will also have helped save energy by designing with standard equipments. With the Industrial Internet of Things in a smart home all the following smart and automatic installations were performed:

- Ability to define intelligent scenarios.
- Load management and energy consumption of buildings.
- Ability to control and execute commands remotely via phone, internet.
- Control of HVAC & R heating, cooling, and air conditioning systems.

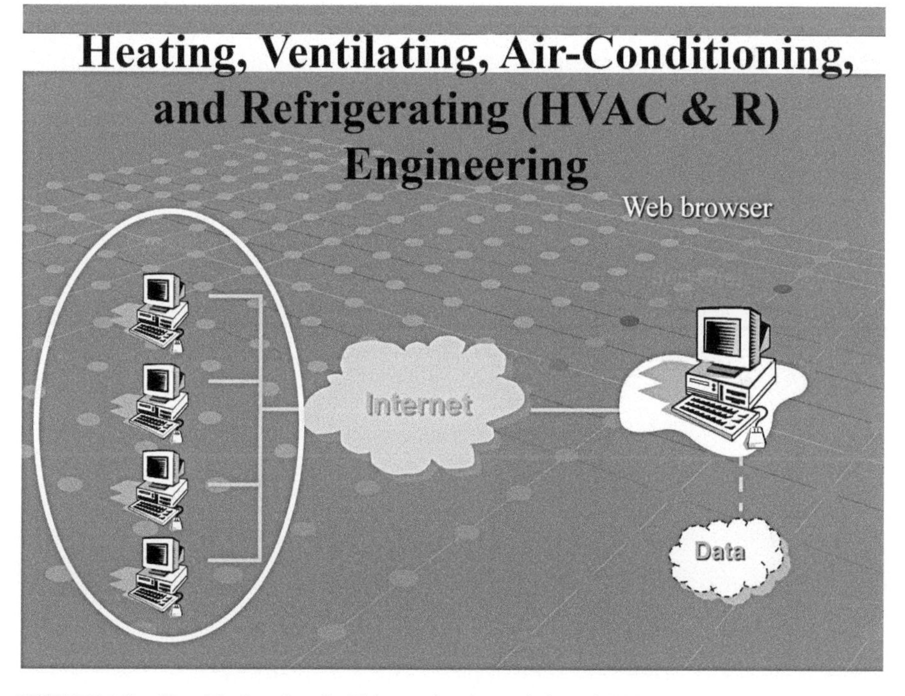

FIGURE 1.2 Troubleshooting facilities under the web-based GIS.

1.2.3 INDUSTRIAL INTERNET OF THINGS (IIOT)

Industrial Internet of Things (IIOT) is one of the most important and widely used areas of IoT deployment in the HVAC&R system. The IOT Industrial Objects Internet means using the technology in industrial fields and using it as a smart industrial network. Using IIOT in industrial units, all objects can be connected online, creating an integrated network for information exchange, control, and monitoring. This technology is one of the five major technologies that will greatly affect the future of industrial automation in all industries as well as in HVAC&R systems.[25–35]

In this work, design factors such as thermal resistance of walls (Table 1.1), temperature factor, and energy consumption unit were investigated. The consumption data, temperature, and power consumption factors on Energy Consumption Intensity (EUI) factors such as Input, Output, and Efficacy were evaluated by the regression model. Factors such as total heat power, occupancy rate, unit income level, and unit energy income can increase energy consumption and temperature factor can lead to energy consumption decrease. In addition, mathematical analysis of the regression model provided a model for managing energy consumption and saving energy through the multi-factor analysis method (Table 1.1).

TABLE 1.1 Minimum Heat Resistance of Non-Light Transient Wall R (M^2. K/W).

Group	Group	Group	Building group in terms of energy savings	
1.5	2.1	2.8	Style	Wall
1	1.4	1.9	Heavy	
0.8	1.1	1.5	Adjacent to uncontrolled space	
2.7	3.7	5	Style	Ceiling
2.2	3	4	Heavy	
1.7	2.3	3.1	Adjacent to uncontrolled space	
1.6	2.2	3	Style	Floor
1.3	1.8	2.4	Heavy	
1	1.3	1.8	Adjacent to uncontrolled space	
2	2.7	3.7	Peripheral insulation	Floor foam
0.9	1.3	1.7	Insulation all below the surface	

1.2.4 RESEARCH TOOLS

The research tools for the first step (determining the share of water loss due to network corrosion and MNF) were included: Meters for consumption data, temperature, and power consumption factors on Energy Consumption Intensity (EUI) factors; GPS for production of Geospatial data; GIS software; Remote Sensing (RS) facilities; advanced modems; data loggers; networked sensors.

The overall heat transfer (1.5–1.9) rate for combined modes is usually expressed in terms of an overall conductance or heat transfer coefficient, U. In that case, the heat transfer rate is

$$\dot{\mathbf{Q}} = h.A.(T2 - T1) \tag{1.5}$$

where

$\dot{Q}$—Overall heat transfer rate.
A—Surface area where the heat transfer takes place, m^2.
T2—Temperature of the surrounding fluid, K.
T1—Temperature of the solid surface, K.
h—Heat transfer coefficient, (W/ (m^2. K)).

$$\mathbf{q} = {}^{d\dot{Q}}\!/\!_{dA} \tag{1.6}$$

where
q: Heat flux, W/ m^2, i.e., thermal power per unit area.
The general definition of the heat transfer coefficient is

$$\mathbf{h} = {}^{q}\!/\!_{\Delta T} \tag{1.7}$$

where
ΔT: Difference in temperature between the solid surface and surrounding fluid area, K.

A simple method for determining an overall heat transfer coefficient that is useful to find the heat transfer between simple elements such as walls in buildings or across heat exchangers is shown below:

$$\mathbf{1}\!/\!_{U.A} = \left({}^{1}\!/\!_{h_1} .A_1 \right) + \left({}^{dx_\omega}\!/\!_{k.A} \right) + \left({}^{1}\!/\!_{h_2} .A_2 \right) \tag{1.8}$$

where

U—The overall heat transfer coefficient (W/(m^2. K)).
A—The contact area for each fluid side (m^2).
k—The thermal conductivity of the material (W/(m·K)).
dx_ω—The wall thickness (m).

$$\mathbf{P} = U.A.T\left({}^{W}\!/\!_{K} \right) \tag{1.9}$$

P—Total thermal power
U—Heat Transfer Coefficient (W/m^2. K).

A—Area (m^2).
T—Temperature (k).
P—Total thermal power (W).

1.2.5 DATA ENVELOPMENT ANALYSIS METHOD (DEA)

A Multi-Factor Productivity Analysis method for evaluating the relative efficiency of an instance building by decision-makers is named Decision Making Units (DMUs). DMU decision units contain a homogeneous set of equipment in which objects are evaluated for performance (1). The purpose of this study is to obtain a building energy efficiency score (Table 1.2) under the name of a DMU. Regression analysis is a statistical technique for examining and modeling the relationship between variables.

This research is a quasi-experimental study and the researcher intends to use Energy Use Intensity (EUI) in the GIS space. The research is conducted by the field method using regression analysis. In addition to using regression analysis or analysis of variance, ANOVA and T-test are defined for the research model. The parameters of the regression model are[36–55]

- Input EUI (kWh/m^2).
- Outputs Percentage of dissatisfied (PPD) DEA Score.
- Efficiency.

The curve of the estimated method, which is the regression test, was thus identified using these data. The model was calibrated using a set of data without changing the parameter values. The research tools used are: Meters for Metering Gas Meters—Meters for Metering Electricity—Timers for Time Logs—CO_2 Measuring Instruments. The statistical population of this study consisted of 12 apartments. Research hypotheses are:[56–68]

1.2.6 HYPOTHESIS ONE

There is a significant relationship between Input (independent variable) and Outputs (dependent variable) (eqs 1.10 and 1.11):

$$Output = f\,(Input) \tag{1.10}$$

(Dependent variable):
Outputs/ Percentage of dissatisfied (PPD)

(Independent variable):
Input /Energy Use Intensity (EUI)

1.2.7 HYPOTHESIS TWO

There is a significant relationship between DEA and Input (independent variables) and Outputs (dependent variable):

$$Outputs = f\,(DEA\ Efficiency,\ Input) \tag{1.11}$$

(Dependent variable):
Outputs/Percentage of dissatisfied (PPD)

(Independent Variables):
Data Envelopment Analysis (DEA)
Input/Energy Use Intensity (EUI)

TABLE 1.2 Results of the Computational Model of Energy Use Intensity (EUI) in the Present Study.

No.	DEA efficiency score	Outputs CO_2 (ppm)	Outputs percentage of dissatisfied (PPD) (%)	Outputs occupancy density (Person/m^2) (%)	Working time (Hours)	Input energy use intensity (EUI) (kWh/m^2)	Building code
1	0.77	799	17.1	0.2	9	78	101
2	1	866	17.5	0.21	11	80	102
3	0.87	777	33.2	0.2	8	80	103
4	0.43	901	34.1	0.24	7	80.8	104
5	0.73	821	20.8	0.18	9	79	105
6	0.88	920	35	0.2	8	81	106
7	0.71	790	19.9	0.19	7	80.4	107
	82%	43%	77%	42%	39%	1.20%	Difference (%)

1.3 RESULTS

In this research, the implementation of GIS and GIS Ready implementation of HVAC&R system maps was performed as follows (Figure 1.3):

- Exchange graphical information from CAD space to GIS space.
- Fixes errors in CAD space.
- Convert graphics data from DWG format to SHP.
- Completing the descriptive and spatial information layers and fixing the errors in the GIS space (descriptive and spatial).
- Eliminate tolls that are in the wrong place.
- Create primary and external keys for the toll table.
- Creating appropriate tolerances and exchanging effects from spaghetti to topology.
- Preparation of conceptual model (Tables 1.3 and 1.4) for modeling network in GIS space.
- Create the right database.
- Ability to track and execute HVAC&R system analyses.
- Speed up the handling of HVAC&R system accidents.
- Systematically store and use HVAC&R system information using GIS.
- Economical savings due to the rapid flow of information in thermal and refrigeration facilities management.
- Create maps and reporting on all of the above.
- Regression analysis of facility systems.

Regression is needed for the estimation and forecasting in almost every field including engineering, physics, economics, management, biological sciences, biology, and social sciences. Regression analysis is one of the most widely used statistical techniques. The use of one variable to perform the prediction for another variable is called regression. Regression using one known and predicted variable predicts the values of another unspecified variable. The rate of change of one variable by the effect of another variable is also called the regression coefficient, which is the amount of change that occurs in the dependent variable by the unit of change in the independent variable. The regression is calculated as one variable and two variables. The door one-variable regression has one independent variable and one function variable, but two-variable regression has one function variable and two independent variables. To begin with, there must be a linear relationship that forms the scatter plot of the original idea. The regression line reflects the trajectory of the total distribution of points in the nominal coordinate system, which can indicate the severity and weakness and the type of correlation between the variables (Tables 1.3–1.28) (Figures 1.4–1.27).

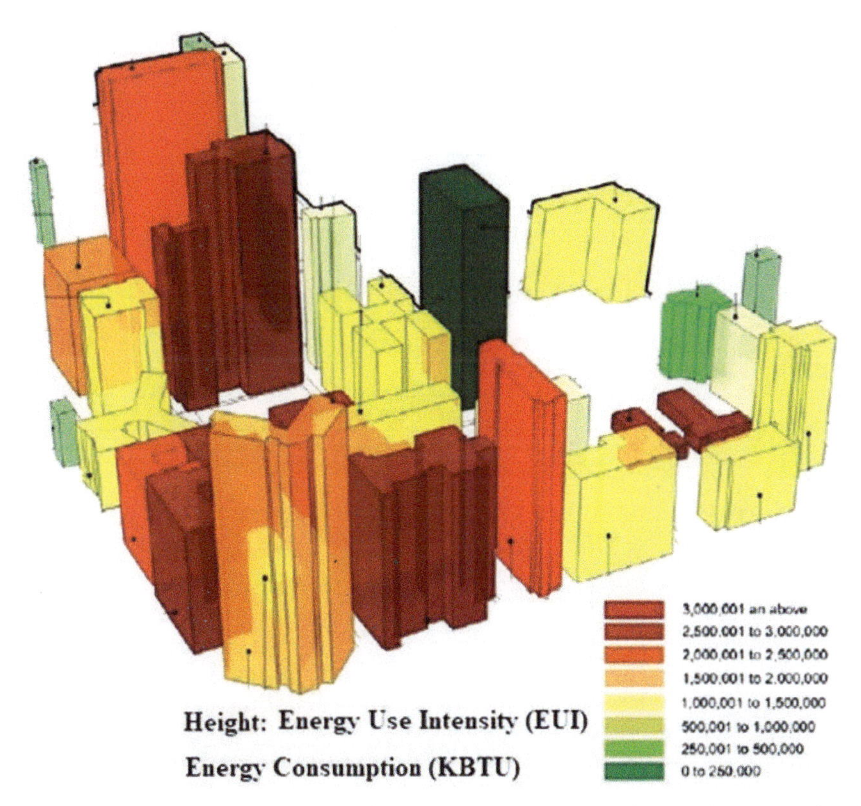

FIGURE 1.3 Comparison of the results of the energy use intensity (EUI) computational model.

TABLE 1.3 Variables Entered/Removed.

Model	Variables entered	Variables removed	Method
1	Energy Use Intensity (EUI) [kWh/m^2][a]		Enter

[a]*All requested variables entered.*

Dependent variable: Percentage of dissatisfied (PPD) [%]

TABLE 1.4 Model Summary.

Model	R	R square	Adjusted R square	Std. error of the estimate
1	.676[a]	.458	.349	6.68048

[a]*Predictors*: (Constant), energy use intensity (EUI) [kWh/m^2].

Dependent variable: Percentage of dissatisfied (PPD) [%].

TABLE 1.5 Case Processing Summary.

Unweighted cases (a)		N	Percent
Selected cases	Included in analysis	7	100.0
	missing cases	0	.0
	Total	7	100.0
Unselected cases		0	.0
Total		7	100.0

[a]If weight is in effect, see classification table for the total number of cases.

TABLE 1.6 Proximity Matrix.

	Euclidean distance						
	1	**2**	**3**	**4**	**5**	**6**	**7**
1	.000	2.053	16.224	17.232	3.833	18.150	3.688
2	2.053	.000	15.701	16.629	3.459	17.529	2.450
3	16.224	15.701	.000	1.282	12.441	2.059	13.307
4	17.232	16.629	1.282	.000	13.425	1.026	14.208
5	3.833	3.459	12.441	13.425	.000	14.341	1.664
6	18.150	17.529	2.059	1.026	14.341	.000	15.113
7	3.688	2.450	13.307	14.208	1.664	15.113	.000

This is a dissimilarity matrix

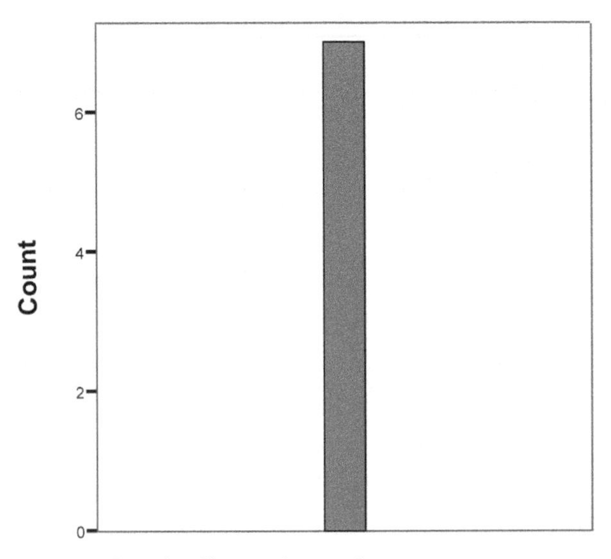

FIGURE 1.4 Sample of interactive graph.

X Graph

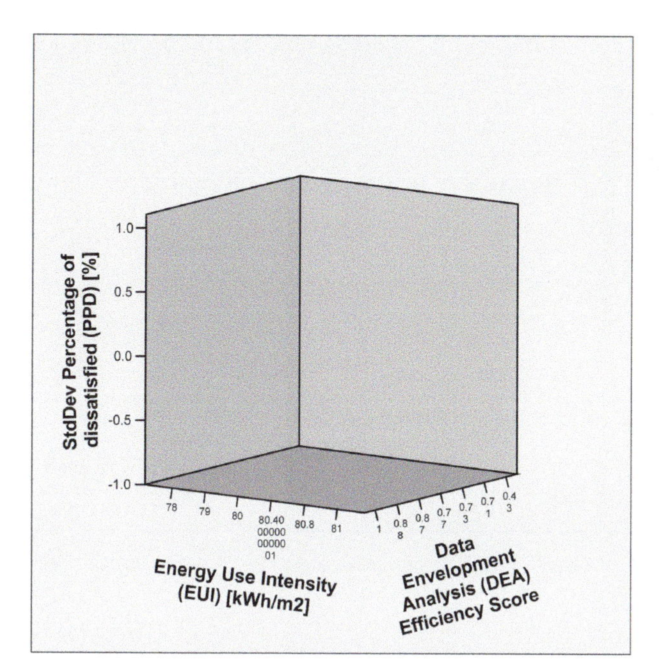

FIGURE 1.5 X Graph: Disadvantage percentage (PPD) based on the regression model.

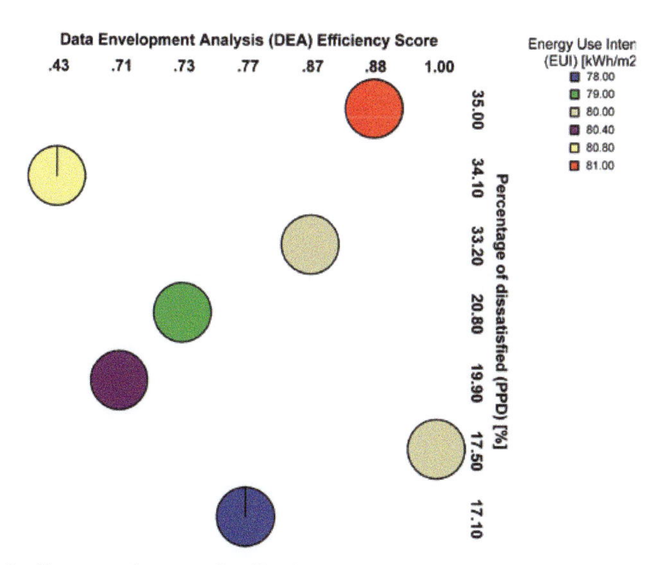

FIGURE 1.6 Data envelopment for disadvantage percentage (PPD) based on the regression model of the present study.

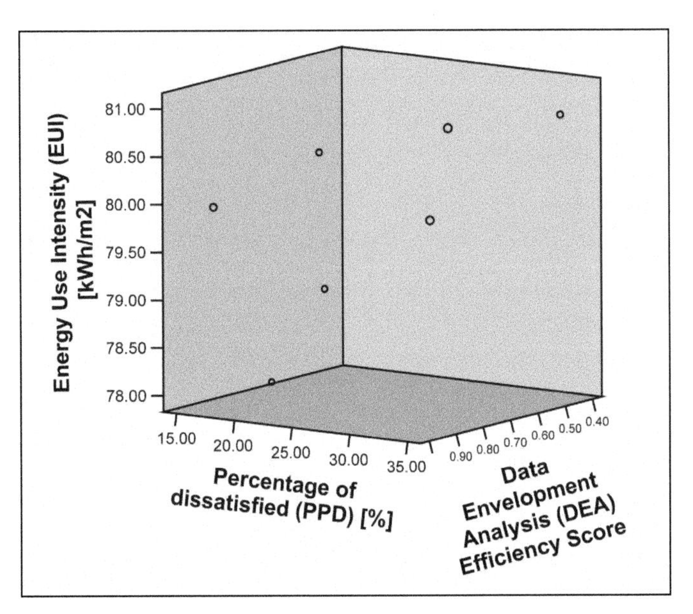

FIGURE 1.7 X Graph: Disadvantage percentage (PPD) based on the regression model of the present study.

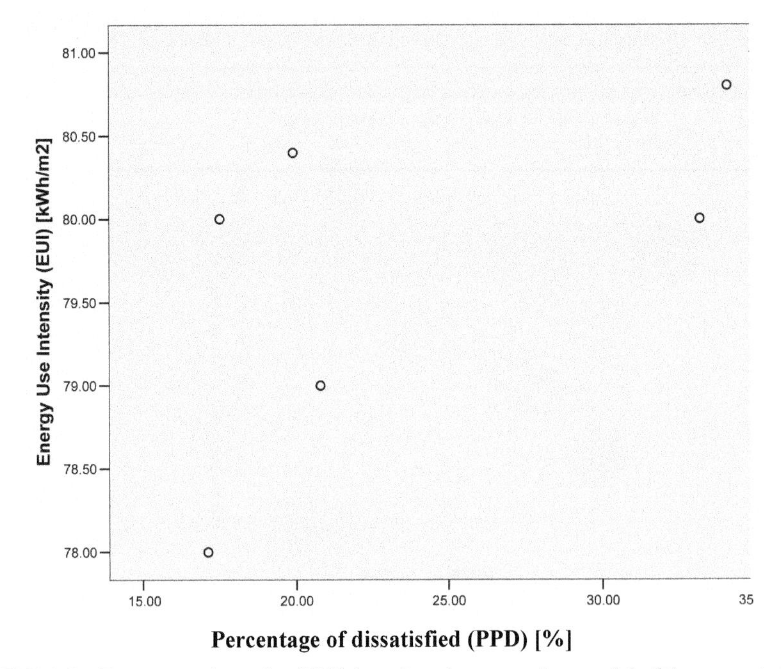

FIGURE 1.8 Energy use intensity (EUI) based on the regression model of the present study.

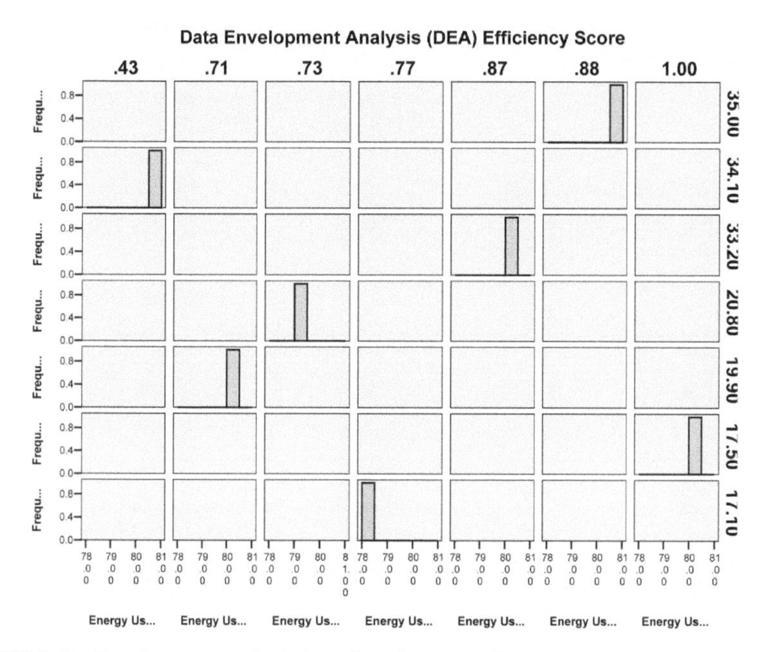

FIGURE 1.9 Envelopment analysis based on the regression.

TABLE 1.7 Model Description.

Model name		**MOD_2**
Series or sequence	1	Energy use intensity (EUI) [kWh/m^2]
	2	Percentage of dissatisfied (PPD) [%]
	3	Data envelopment analysis (DEA) efficiency score
Transformation		Natural logarithm
Non-seasonal differencing		1
Seasonal differencing		0
Length of seasonal period		No periodicity
Standardization		Applied
Distribution	Type	Normal
	Location	Estimated
	Scale	Estimated
Fractional rank estimation method		Blom's
Rank assigned to ties		Mean rank of tied values

Applying the model specifications from MOD_2

TABLE 1.8 Case Processing Summary.

		Energy use intensity (EUI) [kWh/m²]	Percentage of dissatisfied (PPD) [%]	Data envelopment analysis (DEA) Efficiency score
Series or sequence length		7	7	7
Number of missing values in the plot	Negative or zero before log transform	0	0	0
	User-missing	0	0	0
	System-missing	0	0	0

The cases are unweighted.

TABLE 1.9 Estimated Distribution Parameters.

		Energy use intensity (EUI) [kWh/m²]	Percentage of dissatisfied (PPD) [%]	Data envelopment analysis (DEA) Efficiency score
Normal distribution	Location	.0000	.0000	.0000
	Scale	1.00000	1.00000	1.00000

The cases are unweighted.

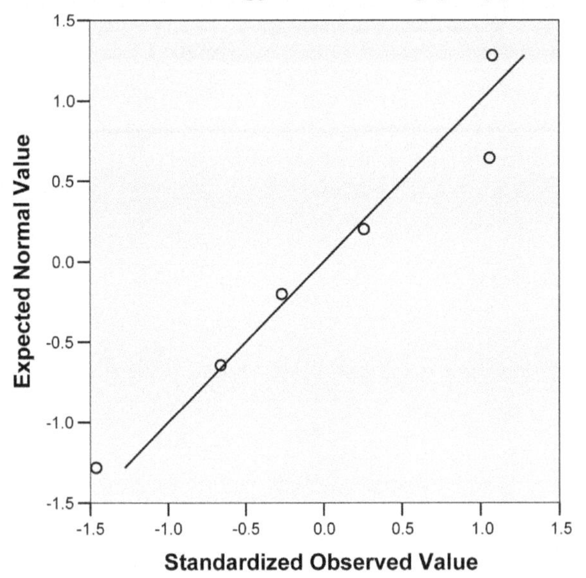

FIGURE 1.10 Normal Q–Q plot of energy use intensity (EUI) [kWh/m²].

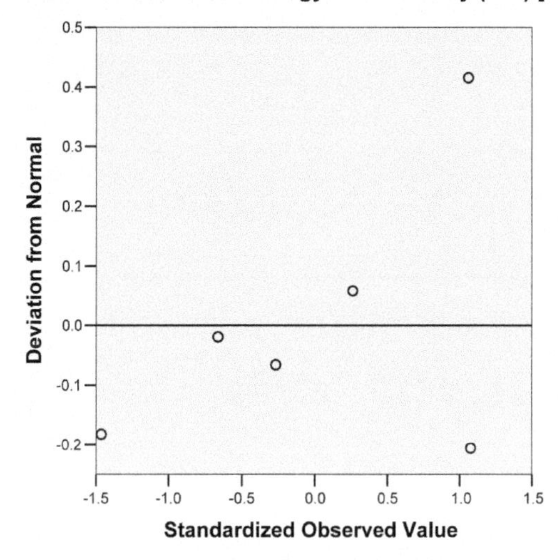

FIGURE 1.11 Normal Q–Q plot of energy use intensity (EUI) [kWh/m^2].

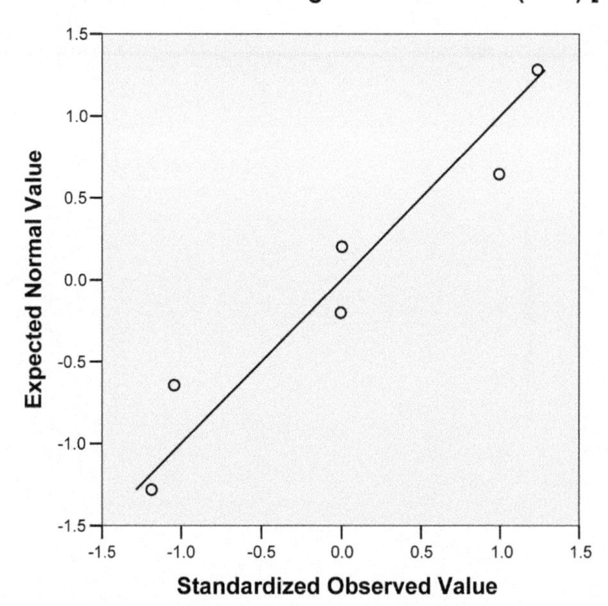

FIGURE 1.12 Normal Q–Q plot of energy use intensity (EUI) [kWh/m^2].

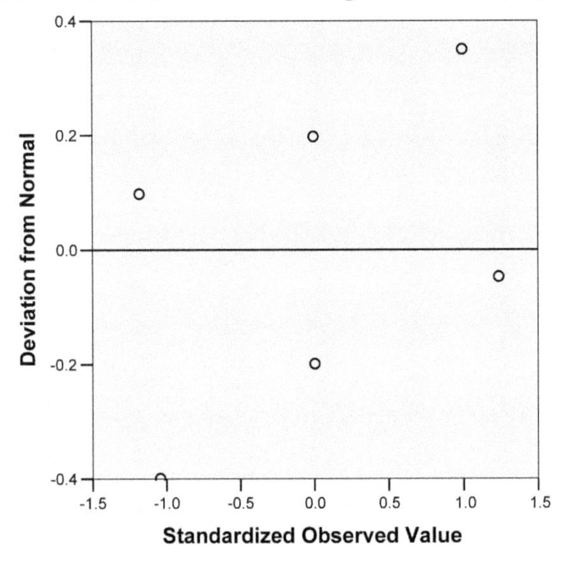

FIGURE 1.13 Normal Q–Q plot of energy use intensity (EUI) [kWh/m^2].

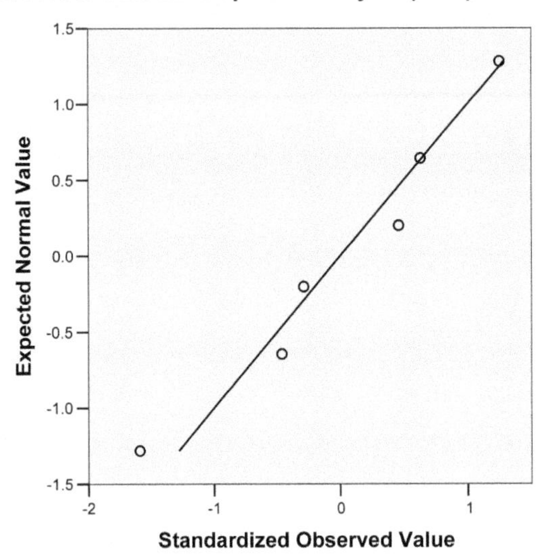

FIGURE 1.14 Normal Q–Q plot of energy use intensity (EUI) [kWh/m^2].

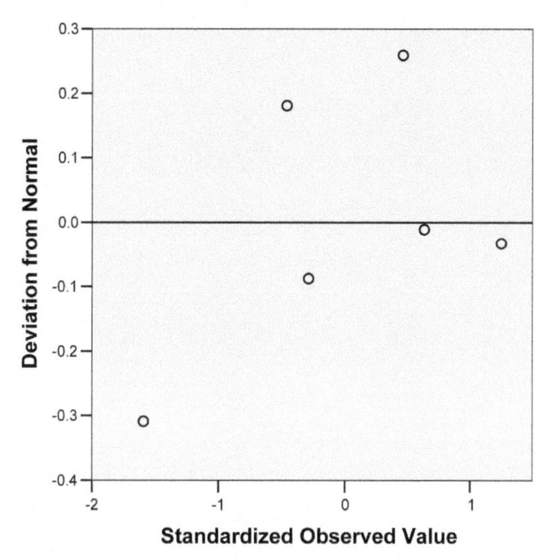

FIGURE 1.15 Normal Q–Q plot of energy use intensity (EUI) [kWh/m²].

TABLE 1.10 Model Description.

Model name		**MOD_3**
Series or sequence	1	Energy use intensity (EUI) [kWh/m²]
	2	Percentage of dissatisfied (PPD) [%]
	3	Data envelopment analysis (DEA) Efficiency score
Transformation		Natural logarithm
Non-seasonal differencing		1
Seasonal differencing		0
Length of seasonal period		No periodicity
Horizontal axis labels		Sequence numbers
Intervention onsets		None
For each observation		Values not joined

Applying the model specifications from MOD_3

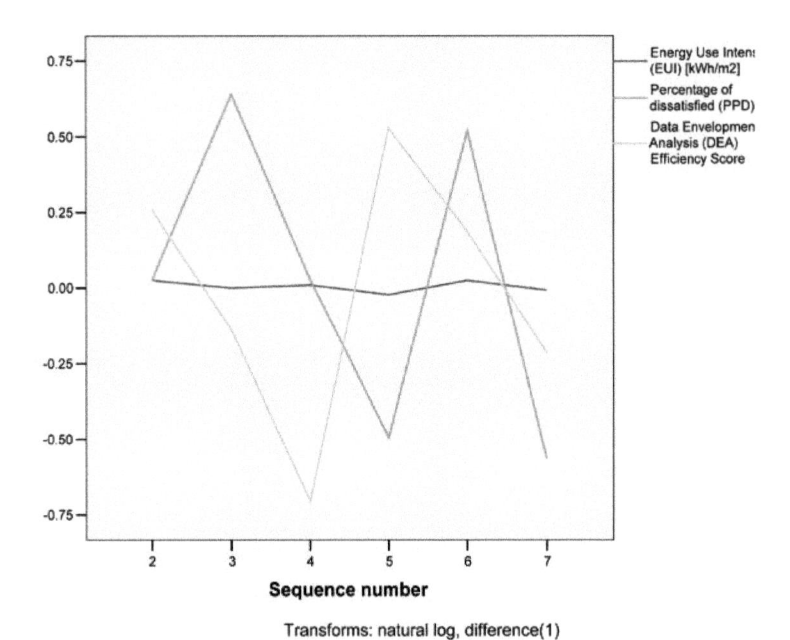

FIGURE 1.16 Data envelopment.

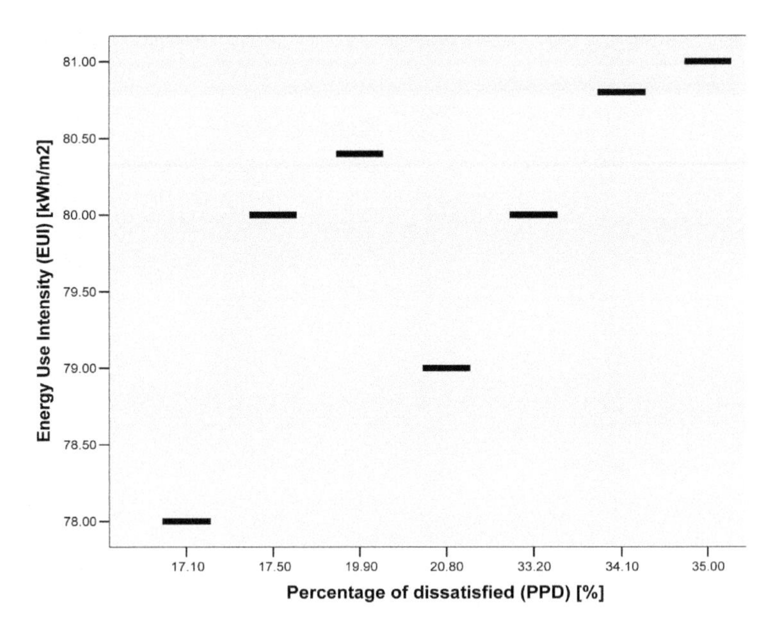

FIGURE 1.17 Plot of energy use intensity (EUI) [kWh/m^2] versus percentage of dissatisfied (PPD) [%].

TABLE 1.11 Case Processing Summary.

		Energy use intensity (EUI) [kWh/m²]	Percentage of dissatisfied (PPD) [%]	Data envelopment analysis (DEA) Efficiency score
Series or sequence length		7	7	7
Number of missing values in the plot	Negative or zero before log transform	0	0	0
	User-missing	0	0	0
	System-missing	0	0	0

TABLE 1.12 Model Description.

Model name		MOD_5
Dependent series		Percentage of dissatisfied (PPD) [%]
Independent series	1	Energy use intensity (EUI) [kWh/m²]
	2	Data envelopment analysis (DEA) efficiency score
Constant		Included
AR		1

Applying the model specifications from MOD_5

TABLE 1.13 Iteration Termination Criteria.

Maximum Parameter Change Less Than	.001
Maximum Marquardt Constant Greater Than	1000000000
Sum of Squares Percentage Change Less Than	.001%
Number of Iterations Equal to	10

TABLE 1.14 Case Processing Summary.

Series length		7
Number of cases skipped due to missing values	At the beginning of the series	0
	At the end of the series	0
Number of cases with missing values within the series		0(a)
Number of forecasted cases		0
Number of new cases added to the current working file		0

[a]*Melard's Algorithm will be used for estimation.*

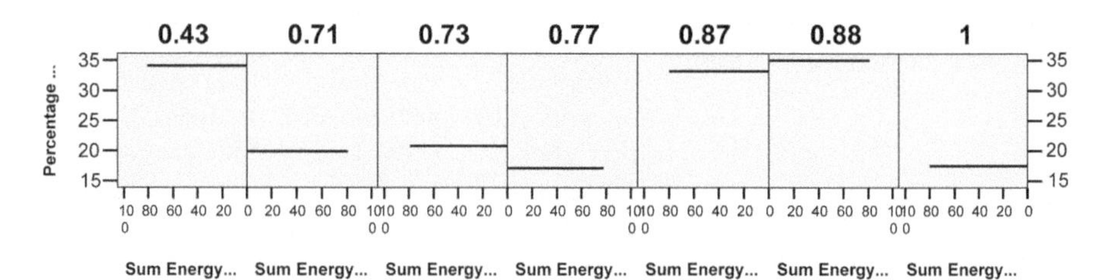

FIGURE 1.18 Data envelopment analysis (DEA) efficiency score.

TABLE 1.15 Requested Initial Configuration.

Rho (AR1)		.000
Regression coefficients	Energy use intensity (EUI) [kWh/m^2]	AUTO(a)
	Data envelopment analysis (DEA) Efficiency score	AUTO(a)
Constant		AUTO(a)

[a]*The prior parameter value is invalid and is reset to 0.1.*

TABLE 1.16 Iteration History.

	Rho (AR1)	Regression coefficients		Constant	Adjusted sum of squares	Marquardt constant
		Energy use intensity (EUI) [kWh/m^2]	Data envelopment analysis (DEA) efficiency score			
0	.000	5.111	−8.574	−376.296	208.983	.001
1	−.759	6.052	−14.935	−446.122	126.690	.001
2	−.763	6.058	−14.943	−446.619	126.687(a)	1.000

Melard's algorithm was used for estimation.

[a]*The estimation terminated at this iteration, because the sum of squares decreased by less than .001%.*

TABLE 1.17 Residual Diagnostics.

Number of residuals	7
Number of parameters	1
Residual df	3
Adjusted residual sum of squares	126.687
Residual sum of squares	208.983
Residual variance	37.283
Model std. error	6.106
Log-likelihood	–20.796
Akaike's information criterion (AIC)	49.593
Schwarz's Bayesian criterion (BIC)	49.377

TABLE 1.18 Parameter Estimates.

		Estimates	Std error	t	Approx sig
Rho (AR1)		−.763	.585	−1.303	.284
Regression coefficients	Energy use intensity (EUI) [kWh/m^2]	6.058	2.567	2.360	.099
	Data envelopment analysis (DEA) Efficiency score	−14.942	11.772	−1.269	.294
Constant		−446.603	208.164	−2.145	.121

Melard's algorithm was used for estimation.

TABLE 1.19 Correlation Matrix.

		Rho (AR1)	Regression coefficients		Constant
			Energy use intensity (EUI) [kWh/m^2]	Data envelopment analysis (DEA) efficiency score	
Rho (AR1)		1.000	0(a)	0(a)	0(a)
Regression coefficients	Energy use intensity (EUI) [kWh/m^2]	0(a)	1.000	.294	−.999
	Data envelopment analysis (DEA) Efficiency score	0(a)	.294	1.000	−.334
Constant		0(a)	−.999	−.334	1.000

Melard's algorithm was used for estimation.

[a]*The Rho (AR1) parameter estimate and the regression parameter estimate are asymptotically uncorrelated.*

TABLE 1.20 Covariance Matrix.

		Rho (AR1)	Regression coefficients		Constant
			Energy use intensity (EUI) [kWh/m^2]	Data envelopment analysis (DEA) Efficiency score	
Rho (AR1)		.343	0(a)	0(a)	0(a)
Regression coefficients	Energy use intensity (EUI) [kWh/m^2]	0(a)	6.588	8.885	−533.829
	Data envelopment analysis (DEA) Efficiency score	0(a)	8.885	138.583z	−817.918
Constant		0(a)	−533.829	−817.918	43332.168

Melard's algorithm was used for estimation.

[a]*The rho (AR1) parameter estimate and the regression parameter estimate are asymptotically uncorrelated.*

TABLE 1.21 Case Processing Summary.

Valid active cases	7
Active cases with missing values	0
Supplementary cases	0
Total	7
Cases used in analysis	7

TABLE 1.22 Model Summary.

Multiple R	R square	Adjusted R square
1.000	.999	.996

Dependent variable: Percentage of dissatisfied (PPD) [%]

Predictors: Energy Use Intensity (EUI) [kWh/m^2] Data Envelopment Analysis (DEA) Efficiency Score

TABLE 1.23 ANOVA.

	Sum of squares	df	Mean square	F	Sig.
Regression	6.995	5	1.399	301.216	.044
Residual	.005	1	.005		
Total	7.000	6			

Dependent variable: Percentage of dissatisfied (PPD) [%]

Predictors: Energy use intensity (EUI) [kWh/m2] Data envelopment analysis (DEA) efficiency score

TABLE 1.24 Coefficients.

	Standardized coefficients		df	F	Sig.
	Beta	Std. Error			
Energy use intensity (EUI) [kWh/m^2]	1.006	.026	2	1505.273	.018
Data envelopment analysis (DEA) Efficiency score	−.107	.026	3	17.100	.175

Dependent variable: Percentage of dissatisfied (PPD) [%]

TABLE 1.25 Correlations and Tolerance.

	Correlations			Importance	Tolerance	
	Zero-order	Partial	Part		After transformation	Before transformation
Energy use intensity (EUI) [kWh/m^2]	.994	1.000	.999	1.000	.987	.998
Data envelopment analysis (DEA) Efficiency score	.006	−.972	−.107	−.001	.987	.998

Dependent variable: Percentage of dissatisfied (PPD) [%]

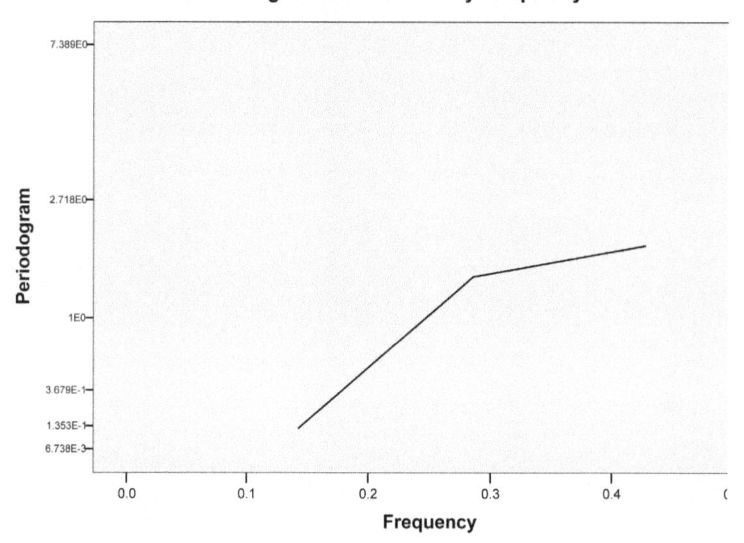

FIGURE 1.19 Data envelopment analysis for energy use intensity (EUI) [kWh/m^2].

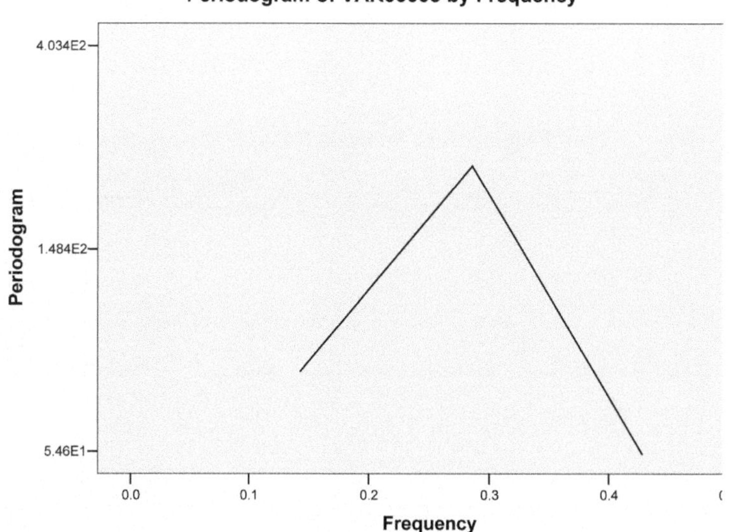

FIGURE 1.20 Data envelopment analysis for percentage of dissatisfied (PPD) [%].

Data Envelopment Analysis (DEA) Efficiency Score

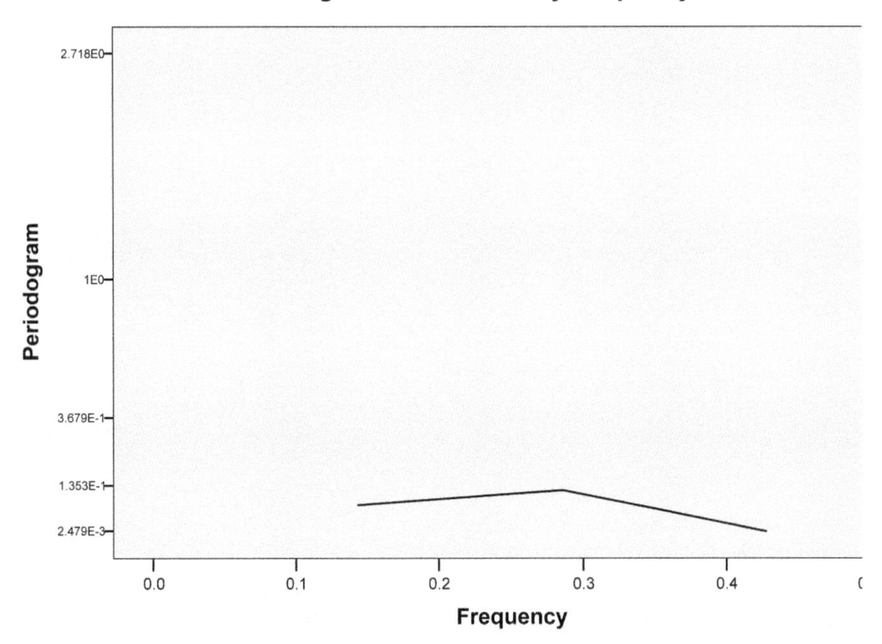

FIGURE 1.21 Data envelopment analysis for percentage of dissatisfied (PPD).

XGraph

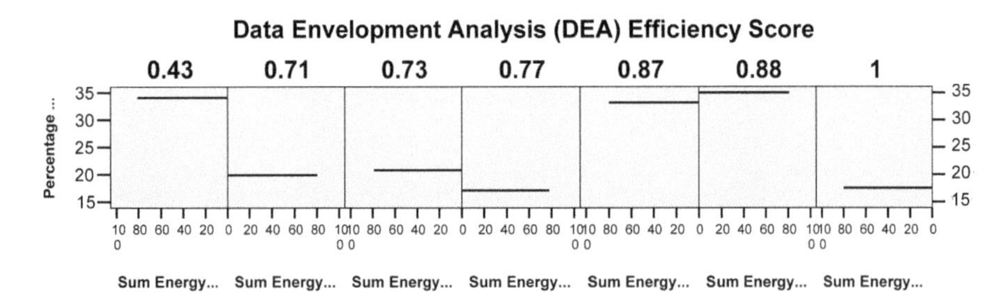

FIGURE 1.22 X graph for data envelopment analysis (DEA) efficiency.

Nominal Regression

TABLE 1.26 Case Processing Summary.

		N	Marginal percentage
Percentage of dissatisfied (PPD) [%]	17.10	1	14.3%
	17.50	1	14.3%
	19.90	1	14.3%
	20.80	1	14.3%
	33.20	1	14.3%
	34.10	1	14.3%
	35.00	1	14.3%
Energy Use Intensity (EUI) [kWh/m^2]	78.00	1	14.3%
	79.00	1	14.3%
	80.00	2	28.6%
	80.40	1	14.3%
	80.80	1	14.3%
	81.00	1	14.3%
Data envelopment analysis (DEA) Efficiency score	.43	1	14.3%
	.71	1	14.3%
	.73	1	14.3%
	.77	1	14.3%
	.87	1	14.3%
	.88	1	14.3%
	1.00	1	14.3%
Valid		7	100.0%
Missing		0	
Total		7	
Subpopulation		7(a)	

[a]*The dependent variable has only one value observed in 7 (100.0%) subpopulations.*

TABLE 1.27 Pseudo R-Square.

Cox and Snell	.980
Nagelkerke	1.000
McFadden	1.000

TABLE 1.28 Likelihood Ratio Tests.

Effect	Model fitting criteria	Likelihood ratio rests		
	-2 log likelihood of reduced model	Chi-square	df	Sig.
Intercept	.000(a)	.000	0	.
VAR00004	.000(a)	.000	0	.
VAR00006	2.773	2.773	6	.837

The Chi-square statistic is the difference in -2 log-likelihoods between the final model and a reduced model. The reduced model is formed by omitting an effect from the final model. The null hypothesis is that all parameters of that effect are 0.

[a]This reduced model is equivalent to the final model because omitting the effect does not increase the degrees of freedom.

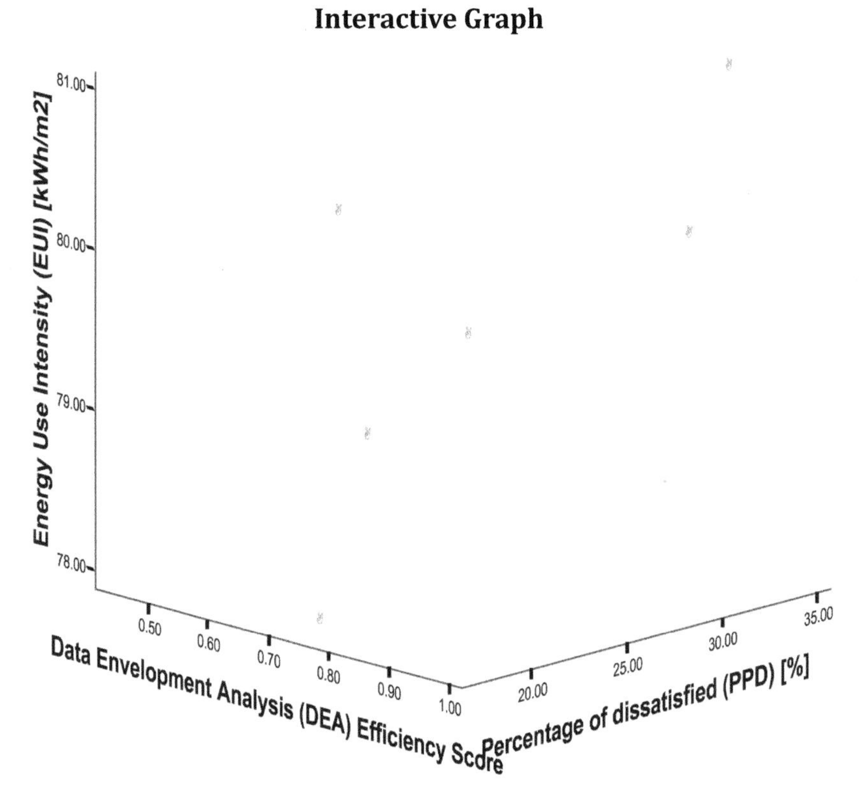

FIGURE 1.23 Interactive graph for energy use intensity (EUI), percentage of dissatisfied (PPD), data envelopment analysis (DEA).

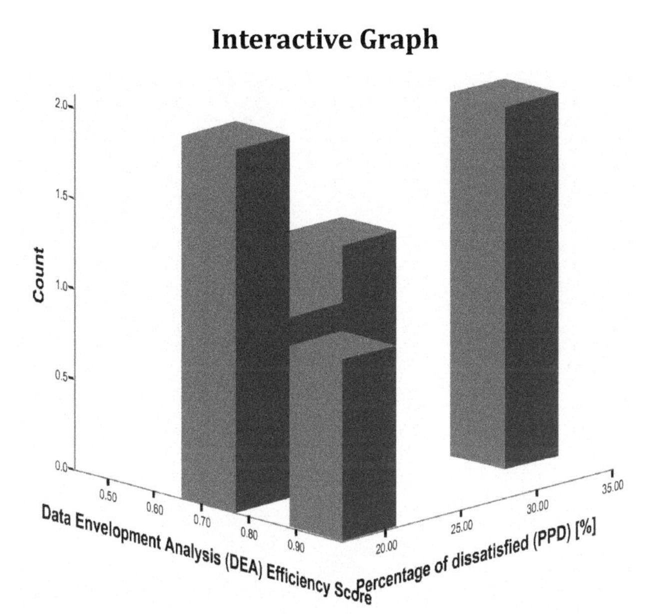

FIGURE 1.24 Interactive graph for energy use intensity (EUI), percentage of dissatisfied (PPD), data envelopment analysis (DEA).

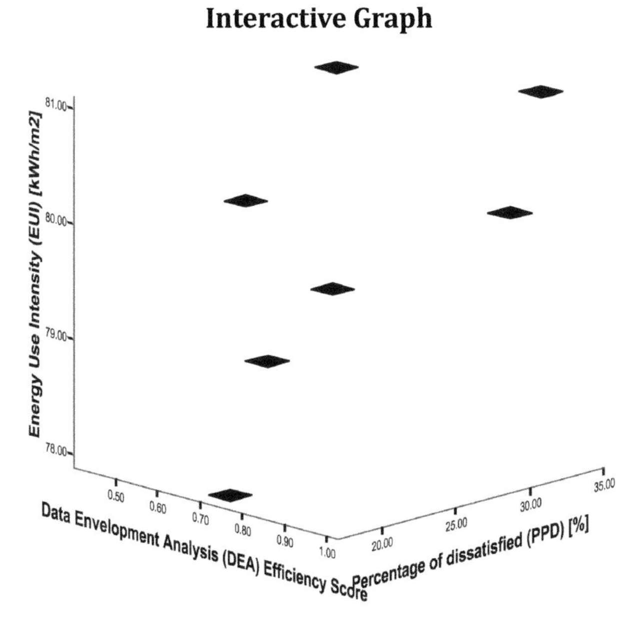

FIGURE 1.25 Interactive graph for energy use intensity (EUI), percentage of dissatisfied (PPD), data envelopment analysis (DEA).

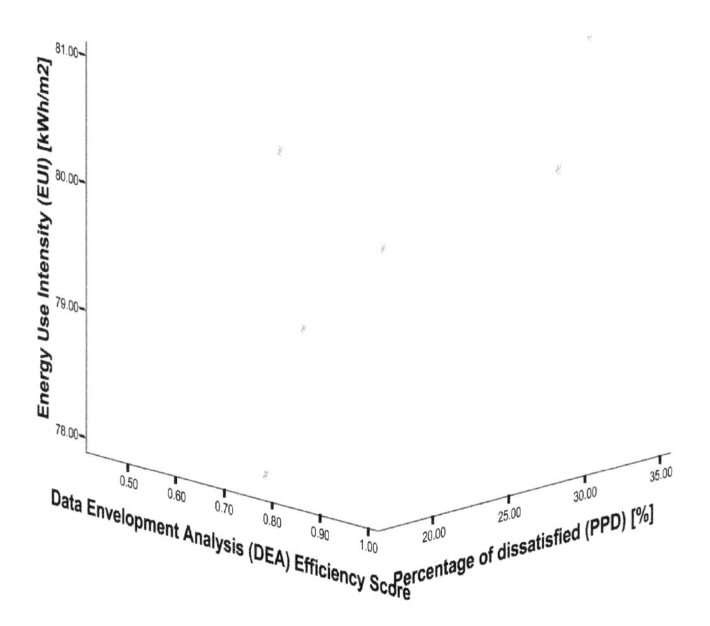

FIGURE 1.26 Interactive graph for energy use intensity (EUI), percentage of dissatisfied (PPD), data envelopment analysis (DEA).

Interactive Graph

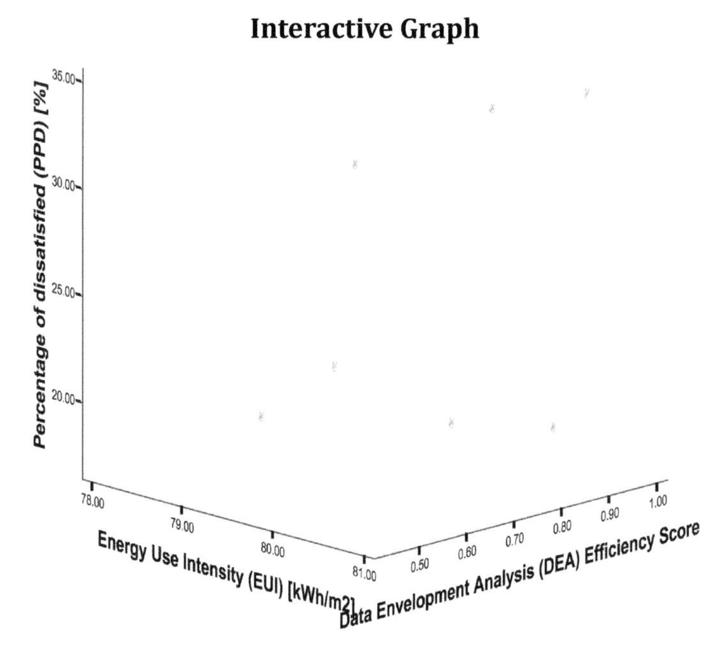

FIGURE 1.27 Interactive graph for energy use intensity (EUI), percentage of dissatisfied (PPD), data envelopment analysis (DEA).

In this research, the regression equations (eqs 1.12–1.33) were used to draw the regression lines. The modeling results (Figures 1.28 and 1.29) were based on the rapid flow of location-based information for the HVAC&R system (Tables 1.29 and 1.30):

- Designing and implementing an energy module to reduce energy consumption, improve the energy rating of the building, and improve the quality of comfort for residents.
- Design and implementation of HVAC module based on simulation of dynamic thermal behavior of building, solar behavior, simulation of central and local control system of mechanical installations.

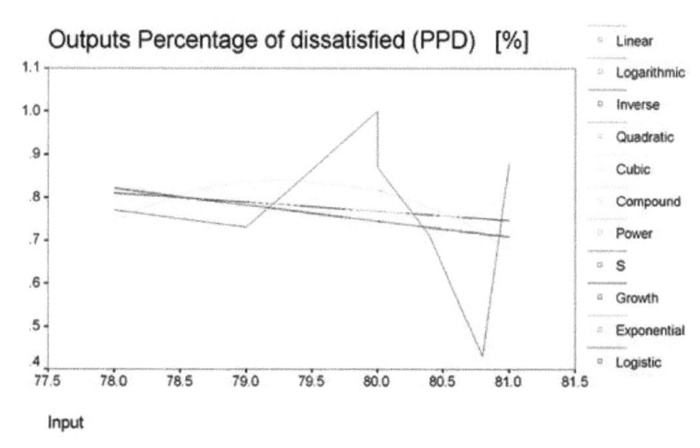

FIGURE 1.28 Disadvantage percentage (PPD) based on the regression model of the present study.

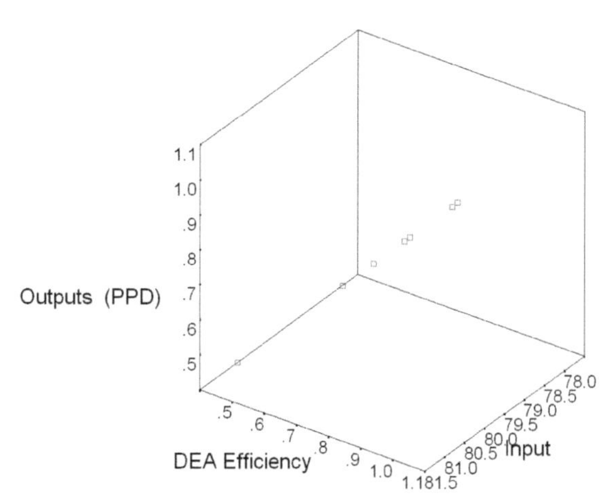

FIGURE 1.29 Energy intensity (EUI) based on the regression model of the present study.

TABLE 1.29 Regression Model in the Present Study Curve Fit.

Mth	Rsq	d.f.	F	Sigf	a_0	a_1	a_2	a_3
Linear function	.016	5	.08	.785	2.5086	−.0218		
Logarithmic function	.016	5	.08	.788	8.2502	−1.7076		
Inverse function	.015	5	.08	.791	−.9070	133.945		
Quadratic function	.017	5	.08	.783	1.6549	−.0001	−.0001	
Cubic function	.105	4	.23	.801	−240.06	4.5557		−.0002
Compound function	.038	5	.20	.677	40.1875	.9514		
Power function	.037	5	.19	.680	2.3E+07	−3.9311		
S function	.036	5	.19	.682	−4.1690	.0499		
Growth function	.038	5	.20	.677	3.6936	−.0499		
Exponential function	.038	5	.20	.677	40.1875	−.0499		
Logistic function	.038	5	.20	.677	.0249	1.0511		

TABLE 1.30 Equations Using the Obtained Regression Models for the Present Research Regression Model.

Equation	Model summary					Parameter estimates			
	R square	F	df1	df2	Sig.	a_0	a_1	a_2	a_3
Linear, (1.12) $\mathbf{y = a_0 + a_1}x$	.016	.08	5	5	.785	2.5086	−.0218		
Logarithmic, (1.13) $\log y = \log(a) - (b)\log x$	.016	.08	5	5	.788	8.2502	−1.7076		
Inverse, (1.14) $\mathbf{y} = f^{-1}(\mathbf{y})$	.015	.08	5	5	.791	−.9070	133.945		
Quadratic, (1.15) $\mathbf{y = a_0 + a_1x + a_2x^2}$	.017	.08	5	5	.783	1.6549	−.0001	−.0001	
Cubic, (1.16) $\mathbf{y = a_0 + a_1x + a_2x^2 + a_3}x^3$	.105	.23	4	4	.801	−240.06	4.5557		−.0002
Compound, (1.17) $\mathbf{A = Ce^{kt}}$	.038	.20	5	5	.677	40.1875	.9514		
Power, (1.18) $\mathbf{y} = c\mathbf{x^p}$	.037	.19	5	5	.680	2.3E+07	−3.9311		
S, (1.19) $\mathbf{y} = f_0\,\mathbf{(T, X, U)}$	.036	.19	5	5	.682	−4.1690	.0499		

TABLE 1.30 *(Continued)*

Equation	Model summary					Parameter estimates			
	R square	F	df1	df2	Sig.	a_0	a_1	a_2	a_3
Growth, (1.20) **(dA/dT) = KA**	.038	.20	5	5	.677	3.6936	−.0499		
Exponential, (1.21) $e^x = \lim_{n\to\infty}\left(1+\frac{1}{n}\right)^n$	.038	.20	5	5	.677	40.1875	−.0499		
Logistic, (1.22) $f(x) = \frac{L}{1+e^{-k(x-x_0)}}$	.038	.20	5	5	.677	.0249	1.0511		

$$\mathbf{A} = 2.5086 - .0218x \quad \text{Linear function,} \quad (1.23)$$

$$\log y = \log(8.2502) + (-1.7076)\log x \quad \text{Logarithmic function,} \quad (1.24)$$

$$\mathbf{y} = -.9070 + 133.945 f^{-1}(\mathbf{y}) \quad \text{Inverse function,} \quad (1.25)$$

$$\mathbf{y} = 1.6549\text{-}.0001\mathbf{x}^{-.0001} \quad \text{Quadratic function,} \quad (1.26)$$

$$\mathbf{y} = .801 + 240.06\mathbf{x} - 4.5557\mathbf{x}^2 - 0x^3 \quad \text{Cubic function,} \quad (1.27)$$

$$\mathbf{A} = 40.1875\mathbf{e}^{t} + .9514 \quad \text{Compound function,} \quad (1.28)$$

$$\mathbf{y} = 2.3\text{E+}07\mathbf{x}^{-3.9311} \quad \text{Power function,} \quad (1.29)$$

$$\mathbf{y} = f_0(-4.1690, \mathbf{X}, .0499) \quad \text{S function,} \quad (1.30)$$

$$(\mathbf{dA/dT}) = 3.6936\mathbf{A} + -.0499 \quad \text{Growth function,} \quad (1.31)$$

$$\mathbf{y} = (40.1875\mathbf{E} - .0499)^{\mathbf{X}} + \mathbf{g} \quad \text{Exponential function,} \quad (1.32)$$

$$\mathbf{y} = .0249\mathbf{x} + 1.0511 \quad \text{Logistic function,} \quad (1.33)$$

Due to the parameter limitations, the power function had a suitable correlation on the scatter diagram and best-fit curve which was used for disadvantage percentage (PPD) based on the regression model shown in (Table 1.31). The percentage of dissatisfied (PPD) varied with the Energy Use Intensity (EUI), as shown in eqs 1.18 and 1.29.

$$\mathbf{f}(\mathbf{x}) = \mathbf{y} = c\mathbf{x}^{\mathbf{p}} \tag{1.18}$$

$$\mathbf{y} = 2.3\text{E}{+}07\mathbf{x}^{-3.9311} \tag{1.29}$$

where

y—Percentage of dissatisfied (PPD),
x—Energy Use Intensity (EUI),

c and *p* are parameters determined by the linear least-squares method using data from the ten parameters tested.

The p-value is the level of marginal significance within a statistical hypothesis test representing the probability of the occurrence of a given event. The p-value is used as an alternative to rejection points to provide the smallest level of significance at which the null hypothesis would be rejected. At this work, the following procedure was done for the p-value evaluation:

- Determine the experiment's expected results.
- Determine the experiment's observed results.
- Determine the experiment's degrees of freedom.
- Compare expected results to observed results with Chi-square.
- Choose a significance level.
- Use a Chi-square distribution table to approximate the p-value.
- Approximate the p-value for experiment, it can be decided whether or not to reject the null hypothesis of the experiment.

Generally, based on the hypothesis and the experimental results, If the p-value is lower than the significance value, it can be shown that the experimental results would be highly unlikely to occur if there was no acceptable relation between the variables which be manipulated (Table 1.31).

In this work, the p-values for three variables include the percentage of dissatisfied (PPD); energy use intensity (EUI) and DEA efficiency are as follows (Table 1.31):

1. P-value for percentage of dissatisfied (PPD):

   $$\text{Chi}^2 = 1.200,\ \text{DF} = 7$$

   The P-value equals .991.

   By conventional criteria, this difference is considered to be not statistically significant.

2. P-Value for energy use intensity (EUI):

$$Chi^2 = 1.200, DF=6$$

The P-value equals .977.

By conventional criteria, this difference is considered to be not statistically significant.

3. P-Value for DEA efficiency:

$$Chi^2 = .000, DF=9$$

The two-tailed P-value equals .962.

TABLE 1.31 Test Statistic and p-Value Calculation.

Parameters	Percentage of dissatisfied (PPD)	Energy use intensity (EUI)	DEA efficiency
Chi-square	1.200	1.200	.000
df	7	6	9
Asymp. Sig.	.991	.977	.962

1.3.1 ARTIFICIAL INTELLIGENCE (AI) PROGRAM

In this work, Artificial Intelligence program was coded in terms of thinking, reasoning, learning, and making decisions with Python. Artificial Intelligence took advantage of existing algorithm and placed it in the structure of the Python language. Therefore, it became possible to learn a set of special conditions, various algorithms, and other things (Table 1.32).

TABLE 1.32 Artificial Intelligence Program was Coded in Terms of Learning Conditions for (DEA), (PPD), and (EUI).

No.	Data Envelopment Analysis (DEA) Efficiency Score	Percentage of dissatisfied (PPD) [%]	Energy Use Intensity (EUI) [kWh/m^2]
1	0.77	17.1	78
2	1	17.5	80
3	0.87	33.2	80
4	0.43	34.1	80.8
5	0.73	20.8	79
6	0.88	35	81
7	0.71	19.9	80.4

1.3.2 HYPOTHESIS ONE

$$y = f(x)$$

Percentage of dissatisfied (PPD) = *f* (Energy Use Intensity (EUI)).

1.3.3 HYPOTHESIS TWO

$$y = f(x1, x2)$$

Percentage of dissatisfied (PPD) = *f* (Data Envelopment Analysis (DEA); Energy Use Intensity (EUI))

If energy use intensity (EUI) ≥ 81, then in order to control the energy use intensity (EUI), pulls were sent to actuator of control valve.

1.3.4 RESEARCH SUGGESTION

In the current research, separate parts of the network that are separated from the distribution network by permanent boundaries, are called independent areas of DMA measurement in order to determine the improvement and development model based on the technology of remote reading and fast information circulation according to GIS considered. This research used the WATERGEMS computer model to solve the water supply network system from the step-by-step algorithm based on the flow and pressure equations. In order to improve the accuracy of the model as much as possible, it was done to fix the above errors. Among the important points that should be considered in relation to the hydraulic model contract is the updating of the model, which should be fully updated while receiving the latest information about the network status and use the results of the model to optimize the network corrections. In general, it can be said that the model will have the necessary efficiency if it is always up-to-date and is used in different stages of network modification, design, and operation. The current research presented the model for the development of the base location in the independent areas of DMA measurement along with the hydraulic analysis of the model to understand the current situation. In order to modify the network, first the existing network with reservoirs and water supply sources was modeled. According to the volume of existing reservoirs, pressure zoning of the network was done. The range of equal pressure lines in the network was determined. According to the volume of the reservoirs, each of these areas was also covered by the reservoirs. Changes were made, including the modification of some of the current network diameters that were inappropriate. Some of the diameters were replaced due to the age of these pipes and the end of their useful life and leakage problems. The leakage problems are produced by the working pressure of fluid. The high pressure of fluid in water transmission line has destructive effects for water piping systems. It can be removed by hydroelectric power through hydro turbines.

1.3.5 CLEAN ENERGY (RENEWABLE ENERGY)

Type of energy: Hydroelectric power through hydro turbines

1.3.6 WHAT'S CLEAN ENERGY?

Any type of energy that is produced from water, wind, or sunlight is called clean energy. This type of energy is also called renewable energy.

There are five main sources of clean energy in the world. These sources, which are known as renewable energy, include:

- Renewable resource: Earth's underground Heat
- Type of energy: Thermal energy
- Renewable resource: Sun
- Type of energy: Solar energy by solar cells
- Renewable resource: running water
- Type of energy: Hydroelectric power through hydro turbines
- Renewable resource: Biomass
- Type of energy: Biomass energy
- Renewable resource: Wind
- Type of energy: Electricity generation through wind turbines

These sources are called renewable sources, and the resulting energy is known as clean energy because these sources regenerate themselves naturally. In other words, light comes from sunlight, plants grow, wind always blows, and rivers always flow.

Solar energy: Solar energy is one of the types of clean energy. Solar energy is actually produced with the help of photovoltaic cells. These cells absorb sunlight and finally convert it into electricity. Solar energy can be used to heat buildings and for hot water, cooking, and even lighting.

Geothermal energy: This type of clean energy is produced from the heat of the earth. Heat that is stored under the earth's crust and requires drilling to access it. Humans have been using this energy naturally for thousands of years. For example, natural hot water springs are one of these common examples that can be found naturally in many cities of our country, Iran.

Wind energy: Wind energy can be used at high altitudes and coastal locations. With the help of wind energy, wind turbines can be moved and in this way electricity can be produced.

Biomass: Biomass power plants use wood waste, sawdust, and combustible waste to produce energy.

Hydroelectric energy: This energy, which is also called hydroelectric power, uses the water flow of rivers, dams, and streams to produce electrical energy. It is interesting to know that it is even possible to generate electricity from the tides in the oceans.

1.3.7 WHY CLEAN ENERGY?

To answer this question, we need to compare it with the cost of providing energy through fossil fuels. The fact is that one-day fossil resources will run out. As a result, the cost of this type of energy will increase. But the cost of producing and using this type of energy will decrease day by day due to the advances made in technology and the necessary tools for its production.

1.3.8 CASE STUDY

- Renewable resource: running water
- Hydrogenerator/benefit and cost modeling.

1.3.9 METHODOLOGY

- Modeling.
- Network analysis.
- Pressure zone design.
- Installation of pressure relief valves.
- Installation of modem and pressure and flow logger.
- Remote reading of pressure and flow facilities.
- Intelligent management of pressure zones.

1.3.10 WATER FACILITIES SPECIFICATIONS

- City water supply from springs and wells.
- Water supply system (gravity and direct pumping).
- The number of the main reservoir and the distribution network.
- The working pressure of the main water supply lines.
- Heterogeneous topography of the city and pressure zone in the context of GIS.

1.3.11 NETWORK ANALYSIS

1.3.11.1 RESEARCH MATERIALS AND METHODS

In this research, for the establishment of GIS and rapid information circulation, firstly, the preparation of GIS Ready maps was carried out as follows:

- Exchange graphic information from CAD space to GIS space
- Fixed errors in the CAD space
- Convert graphic information from DWG format to SHP
- For the coordinate system of SHAPE files, the coordinate system of the original map was followed.
- CODEPAGE font type used in descriptive information table is UNICODE type.
- The operating system used to generate data is WINDOWSXP
- At the end of the project, point, line, and surface folders called POINT, LINE, AREA were created and maps related to each were placed in the relevant folder.
- Useful features were preserved and features specific to the CAD space were removed.
- Fixing descriptive and spatial errors in the GIS space.
- Eliminate overshoot errors
- Eliminate undershoot errors
- Eliminate multipart errors
- Eliminate duplicate errors
- Removal of SILVER-GAP errors
- Removal of PATTERN errors
- Removing the error of polygons not being closed
- Snap implementation of separated complications with proper tolerance.
- Removing toll errors that were in the wrong place.

- Create primary and foreign keys for the tax table.
- Establishing proper tolerance and exchange of complications from the topology space.
- Preparing a conceptual model for transmission line modeling in GIS space
- The ERD (Entity-Relationship Diagram) model is desired and in this model information is expressed by the three basic concepts of entity, attribute, and relationship (including the type and degree of relationship).
- Complete the characteristics of each complication.
- Preparation of a suitable name.
- Completing the fee tables based on the prepared forms.
- Create a proper land database.

The proposed ground database includes a database that is supported by ArcGIS and SQLSERVER and is programmable. Database features include the following:

- Creating domains for descriptions (Attribute domains) necessary to avoid entering false descriptions into the database (by giving description codes).
- Creating different categories (datasets) for complications for different management.
- Automatic updating of sizes and areas after each change (dimension update).
- Creating a spatial reference (spatial reference) for complications and also creating spatial domains for complications in order not to enter extraneous spatial data.
- Creation of relationship classes (relationship classes) between spatial and nonspatial data for the spatial relationship of complications so that each complication is related according to the type of relationship (simple or compound) and its degree (cardinality). With these classes, the leakage points of the transmission line can be reached.
- Creating subgroups for complications to create validation rules (Validate rules) and also prevent additional queries in the database.
- Creating an index for descriptive tables to increase the accuracy and speed of queries.
- Creating topology rules to prevent wrong edits, such as separation of equipment from pipes, such as separation of a valve from a pipe or a manhole from a pipe, etc.

- Creation of Annotation classes for immediate calling of descriptions in order to print them.
- Determining the equipment needed to solve the accident, by having available the type of valve or pipe involved in the accident.
- Creating a smart network of water transmission lines by introducing charges in the network (Networking).
- Create tracking capability and perform network analysis such as:
- Descriptive information about the location of the valves that must be closed to cut off the water in the affected area.
- Descriptive information about the location of pipes that become waterless due to closing a transmission line valve.
- Descriptive information about the location of common valves of pipes.
- Descriptive information about the location of the pipes downstream of the Manhol accident.
- Descriptive information about the location of the pipes above Manhole that have had an accident (caught pipes) and...
- Increasing the speed of dealing with incidents and reducing physical losses of water.
- Systematizing the storage and use of distribution network equipment information using GIS.
- Economic savings due to reducing the follow-up of events and...
- Creating maps and reporting in all the above cases.
- Combination of the above analyzes for optimal and immediate management of facilities.

In this research, GIS Ready maps were prepared in order to achieve the following goals:

- Estimation and evaluation of water distribution system.
- Water quality studies.
- Studies to prevent wastage of water and energy.
- Improvement and development of water distribution system.
- Connecting with nearby networks.
- Update existing facilities.

The process of choosing the model and the type of software to prepare the hydraulic model is as follows:

- Hydraulic analysis
- Water quality analysis
- Simulation features and specifications

- Graphic features
- Hardware requirements
- Software requirements
- Software capabilities
- Required data
- Cost

The information needed in the preparation of the hydraulic model of the water distribution system includes information about pipes, faucets, pumps, nodes, operating conditions, water consumption and needs, consumption fluctuations and water consumption patterns in the city, and loss coefficient. Therefore, quantitative information required for hydraulic modeling and analysis was prepared and studied under three branches as described below:[3-6]

- Network information
- Information related to exploitation
- Information about consumption
- Information about the network including pipes is as follows:
- Pipe code
- Flow direction in the pipe
- Node code of the beginning and end of the pipe
- Diameter
- Material
- The length
- Roughness coefficient
- Coefficient C
- Partial loss coefficients
- Working pressure and design pressure
- Node code

1.3.11.2 PRESSURE ZONE DESIGN

The water distribution network in this research is designed as a combination of gravity system and direct water pumping. The hydraulic analysis of the base location for the current state of the network indicates that the network should be designed in the form of independent pressure zones and independent consumption zones, which is done by closing the faucets and placing small volume tanks in each zone. Without taking into account separate areas of pressure and if the reservoirs with sufficient volume are watered, the network is in a stable state; therefore, the working pressure of the network has an upward trend, which causes an increase in the number of accidents and, as a result, large losses in the network, as well as an increase in the amount of water becomes without income. The water of the city is supplied through a series of wells that are dug around the river after the purification process is done in the refinery, because of the special topographical situation of the city, the output water of the refinery is pumped to the intermediate reservoir in the first stage and to the main reservoir of the city in the second stage. From the main tank with a higher height, the water enters the smaller tanks in pressure zones by gravity and then feeds the network. The large height difference of the reservoirs indicates the situation of static head and dynamics of the city's water distribution facilities. To design the pressure zone and the independent measurement zone of DMA, first, the network is divided into separate areas, then for the entrance and exit of each area, meter and control valve installed (Figures 1.30 and 1.31). If the flow rate of entry and exit is continuous in each size area be taken. After measuring for about a month flow diagram of input and output continuously in each size area is taken. After measuring the entry and exit of each area consumption graph was drawn at different hours for a month.

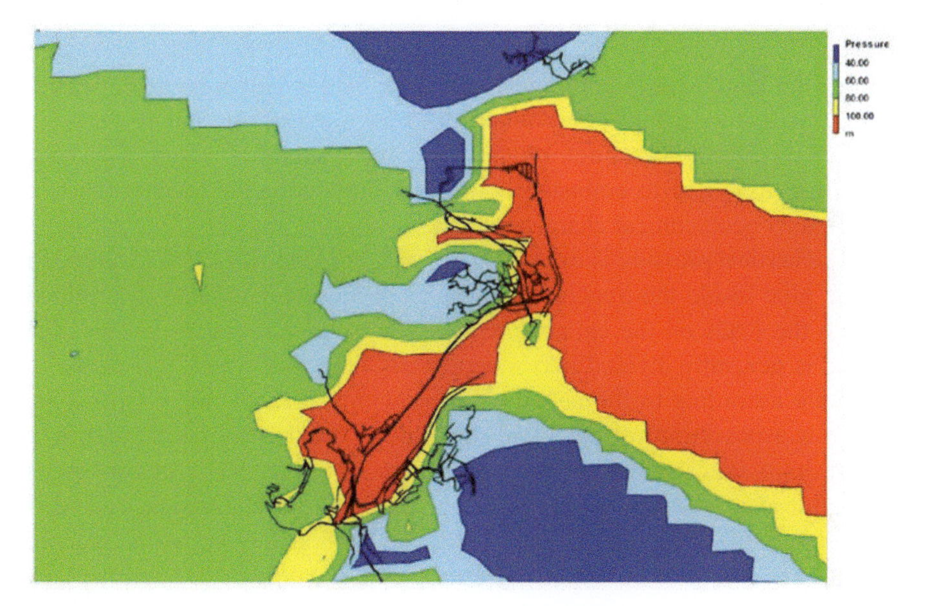

FIGURE 1.30 Pressure distribution at minimum consumption.

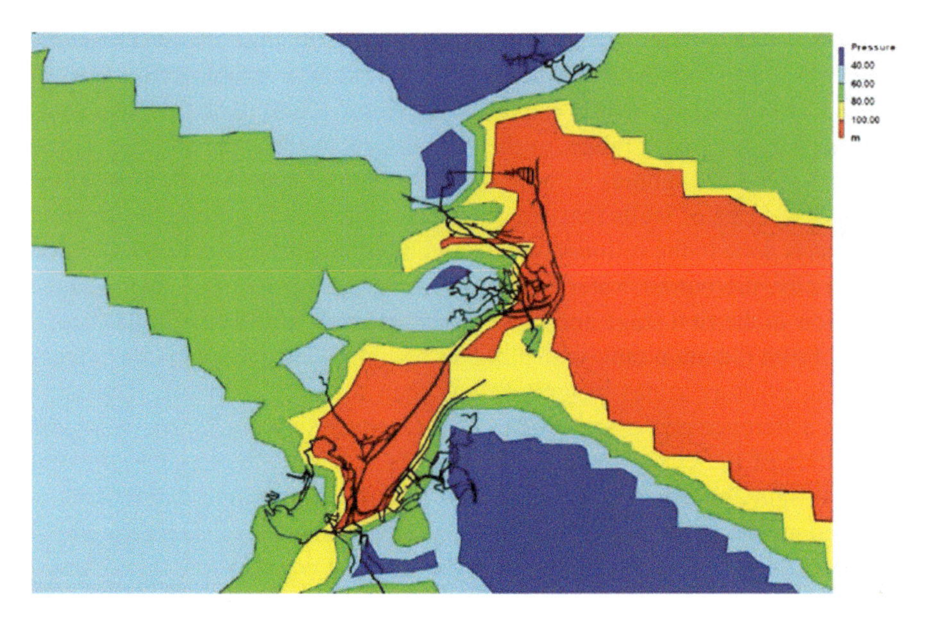

FIGURE 1.31 Pressure distribution at maximum consumption.

1.3.11.3 PRESSURE RELIEF VALVES

Several factors are considered in the design and selection of pipes. Among these, the high working pressure of fluid, water hammer, geotechnical characteristics of the soil bed trench can play a major role in choosing the type of pipe. In a lateral deformation pipeline, the gasket collapses sealing and ring stress due to embankment load is among the parameters that are affected by geotechnical characteristics.

The vertical pressure is caused by the weight of the soil in the installation of the pipe inside the trench. The pipe must be calculated and determined due to the loads on the embankment surface. The land where the pipeline is laid is generally divided into several parts. The land where is possible to excavate according to the height, digging with a vertical slope is possible. There is a possibility of cracks in the wall of the trench or the possibility of crushing after excavation can be imagined. In watery and muddy lands that after excavating, without immediate protection of the trench wall, it is impossible. In watery and muddy fields, the underlayment of pipes is more important. In case of failure to pay attention to the type of soil and the lack of appropriateness and coordination of the chosen pipe with those problems such as soil and pipe settlement and water, the band will include the tube exit.

The impact of the water hammer caused by the sudden changes in speed may cause major damages to the pipes and as a result heavy leaks happen. Pressure fluctuations in the pipes can damage the pipes and if the pipe is placed on a stone or any kind of hard and unsuitable substrate, it may burst. The impact of the ram causes the pipes to settle and the pipe gaskets to collapse. The compression of the pipe gasket creates an open space at the opposite point of the compression, which causes water leakage. The pipes are not sealed and water leaks from the joint of the two pipes (Figures 1.32 and 1.33).

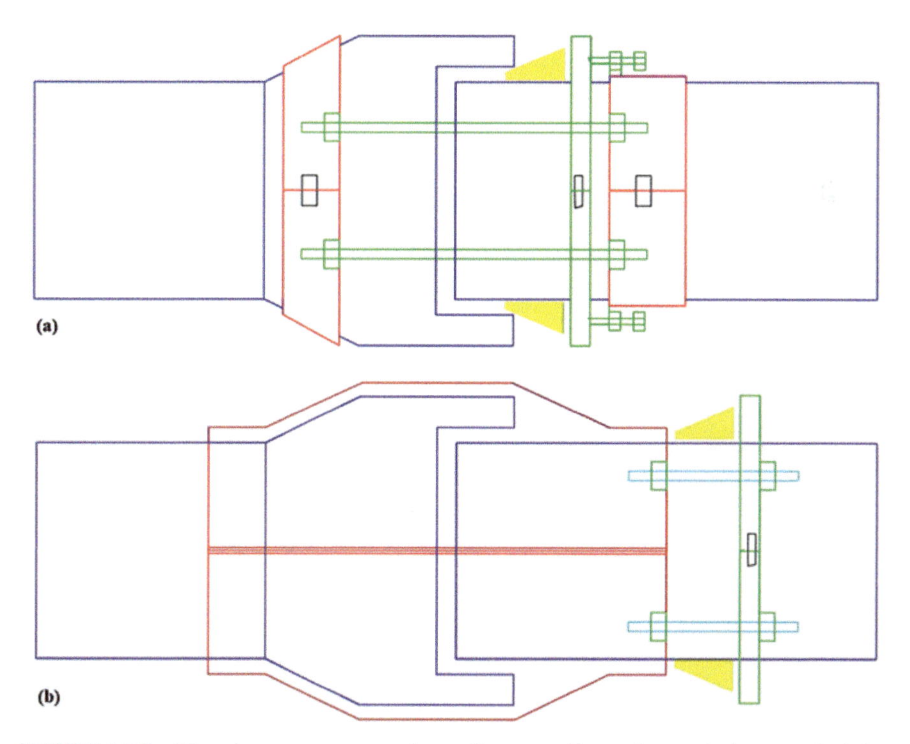

FIGURE 1.32 Water hammer causes gasket collapses sealing and water leakage on the joint of the two pipes: (a) normal gasket; (b) collapse gasket due to water hammer.

The phenomenon of water hammer is caused by a sudden change in the speed of water flow in the pipes. One of the most important causes of water hammer is high-pressure flow due to interruption in electric current to failure of the pump. If a pump fails intentionally or as an emergency, the amount of liquid flows into the pipeline. The water flow in pipe decreased rapidly, but due to the momentum of the liquid column, the fluid movement continues and causes creation of pressure reduction behind the pressure wave. This pressure may be reduced to the vapor pressure. The flow rate decreases after a while due to friction and static pressure of the system and reaches zero. So from this moment, flow happens in reverse direction and liquid flows toward the pump. At this moment, the one-way valve in the thrust line of the pump is closed, the pressure in the place of cavitation is increased and causes the liquid vapor to evaporate. Two liquid columns collide strongly with each other. This causes two columns of liquid to rise. The pressures rise very high and cause serious damage to the pipeline and pump station and related equipment. The impact of the water hammer happens instantaneously and as a result, the flow speed must be low enough to protect the system against damage. In pump stations, water hammer impact may occur for the following reasons (Figures 1.32 and 1.33).

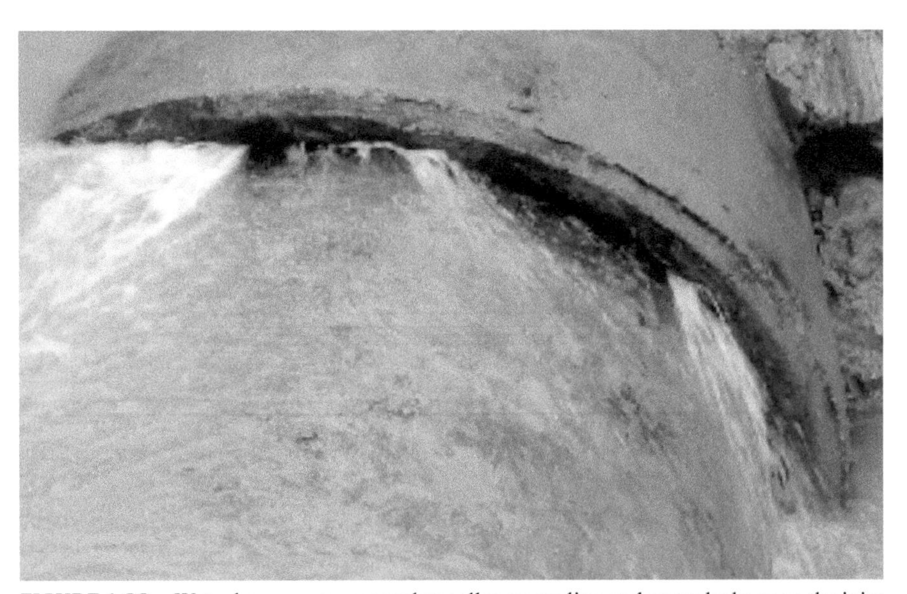

FIGURE 1.33 Water hammer causes gasket collapses sealing and water leakage on the joint of the two pipes.

- Changing the settings of the valves or sudden closing of the valve turning off one or more pump turning on one or more pumps.
- Apart from the last case, by taking measures such as choosing suitable one-way valves, closing or opening the valves at the pump outlet, sudden failure of one or more pumps, use of inappropriate one-way valves, improper filling of the pipeline, changing the rotational speed of the pump (in variable speed systems).

It is possible to reduce pressure changes to the minimum possible time in a completely calm manner. One of the reasons why it is strongly recommended that before to turn on a pump, the outlet valve should be closed and opened slowly after starting and also before turning off the pump. First, the outlet valve should be closed slowly and then the pump should be turned off, that is the issue. But in the case of sudden interruption, these measures will not be applicable and the phenomenon of battering water hammer will occur with great intensity and in these conditions surge tank as a systems protection can reduce its destructive effect to an acceptable level.

1.3.11.4 INSTALLATION OF SURGE TANK

Among the equipment to deal with ram impact, the pressurized shock absorber tank provides the most facilities against the phenomenon of ram impact. This device works both in the negative pressure phase and in the positive pressure phase, which means that it prevents the column from breaking. Water prevents and absorbs the increase in pressure. It is much more with other systems and also requires experienced operators to use.

This system consists of a reservoir, which is approximately half water and half compressed air when the pressure is reduced. The air expands and sends water into the pipeline, and in the phase of positive pressure, water flows from the pipeline to the tank. It re-compresses the air that has been expanded and thus prevents both the decrease and the unauthorized increase in pressure in the pipeline (Figure 1.34).

1.3.11.5 HYDRAULIC MODEL AND PRESSURE AND FLOW MONITORING

Based on the principles of designing water transmission systems, if in some cases because of the distribution network being integrated and also the reservoirs' feeding range being unclear, the network condition seems unfavorable by performing the calculations of the theoretical model; the only basic solution to solve the problems of the network is to build reservoirs with sufficient volume. Appropriate height and isolation of pressure and consumption areas are covered by these tanks. This operation was carried out for the present research in the form of a plan to improve and optimize the water transmission and distribution system. The base location model and intelligent pressure management were investigated after the modification and optimization of the water transmission and distribution system.

FIGURE 1.34 Installation of surge tank against damage of pipeline.

In the preparation of the network model, error sources include the following:

- The error related to the connection of pipes, the diameter and type of pipes, and the topology of the network in general.
- The error related to the network not being isolated.
- Existence of leakage in many areas, which causes mismatch between model results and measured parameters.
- An error caused by the presence of valves with an unknown status in the network, such as half or open valves, closed valves on some pipes of a ring.
- The existence of incidents and events in the network that cause errors in the measured parameters.

- Incompatibility of model calculations with field information.
- The error related to the topographical maps of the region, which causes the error in the number of network nodes and the calculation of the length of the pipes.
- The absence of an accurate map of subscribers (and as a result, the preparation of a model based on the connection of all subscribers to the network and considering the consumption pattern related to different uses such as residential, commercial and industrial, and considering the diameter of the main and secondary pipes and branches in the network has not been done).
- The errors related to the roughness coefficient of the pipe are created due to the lack of information about the age of the pipe and the type of pipes.

1.3.11.6 PRESSURE ZONE CHECK

By creating pressure zones in the network and installing pressure relief valves in appropriate geographical coordinates, pressure management can be applied to the network. In order to record flow and pressure parameters and in the pressure zones of the city, pressure and flow monitoring equipments are installed including pressure gauges with remote reading and ultrasonic flowmeter with remote reading.

1.3.11.7 REMOTE READING BAROMETER AND CHECKING PRESSURE ALIGNMENT LINES

The change of attitude from dependence on more production, which resulted in water losses without more income, toward reducing production in line with resource management and fair distribution with the aim of scientific supply and demand to increase income and reduce costs is one of the general results of this research. The operations performed are as follows:

- Investigating and analyzing the city's water supply system with a scientific point of view, creating a pressure zone with a policy of reducing water pressure in the framework of the water supply network's tolerance against the standards defined in the water supply system, and creating a water supply network loop and installing valves in all important routes.
- Setting up and using all the storage capacities in the network.

- Creating a ring in the network to balance pressure in low-pressure and high-pressure areas.
- Increasing income in the framework of increasing sales volume.
- Cost reduction by reducing water losses caused by unconventional pressure in the network.
  - Using a portable ultrasonic meter for flow measurement of production sources.
  - Balance of production and distribution, as well as analysis of weaknesses and strengths of the network.
  - Preventing water losses without income.
  - Repairing the floating valve of the tanks to control the water from the secondary tanks to the main tank of the city.

The positive effects of this research can be seen in the comparison of production and income increase with the same time of the previous year. It is clear that any research results in compliance with water supply knowledge and expertise have had very favorable results. The result is that replacing the damaged contour and putting contour on the branch without contour, before creating the water supply engineering platform (adjustment of pressure based on water supply knowledge) will cause harm. In the executive movement without scientific calculation, temporary income may be created, but sustainable income depends on following the scientific path. The results of the current scientific activities have led to a decrease in production and an increase in income.

1.3.11.8 INTELLIGENT MANAGEMENT OF PRESSURE ZONES

By using area analysis, pressure distribution analysis diagrams showed the high speed of flow in the network. This state of speed in the network increased the water retention time in the network and increased the possibility of corrosion in the network.

Considering the static head of the tanks, it should be said that the small local tanks act as a pressure breaker and if these tanks are removed from the circuit, the high pressure applied to the pipes will cause a lot of damage. The prepared model showed very high pressure in some areas of the network. The graphs of the statistical distribution of pressure in the network nodes are also indicative that during the time of minimum consumption in the network, most of the network nodes have high pressure. There is no appreciable difference in the pressure condition during the hours of maximum consumption.

The working pressure of high-pressure fluid in water transmission line has destructive effects for water piping systems. It is necessary to be removed in water transmission line. The high pressure in this work must be reduced by Pressure Reducing Valve (PRV) or in Pressure Relief Valve (PRV). As a renewable resource and advanced technology high-pressure fluid has the potential energy for handling of hydro-generator through hydro turbines, the energy loss in Pressure Reducing Valve (PRV) or in Pressure Relief Valve (PRV) can be saved.

1.3.11.9 REMOTE READING OF PRESSURE AND FLOW FACILITIES

According to the characteristics of the network, it can be concluded that a DMA is a separate area with several water supply locations. That is, a DMA is not connected to the adjacent DMA. The flow entering the DMA is measured by the meter and the water transfer from one DMA to the adjacent DMA is in the form of a cascade. The DMA designer first divides the network into separate areas. Then a meter and control valve are installed for the entry and exit of each area. Inlet and outlet flow should be measured continuously in each area. After a month of measuring the inlet and outlet and pressure in each area, the flow chart of inlet and outlet and pressure curves are drawn at different hours. In case of any malfunction in the network, the malfunction can be identified through changes in the drawn diagram. In DMA design, it is in the form of separate areas. The urban facilities can be controlled by remote reading of pressure and flow in isolated DMA and it prevents the disruption of network parts.

For quick intercommunication, information in all types of complications, classes and sub-classes related to water supply, storage, transmission and distribution facilities, creating a GIS-based location platform is required. Therefore, urban facilities can be guided and controlled in the best way by Networked Sensors, Remote Sensing (RS) and Internet of Things (IoT) and GIS (Figures 1.35 and 1.36).

1.3.11.10 INSTALLATION OF MODEM AND PRESSURE AND FLOW LOGGER

In order to intelligently control water losses, performing per capita water consumption calculations in the theoretical model is the only basic solution to reduce the amount of water without income in the existing network. Isolating

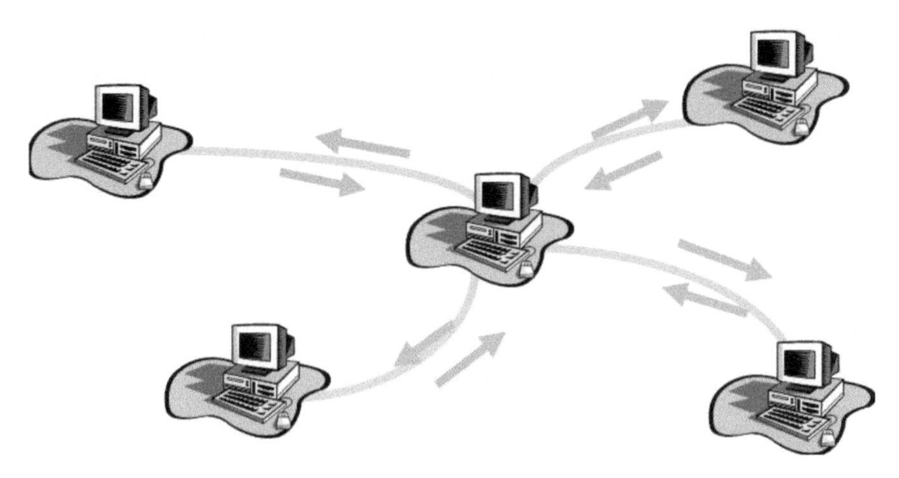

FIGURE 1.35 Internet of things (IoT) based on (GIS).

FIGURE 1.36 DMA control by networked sensors, remote sensing (RS).

the pressure and consumption areas covered by reservoirs in the network and determining the per capita consumption of water are important factors for the design of the water supply system. In the design of the network, it is necessary to determine the water consumption per capita as a first step. In order to control water losses in the distribution network, it is necessary to create separate pressure zones in the network. In order to create an isolated area, the water distribution network under the cover of each reservoir is divided into separate areas with smaller networks so that the water enters and leaves it through one or more specific and limited routes and from all its adjacent areas through closing the shut-off and connection valves are separated in a separate area. According to the definition, each isolated area has certain characteristics in terms of population, number of branches, number of subscribers, length of main and sub-network pipes, ratio of pipe length to branches, number and type of major consumers at night, infrastructure conditions, age of the network, and number of accidents. The installation of modems, pressure, and flow data loggers on the main pipelines and branches forms the intelligent management of pressure (Figure 1.37).

FIGURE 1.37 Modem and pressure and flow logger.

1.3.11.11 CALCULATIONS OF POWER AND PRODUCTION ENERGY OF HYDROGENATOR

$$Pe = \Delta Z.Q.\rho.g.\eta t.\eta e.\eta f.1/1000 \tag{1.34}$$

*ΔH = Elevatoin difference of highest reservoirs (m)
*ΔZ = Elevatoin difference of lowest reservoirs (m)
Q = Pump station discharge (m^3/sec)
ρ = Water density (1000) (kg/m^3)
g = Acceleration of gravity 9.81 (m/sec^2)
ηt = Turbin efficiency (0.73)
ηe = Hydrogenator efficiency (0.78)
ηf = Transformer efficiency (0.93)
Pe = Power (KW)

Note: The 110 meters of the total 210 meters difference height between the highest number of the network and the lowest number of the network ΔH can be used to produce electric energy by the hydrogenator installed on the main water supply lines (Tables 1.32–1.34) (Figures 1.38–1.49).

$$Ee = Pe(KW)8760(hrs/Year).\ Ra \tag{1.35}$$

Pe = power (kW)
Ra = a viability ratio (up time/total time)
Ee = energy (KWh)
Ee = energy (KWh/year)

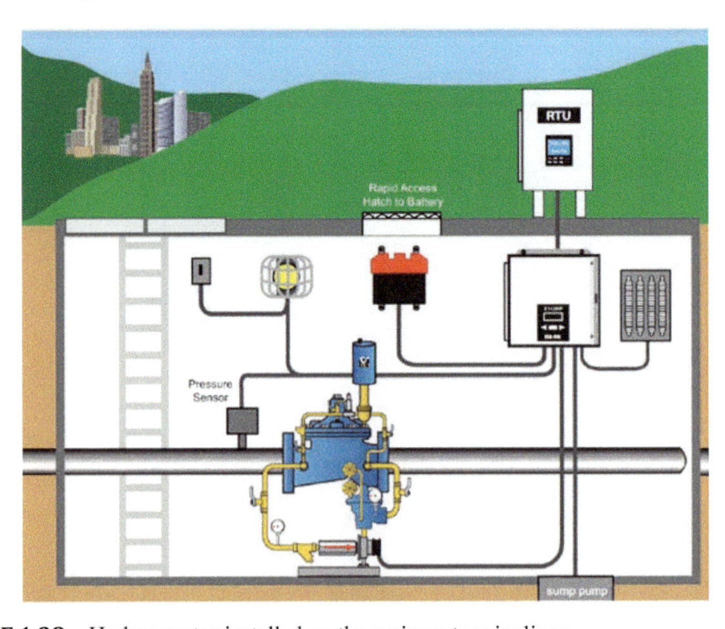

FIGURE 1.38 Hydrogenator installed on the main water pipelines.

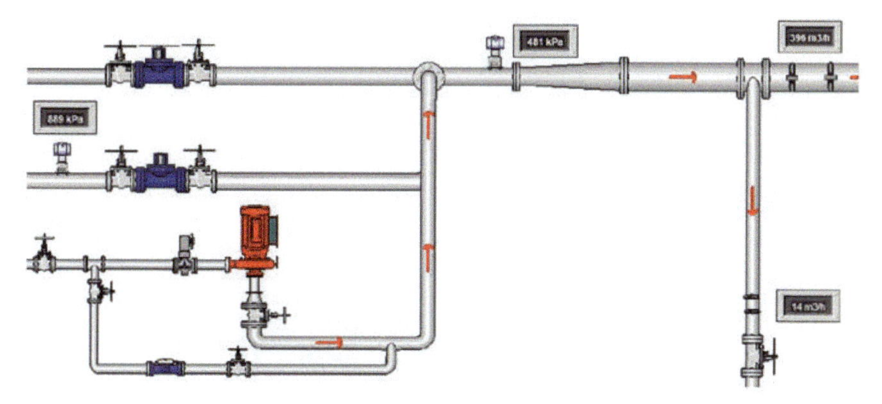

FIGURE 1.39 Working specifications of the hydrogenerator of pump stations.

FIGURE 1.40 Installation of hydrogenerators in pump stations of water supply.

FIGURE 1.41 Installation of hydrogenerators in pump stations of water supply facilities.

FIGURE 1.42 Installation of hydrogenerators in pump stations of water supply facilities.

FIGURE 1.43 Installation of hydrogenerators in pump stations of water supply facilities.

FIGURE 1.44 Installation of hydrogenerator on the main water supply lines.

TABLE 1.32 Differential Pressure-Flow Data of Hydrogenator for Water Pipelines.

Flow (watts)	Differential pressure (bar)
6	2
9	4
13	8
13.1	9
13.1	10
13.5	13
13.6	14

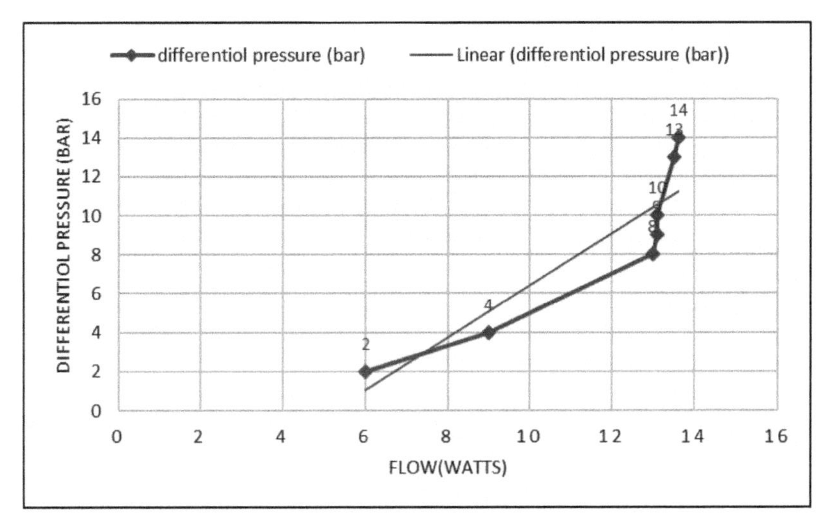

FIGURE 1.45 The pressure–flow relationship of the main water supply lines in the hydrogenator.

TABLE 1.33 Power(watts)-Differential Pressure (bar) Data of Hydrogenator for Water Pipelines.

Power (watts)	Differential pressure (bar)
22	25
110	40
130	44
170	50
220	60

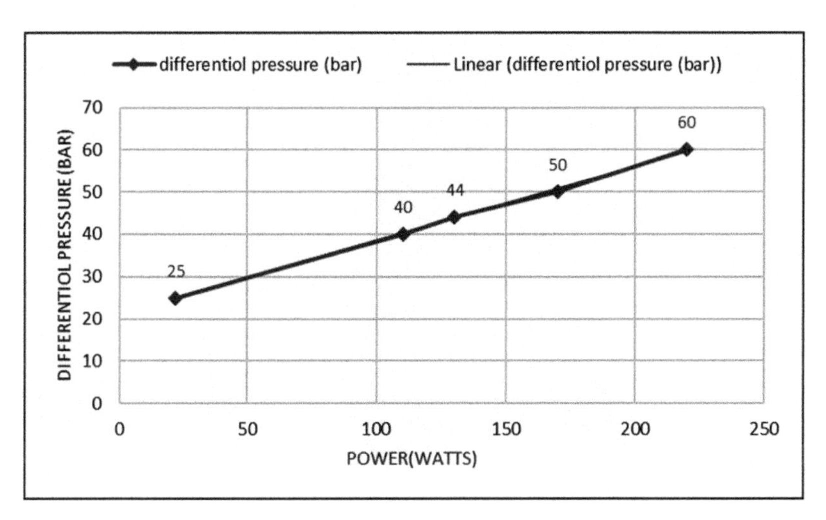

FIGURE 1.46 Pressure–power relationship in the main water supply lines in the hydrogenator.

TABLE 1.34 Differential Pressure-Flow Data of Hydrogenator for Water Pipelines.

Flow (watts)	Differential pressure (bar)
0	25
17	32
19.5	40
22.2	50
24	60

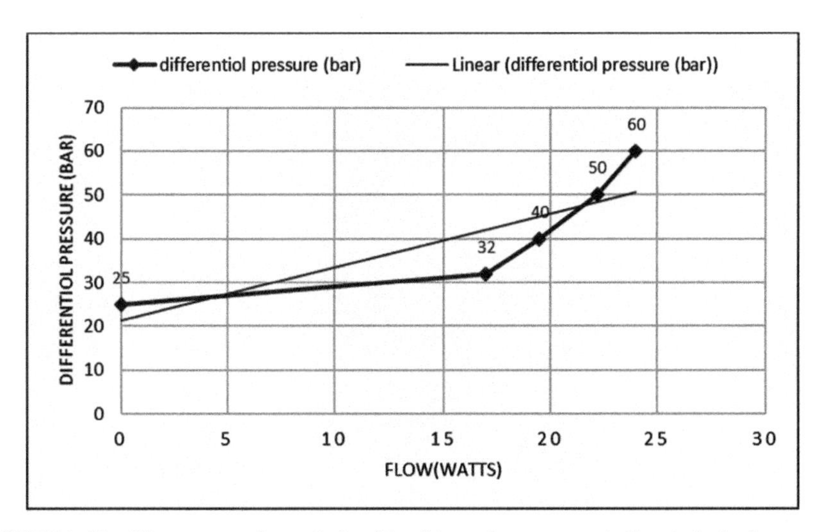

FIGURE 1.47 The pressure–flow relationship of the main water supply lines in the hydrogenator.

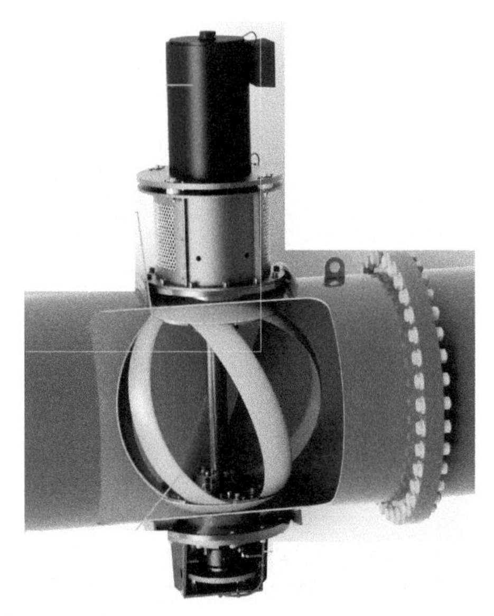

FIGURE 1.48 Installation of hydrogenerator on the main water supply lines.

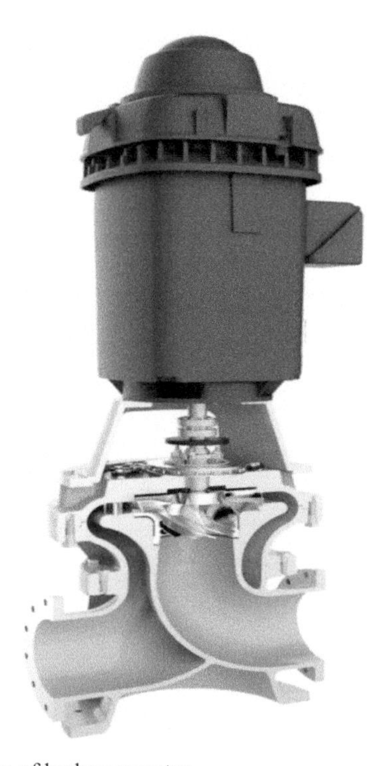

FIGURE 1.49 Installation of hydrogenerator.

Working specifications of the hydrogenator

- Easily installable.
- Mechanical technology.
- No live electricity needed.
- Three-phase AC generator.
- No external power source needed.

Pipe diameters:

HG300 – 3″ (80 mm)

HG400 – 4″ (100 mm)

HG600 – 6″ (150 mm)

HG800 – 8″ (200 mm)

1.3.11.12 COST-BENEFIT OF PRODUCING ELECTRIC ENERGY FROM WATER SUPPLY LINES

- One-year investment return period.
- Reduction of accidents and wastage of city water.
- Installation of a 50 kW hydrogenator in the water supply line.
- The cost of preparing and installing a hydrogenator.
- Elimination of the cost of electricity consumed in the annual production of city water.
- Exchange of electric energy with the electricity distribution company during the peak load of the network.
- The average value of electricity produced by a hydrogenator is 50 kilowatts.

1.3.11.13 RESULTS OF HYDROGENERATOR

The results of the present research emphasized the following actions:

- The necessity of remote reading of pressure and flow parameters in pressure zones.
- Using intelligent pressure management technology in GIS platform.
- Completing the operation of isolation of pressure zones.
- Network modification and branch standardization.
- Pressure zone management in GIS platform.

- Reduction of water incidents.
- Free electrical energy production.
- The final conclusion.
- Management of water consumption and demand.
- Identification of large and high consumption unauthorized branches.
- Identifying the location of water harvesting for green spaces and separating green space irrigation systems from drinking water.
- In case of not separating irrigation systems of green areas from drinking water, irrigation should be done during minimum consumption hours.
- Identifying subscribers with average consumption higher than half a liter per second and negotiating with them to shift their consumption to minimum consumption hours.
- Reservoirs should be used as a tool to control and move the network peak.
- Avoid water cuts and rationing as much as possible and keep the network condition stable by reducing network pressure.
- Network pressure management by tanks, pressure relief valves, and flow control valves and installation of logger pressure gauges and zoning.
- Installation of devices to improve consumption in public places.
- Process costs such as the costs of treatment plant units, network washing, filters should be controlled.
- By carrying out cultural measures, people will be sensitized to correct consumption, not taking unauthorized branches, reporting fractures and wastage of water.
- Installing a meter at the entrance of the network to find the leak that can be removed and take the necessary measures to reduce it.
- Installing a timer in the control circuit of the electric pump of wells and pumping stations to prevent overflow of tanks.
- Intelligent management on the pressure relief valves in order to manage the network pressure at the time of minimum consumption.

1.4 CONCLUSION

Currently, the building industry accounts for the largest amount of energy consumption, and heating and cooling systems are the largest energy consumers in the building. Therefore, there is a need to improve energy efficiency or optimize energy consumption. In the present work, the role of Networked Sensors of HVAC&R facilities and energy audit of buildings was investigated. In this regard, consumption patterns in energy were determined.

The research method in this work was documentary-analytical and the type of research was based on the presented analytical-practical solutions. Based on the findings and results of the work, the use of remote sensing technology and the necessity of applying GIS in the management of various industries such as thermal, refrigeration, and air conditioning industry for energy audit and optimization of energy consumption pattern were emphasized. Therefore, the achievements of utilizing the remote sensing of HVAC&R facilities and the energy audit of buildings in the present research for the facility industry are as follows:

1. Scientific management by analyzing the received data on a variety of hydraulic and thermodynamic parameters by remote reading in the facility systems and creating the following capabilities:
   - Hydraulic analysis
   - Simulation properties and specifications
   - Graphic properties
   - Hardware requirements
   - Software requirements
   - Software capabilities
   - Required data
   - Cost
2. By identifying any qualitative and quantitative changes in the facility setup in least possible time, it is possible to analyze the related data at the system outlet.
3. Develop the ability to cope with a variety of hydraulic and thermodynamic instability factors at the facility.
4. By managing HVAC&R ON-LINE Thermal, Refrigeration and Air Conditioning facilities while utilizing remote read and fast flow information technology in accordance with GIS.
5. Scientific leadership and upgrading of HVAC&R installation systems' technical and health safety coefficient and the following capabilities:
   - Eliminating unwanted usage.
   - Fixing and controlling HVAC&R thermal, refrigeration, and air conditioning facilities at different times of the day.
   - Reducing depreciation and increasing efficiency of HVAC&R thermal, refrigeration, and air conditioning installations.
   - Alerting alarms for periodic equipment reviews.
   - Controling the number of HVAC&R thermal, refrigeration, and air conditioning equipment such as in-service burners tailored to the building's thermal load demand and so on.

- Deactivating HVAC&R thermal, refrigeration, and air conditioning installations in office buildings in accordance with work schedules or in accordance with outdoor temperatures.
- Ability to remotely control and monitor the status of HVAC&R thermal, refrigerating, and air conditioning installations.
- Alarm system and alarm and event logging.
- Accurate statistical reporting of the performance of various components of the HVAC&R building's thermal, refrigeration, and air conditioning installations.
- Cleverly prioritizing emergency and peak consumption.

KEYWORDS

- **networked sensors**
- **geographic information system**
- **internet of things**
- **HVAC&R**
- **energy loss**
- **hydrogenerator**

REFERENCE

1. Kevin, A. Internet of Things. *RFID J.* **2009**.
2. Parker, D. S.; Merrill, M. S. Oxygen and Air Activated Sludge: Another View. *J. Water Pollut. Control Fed.* **1976,** *48* (11), 2511–2528.
3. Hariri Asli, K., Nonlinear Dynamics Modeling by Geospatial Information System (GIS), *Mapp. Geospatial Inform. J. Guilan* **2019,** *3* (1), Iran, www.mpogl.ir
4. https://en.wikipedia.org/wiki/Artificial_intelligence
5. Saraei, N.; Khanal, M.; Tizghadam, M. Removing Acidic Yellow Dye from Wastewater Using Moringa Peregrina. *Transac. Civil Environ. Eng.* **2022,** *8* (3) 2483, 1–8.
6. https://fa.wikipedia.org/wiki/ Geographic Information System
7. Wylie, E. B.; Streeter, V. L. *Fluid Transients*, Feb Press: Ann Arbor, MI, 1983; pp 166–171, corrected copy 1982.
8. Lee, T. S.; Pejovic, S. Air Influence on Similarity of Hydraulic Transients and Vibrations. *ASME J. Fluid Eng.* **1996,** *4,* 706–709.
9. Yoon, S. H.; Park, C. S. Objective Building Energy Performance Benchmarking Using Data Envelopment Analysis and Monte Carlo Sampling. *Sustainability* **2017,** *9,* 780. DOI: 10.3390/su9050780.

10. Hariri Asli, H.; Nazari, S. Water Age and Leakage in Reservoirs; Some Computational Aspects and Practical Hints. *Larhyss J.* **2021,** *48,* 151–167.
11. Hariri Asli, H.; Hozori, A. Non-Revenue Water (NRW) and 3d Hierarchical Model for Landslide. *Larhyss J.* **2021,** *48,* 189–210.
12. Kensek, K. *Buildings* **2015,** *5,* 899–916. DOI: 10.3390 /buildings5030899
13. Wang, S. K. *Handbook of Air-Conditioning and Refrigeration*; McGraw-Hill, Inc., 1994. ISBN 0-07-068138-4.
14. https://fa.wikipedia.org/wiki/ Internet of Things _IoT
15. Ferras, D.; Manso, A. P. A.; Schleiss, A. J. Covas, D. I. C. One-Dimensional Fluid-structure Interaction Models in Pressurized Fluid-filled Pipes: A Review. *Appl. Sci.* **2018,** *8,* 1844–1856.
16. Vardy, A. E.; Brown, J. M. B.; He, S.; Ariyaratne, C.; Gorji, S. Applicability of Frozen-Viscosity Models of Unsteady Wall Shear Stress. *J. Hydraul. Eng,* **2015,** *141,* 1–13.
17. Asli, K. H.; Aliyev, S. A. O.; Thomas, S.; Deepu, A. *Handbook of Research for Fluid and Solid Mechanics: Theory, Simulation, and Experiment*, 1st ed.; Apple Academic Press, 2017. DOI: 10.1201/9781315365701
18. Asli, K. H.; Naghiyev, F. B. O.; Haghi, R. K.; Asli, H. H. *Advances in Control and Automation of Water Systems*, 1st ed.; Apple Academic Press, 2012. DOI: 10.1201/b13114
19. Lema, M.; Peña, F. L.; Buchlin, J. M.; Rambaud, P.; Steelant, J. Analysis of Fluid Hammer Occurrence with Phase Change and Column Separation due to Fast Valve Opening by Means of Flow Visualization. *Exp. Therm. Fluid Sci.* **2016,** *79,* 143–153.
20. Zecchin, A. C.; Maier, H. R.; Simpson, A. R.; Leonard, M.; Nixon, J. B. Parametric Study for an Ant Algorithm Applied to Water Distribution System Optimization. *IEEE Trans. Evol. Comput.* **2005,** *9,* 175–191.
21. Andrade, D. M.; de Freitas Rachid, F. B. A Versatile Friction Model for Newtonian Liquids Flowing Under Unsteady Regimes in Pipes. *Meccanica* **2022,** *57,* 43–72.
22. Sarkardeh, H.; Zarrati, A. R.; Jabbari, E.; Marosi, M. Numerical Simulation and Analysis of Flow in a Reservoir in the Presence of Vortex. *J. Eng. Appl. Comput. Fluid Mech.* **2014,** *8* (4), 598–608.
23. Zhao, L.; Yang, Y.; Wang, T.; Han, W.; Wu, R.; Wang, P.; Wang, Q.; Zhou, L. An Experimental Study on the Water Hammer with Cavity Collapse under Multiple Interruptions. *Water* **2020,** *12,* 2566.
24. Duan, H. F.; Pan, B.; Wang, M.; Chen, L.; Zheng, F.; Zhang, Y. State-of-the-Art Review on the Transient Flow Modeling and Utilization for Urban Water Supply System (UWSS) Management. *J. Water Supply Res. Trans.* **2020,** *69,* 858–893.
25. Zhou, L.; Liu, D.; Karney, B. Investigation on Hydraulic Transients of Two Entrapped Air Pockets in a Water Pipeline. *J. Hydraul. Eng.* **2013.** DOI: 10.1061/(ASCE)HY.1943-7900.0000750
26. Möller, G.; Detert, M.; Boes, R. M. Vortex-Induced Air Entrainment Rates at Intakes. *J. Hydraul. Eng.* **2015,** DOI: 10.1061/(ASCE)HY.1943-7900.0001036
27. Hariri Asli, K.; Nagiyev, F. B.; Haghi, A. K. Interpenetration of Two Fluids at Parallel Between Plates and Turbulent Moving in Pipe. In *Computational Methods in Applied Science and Engineering*; (Nova Science publisher: USA, 2009; Chapter 7, pp 115–128. https://www.novapublishers.com/catalog/product_ info.php?products_id=10681
28. Hariri Asli, K.; Nagiyev, F. B.; Haghi, A. K. Some Aspects of Physical and Numerical Modeling of water hammer in pipelines. *Int. J. Nonlin. Dynam. Chaos Eng. Syst.* **2009.**

ISSN: 0924-090X (print version) ISSN: 1573-269X (electronic version) Journal no. 11071 Springer: USA, http://nody.edmgr.com/

29. Hariri Asli, H.; Arabani, M.; Golpour Y. Reclaimed Asphalt Pavement (RAP) based on a Geospatial Information System (GIS). *Slovak J. Civil Eng.* **2020,** *28* (2), 36–42.
30. Hariri Asli, H. Investigation of the Factors Affecting Pedestrian Accidents in Urban Roundabouts. *Transac. Civil Environ. Eng.* **2022,** *8*, 2226, 1–4, Special Issue: NCTT **2021**. DOI: 10.52547/crpase.8.1.2255
31. Hariri Asli, H.; Arabani, M. Analysis of Strain and Failure of Asphalt Pavement. *Comput. Res. Prog. Appl. Sci. Eng. Transac. Civil Environ. Eng.* **2022,** *8,* 1–11, 2250.
32. Lee, T. S.; Pejovic, S. Air Influence on Similarity of Hydraulic Transients and Vibrations. *ASME J. Fluid Eng.* **1996,** *118* .(4), 706–709.
33. Monzavi, M. T. *Wastewater Treatment*; Tehran, Iran, 1991; pp 7–162.
34. Filion, Y.; Karney, B. W. In *A Numerical Exploration of Transient Decay Mechanisms in Water Distribution Systems*, Proceedings of the ASCE Environmental Water Resources Institute Conference, American society of civil engineers: Roanoke, Virginia, 2002.
35. Shirzad, S.; Hassan, M. M.; Aguirre, M. A.; Mohammad, L. N.; Daly, W. H. Evaluation of Sunflower Oil as a Rejuvenator and Its Microencapsulation as a Healing Agent. *J. Mater. Civil Eng.* **2016,** *28* (11), 04016116.
36. Chen, M.; Leng, B.; Wu, S.; Sang, Y. Physical, Chemical, and Rheological Properties of Waste Edible Vegetable Oil Rejuvenated Asphalt Binders. *J. Construct. Build. Mater.* **2014,** *66* (15 Sep), 286–298.
37. Fini, E. H.; Kalberer, E. W.; Shahbazi, A.; Basti, M.; You, Z.; Ozer, H. Chemical Characterization of Bio-Binder from Swine Manure: SUSTAINABLE MODIFIER for Asphalt Binder. *J. Mater. Civil Eng.* **2011,** *23* (11), 1506–1513.
38. Hassan, M.; Essam, T.; Yassin, A. S.; Salama, V. Optimization of Rhamnolipid Production by Biodegrading Bacterial Isolates Using Plackett–Burman Design. *Int. J. Biol. Macromol.* **2016,** *82* (4), 573–579.
39. https://en.wikipedia.org/wiki/Computational_physics
40. NZS 3106, *Design of Concrete Structures for the Storage of Liquids, Cement and Concrete Association of New Zealand*; Standards Association of New Zealand, 2009.
41. *Recommended Standards for Water Works*, Great Lakes-Upper Mississippi River Board of State and Provincial Public Health and Environmental Managers, Health Research Inc.; Health Education Services Division, POBox 7126: Albany, NY 12224, 2012; 102-SA Water TG
42. Planning and Infrastructure, Above Ground Circular Concrete Tank, Manager Engineering, 2008. 103-Selected Earthquake Engineering Papers of George W. Housner, American Society of Civil Engineers, 1990.
43. *Specification for Construction of Concrete Reservoirs, Standard Specifications and Drawings Specification for Construction of Concrete Reservoirs*, Gold Coast, Australia, SS 11, 2004.
44. TM 5-809-10, *Seismic Design Guidelines for Essential Buildings*; US Department of the ARMY: Washington, D.C., 1986.
45. *USEPA Finished Water Storage Facilities*, U.S. Environmental Protection Agency, 2002.
46. Zecchin, A. C.; Maier, H. R.; Simpson, A. R.; Leonard, M.; Nixon, J. B. Parametric Study for an Ant Algorithm Applied to Water Distribution System Optimization. *IEEE Trans. Evol. Comput.* **2005,** *9* (2), 175–191.

47. Walski, T. M. Hydraulic Design of Water Distribution Storage Tanks. In *Water Distribution Systems Handbook*; Mays, L. W., Ed.; McGraw-Hill: New York, 2000; Chapter 10.
48. Farmani, R.; Walters, G.; Savic, D. Evolutionary Multi-objective Optimization of the Design and Operation of Water Distribution Network: Total Cost vs. Reliability vs. Water Quality. *J. Hydroinform.* **2006,** *8* (3), 165–179.
49. Farmani, R.; Savic, D. A.; Walters, G. A. The Simultaneous Multi-Objective Optimization of Anytown Pipe Rehabilitation, Tank Sizing, Tank Siting and Pump Operation Schedules. In *Critical Transitions in Water and Environmental Resources Management*; ASCE, American Society of Civil Engineers, Salt Lake City, 2004; pp 4663–4672.
50. Haarhoff, J.; Van Zyl, J. E.; Nel, D.; Van der Walt, J. J. Water Supply and Water Quality in Rural Areas: Stochastic Analysis of Rural Water Supply Systems. *Water Supply*. **2000,** *18* (1), 467–470.
51. Maier, H. R.; Simpson, A. R.; Zecchin, A. C.; Foong, W. K.; Phang, K. Y.; Seah, H. Y.; Tan, C. L., In *Ant Colony Optimization for the Design of Water Distribution Systems*, Proceedings of World Water and Environmental Resources Congress 2001, ASCE, The Rosen Plaza Hotel: Orlando, Florida, United States, 2001.
52. Maier, H. R.; Simpson, A. R.; Zecchin, A. C.; Foong, W. K.; Phang, K. Y.; Seah, H. Y.; Tan, C. L. Ant Colony Optimization for Design of Water Distribution Systems. *J. Water Resour. Plann. Manag.* **2003,** *129* (3), 200–209.
53. Walski, T. M.; Chase, D. V.; Savic, D. A.; Grayman, W.; Beckwith, S.; Koelle, E. *Advanced Water Distribution Modeling and Management*; Watertown, Conn: Haestad Press, 2003.
54. Hashemi, S. S. Optimization of Water Networks with Minimization of Pumping Energy. Master Thesis, University of Tehran, Iran, 2010.
55. Mau, R. E.; Boulos, P. F.; Clark, R. M.; Grayman, W. M.; Tekippe, R. J.; Trusell, R. R. Explicit Mathematical Models of Distribution Storage Water Quality. *ASCE J. Hydraul. Eng.* **1995,** *121* (10), 669–709.
56. Rossmann, L. A. *EPANET 2 User's Manual. Water Supply and Water Resources Divison*; U.S. Environmental Protection Agency: Cincinnati, OH. Available at: http://www.epa.gov/ORD/NRMRL/wswrd/epanet.html
57. Rossmann, L. A.; Grayman, W. M. Scale-Model Studies of Mixing in Drinking Water Storage Tanks. *J. Environ. Eng.* **2000,** *125* (8), 755–761.
58. Yeung, H. Modeling of Service Reservoirs. *J. Hydroinformatics* **2001,** *3* (3), 165–172.
59. Martínez-Solano, F. J.; Lopez-Jimenez, P. A.; Iglesias-Rey, P.; Perez, R. Utilization of fluid dynamics co, 2006.
60. Mahmood, F.; Pimblett, J. G.; Grace, N. O.; Grayman, W. M. Evaluation of Water Mixing Characteristics in Distribution System Storage Tanks. *J. Am. Water Works Assoc.* **2007,** *97* (3), 74–88.
61. Marek, M.; Stoesser, T.; Roberts, P. J. W.; Weitbrecht, V.; Jirka, G. H. *CFD Modeling of Turbulent Jet Mixing in a Water Storage Tank*; XXXII IAHR Congress: Venice (Italy), July 1/6, 2007.
62. Van Zyl, J. E.; Haarhoff, J. Reliability Analysis of Municipal Storage Reservoirs Using Stochastic Analysis. *J. S. Afr. Inst. Civil Eng.* **2007,** *49* (3), 27–32.
63. Arturo Leon, S. Improved Modeling of Unsteady Free Surface, Pressurized and Mixed Flows in Storm-Sewer Systems. *Submitted in Partial Fulfillment of the Requirements*

for the Degree of Doctor of Philosophy in Civil Engineering in the Graduate College of the University of Illinois at Urbana-Champaign, 2007; pp 57–58.
64. Hariri Asli, K.; Nagiyev, F. B.; Haghi. A. K. Physical Modeling of Fluid Movement in Pipelines. In *Nanomaterials Yearbook*, USA, 2009. https://www.novapublishers.com/catalog/product_info.php?products_id=11587
65. Leon A. S.; Ghidaoui, M. S.; Schmidt, A. R.; Garcia, M. H. Godunov Type Solutions for Transient Flows in Sewers. **2005,** 20–44.
66. Asli, K. H.; Aliyev, S. A. O.; Thomas, S.; Deepu, A. *Handbook of Research for Fluid and Solid Mechanics: Theory, Simulation, and Experiment*, 1st ed.; Apple Academic Press, 2017. DOI: 10.1201/9781315365701
67. Asli, K. H.; Naghiyev, F. B. O.; Haghi, R. K.;Asli, H. H. *Advances in Control and Automation of Water Systems*, 1st ed.; Apple Academic Press, 2012. DOI: 10.1201/b13114

CHAPTER 2

Energy Applications of Metal and Metal Oxide Nanoparticles

ABHIJITH SHARMA[1], DIVYA NERAVATHU GOPI[2], SAJU M. MOHAMMED[2], and RIJU K. THOMAS[2]

[1]*Cochin University of Science and Technology, South Kalamassery, Cochin, India*

[2]*Bharata Mata College, Thrikkakara, Edappally, Cochin, Kerala, India*

ABSTRACT

Nanotechnology is offering true elucidations to ensure supportable energy for the upcoming energy requirements. Due to the unique structural properties of metal and metal oxide nanomaterials, they are regarded as the most promising nanomaterials in the field of nanotechnology, especially for energy applications. To invoke the complete potential of nanomaterials for diverse applications, different nanostructures, such as nanorods, nanofibers, nanotubes, and so on have been synthesized and investigated. As a result, these nanostructures are identified as the most promising morphologies for energy storage device applications and energy conversion applications due to their eminent properties, such as large surface areas, efficient charge transport kinetics, photoconductivity, optical, thermoelectric, electronic, and catalytic properties, and porosities. Moreover, these materials can be produced on large scale for industrial applications by employing both top-down and bottom-up synthesis procedures that also permit the deposition of these nanostructures on suitable substrates which makes them superior in industrial applications. The devices such as photovoltaic cells and lithium–ion batteries constructed on metal oxide nanofibers exhibit good power

Technological Advancement in Clean Energy Production: Constraints and Solutions for Energy and Electricity Development. Amritanshu Shukla, Kian Hariri Asli, Neha Kanwar Rawat, Ann Rose Abraham, & A. K. Haghi (Eds.)

conversion efficiency, better reversible ability, and electrochemical stability. In this chapter, the possibilities of metal and metal oxide nanomaterials for energy applications are elaborated.

2.1 INTRODUCTION

Metal oxide nanoparticles, are a growing field of interest worldwide in material science. They manifest excellent properties and are of immense application in several areas, including catalysis, sensors, optoelectronic materials, fertilizers, engineering, pharmaceuticals. There are numerous methods for preparing nano metal oxides, such as sol–gel synthesis, microwave synthesis, and CVD. Scientists worldwide are interested in nanomaterials because of their numerous practical applications.[3] They also have improved physical and chemical properties. Nanotechnology is the technology that makes synthesis possible.[7] Richard P. Feynman is considered the father of nanotechnology due to his speech in 1959 entitled "There is plenty of room at the bottom." Norio Taniguchi first used the term nanotechnology in 1974. Successful application of metal oxide nanoparticles requires controlled synthesis, and solution-phase approaches offer a high degree of control over the synthesis products. In simple terms, nanotechnology mainly consists of the processing of separation, consolidation, and deformation of materials by one atom or one molecule. The technology aims to reduce the size and increase materials' efficiency.

The particles produced by nanotechnology range between 1 and 100 nm (1 nm = 10^{-9} m) in each spatial dimension.[22] The technology proved that by making smaller, faster, lighter, and cheaper devices with greater functionality while using less raw materials and consuming less energy, thereby achieving sustainable usage of materials and energy. Nanoscale particles can have different physicochemical properties concerning micro scale or macro scale particles of the same material.[9]

2.2 SYNTHESIS METHODS

There are a large number of effective and potential methods for the synthesis of nanoparticles.[18] Various forms, such as colloids, clusters, powder, rods, wires, and thin films can be synthesized using these methods. All these syntheses can be categorized into two approaches, "top-down" and "bottom-up." The schematic representation of both these approaches for the synthesis of nanomaterials is illustrated in Figure 2.1. The two can be contrasted in that the bottom-up approach draws its inspiration from scientific research, including nanoscience, but the top-down approach does not. The bottom-up approach involves creating nanomaterials and objects within the same nanosphere based on atoms, molecules, and aggregating grouping. This method allows an increased functionality of structure of such materials. While the top-down approach, which is sourced from microelectronics, has to do with breaking down systems in their present state by making technologies more efficient. This reduces the size of the devices in nanoscale aspects. Regarding the size range of objects, both techniques frequently converge. The former method, however, tends to be more abundant in terms of material type, design variety, and nanometric control. In contrast, the latter method emphasizes the importance of material acquisition with potentially weaker power. The top-down approach is beneficial in producing technological structures and for connecting macroscopic devices; the bottom-up is suitable for the arrangement and synthesis of short range order at nanoscales. The optimal equipment integration for nano-based fabrication is anticipated to result from combining the two approaches. The standard form of the top-down approach is the lithographic technique that makes use of intensified visual sources of a short wavelength. One key advantage of using the top-down approach to design joint circuits is that all pieces are built and organized in a way that eliminates the need for additional assembly.

Using these shorter wavelengths such as UV and X-ray, nanoparticles of the range 10–100 nanometers can be synthesized (electron lithography can goup to 20 nanometers). While nanoscale imprinting, stamping, and molding techniques produce nanoparticles ranging from 20 to 40 nanometers.

2.3 TOP-DOWN APPROACHES

2.3.1 SPUTTERING

One of the most widely utilized synthesis methods involves the deposition of nanoparticles as a thin layer over the substrate produced by ion collisions, followed by annealing. This procedure is known as the physical vapor deposition (PVD) procedure.[5] Sputtering efficiency depends on layer thickness, annealing duration, substrate type, and temperature, which directly affect the nanoparticle size and shape.[13,19]

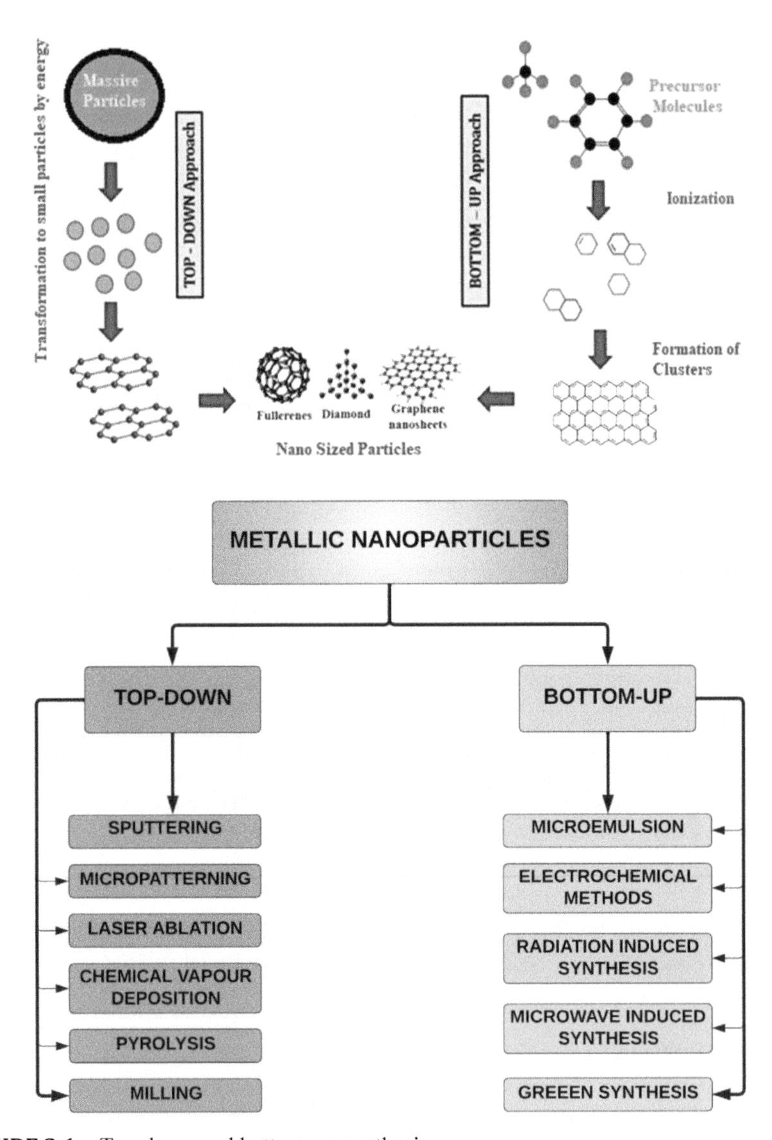

FIGURE 2.1 Top-down and bottom-up synthesis.

2.3.2 MICROPATTERNING

It is a widely used technique employed in microarrays, biosensors, and cellular studies; it is also practiced for producing metallic nanoparticles. To synthesize nanostructured arrays from a suitable precursor, this technique is often analogous to a printing process in which a material is cut or molded into the necessary shape and size either with a light or electron beam. This low-temperature, the non-vacuum approach uses laser sintering to create metallic nanoparticles from metallic nanoparticles ink using photolithography. Numerous lithography methods have been developed in addition to photolithography, which include soft nanoimprinting, colloidal, nanosphere, and E-beam lithography.[2]

2.3.3 LASER ABLATION

One technique that is thought to be a good substitute for traditional chemical methods is the laser ablation due to its quick processing time, its superior control over particle size and form, high yields, and improved long-term stability.[5,15] A solid surface (often a plate of pure metal) is exposed to a laser beam during a laser ablation procedure. This creates a low-flux plasma plume that evaporates or sublimates to form nanoparticles.[25] The materials are transformed into a higher flux. The use of the laser ablation method in biomedical applications, such as the in situ conjugation of biomolecules with metallic nanoparticles, which has been demonstrated to be more effective than standard techniques, has been made possible by the lack of necessity to remove excess reagents as well as the capability of metal nanoparticle synthesis in both aqueous and organic solvents.

2.3.4 CHEMICAL VAPOR DEPOSITION

Here the gaseous reactant is deposited as a thin film onto a substrate. The process is also referred to as the vacuum deposition method. Along with amixture of additional gas molecules that help the substrate superheat, the substrate interacts with the mixed gases during the process, which causes anion reduction. The product of this reaction is typically a film from which the nanoparticles must be scraped. The technique creates extremely pure, uniform, and nonporous nanoparticles; as a result, it has grown in importance in the semiconductor and electronics industries. Despite these significant benefits, this process has some notable drawbacks, including the need for specialized equipment to create the reaction chambers and films and the extremely poisonous nature of the reaction's gaseous by-products.[12]

2.3.5 PYROLYSIS

Another crucial method frequently employed alone or in conjunction with other physical techniques for metallic nanoparticle production is the thermal breakdown.[6] It is an endothermic chemical decomposition method that breaks chemical bonds in the molecule using heat, propelling the precursor into a chemical reaction that yields nanoparticles as well as other by-products in the form of ash. Nanoparticles are recovered through additional processing of the resultant solid ash. Noble metallic nanoparticles are typically made using the process of pyrolysis. One of this method's most significant limitations is its high energy usage.

2.3.6 MILLING

Since milling entails the direct breakdown of bulk materials into micro- and nanostructures, it is frequently referred to as the outward manifestation of top-down processes. The kinetic energy of the rollers or balls is transferred to the bulk material during mechanical milling, reducing the grain size. The form and size of the nanoparticles are greatly influenced by factors such as the type of mill, milling environment, milling media, intensity, time, and temperature.[23] Several techniques have been developed to overcome these difficulties, such as shaker mills, vibratory mills, planetary mills, etc.

2.4 BOTTOM-UP APPROACHES

2.4.1 MICROEMULSION

A growing area of research is the production of metal nanoparticles based on microemulsions. This technique is efficient and offers superior control over the physical properties of the synthesized nanoparticles, such as size and form. When a surfactant is present, microemulsions are often only combinations of two immiscible liquids. Thermodynamic stability, a wide interfacial area, and extremely low interfacial tension are typical characteristics of these systems.[17] The team of Muoz-Flores et al. published the first microemulsion-based synthesis of metallic nanoparticles using platinum, palladium, and rhodium nanoparticles as examples. Two distinct microemulsions are made for the microemulsion-based nanoparticle synthesis, one containing the ionic salt and the other having the reducing agent created in an amphiphilic environment. The salt ions are converted to neutral atoms by colliding the two emulsions, which causes the reactants to mix and create nanoparticles (Azharuddin et al., 2019b). Metal nanoparticles are typically synthesized using water-in-oil systems, and as these nanoparticles are made as emulsions, they are typically thermodynamically stable. Depending on the need, the process can be altered to synthesize a particular type of nanoparticle by changing the ratio of the surfactant to oil, which brings control over the size and shape of the particles.

2.4.2 ELECTROCHEMICAL METHODS

Since metallic nanoparticles and nanocomposites are primarily used for their catalytic capabilities and have lately been used in biological applications as biosensors, electrochemical techniques are frequently used for their manufacture. In order to achieve the deposition of metal salt on the cathode of an electrochemical cell in the presence of an electrolyte to make nanoparticles (Azharuddin et al., 2019b), Reetz and Helbig in 1994 devised the electrochemical technique. The type of the reducing agent, the purity of the metal and stabilizer, the choice of the electrolyte, the concentration ratio, and temperature are some of the factors that affect the physical characteristics of the nanoparticles and affect how effective this process is. At present, the synthesis of nanocomposites (especially those with graphene) using electrochemical methods is preferred to the synthesis of nanoparticles.[28]

2.4.3 RADIATION-INDUCED SYNTHESIS METHODS

Ionizing radiation, particularly gamma radiation, as well as X-rays and UV light, are used in this process to create metal nanoparticles. As it produces entirely reduced, extremely pure (by-product-free), metal nanoparticles, it has been demonstrated to be significantly more efficient than the traditional methods of nanoparticle synthesis. In this procedure, radiolysis caused by exposure to radiation results in the creation of nanoparticles from an aqueous solution containing a reducing and stabilizing agent. The water molecules split apart as a result of the radiation exposure, producing momentary products that have strong oxidizing or reducing properties. They also decrease metal ions to neutral metal atoms, which further nucleate to produce nanoparticles. Real-time tracking[27] of the growth trajectories of colloidal nanoparticles was made possible using synchrotron X-ray techniques. The radiation dose, the pH of the system, and the kind of solvent employed in the synthesis are the physical parameters important for the synthesis of nanoparticles. Recently, tween 80 stabilized Ag nanoparticles for antibacterial applications were made using radiation-induced synthesis.

2.4.4 MICROWAVE-INDUCED GREEN SYNTHESIS METHODS

This method is also known as one-pot synthesis and uses salts and surfactant solutions for the synthesis of nanoparticles. The method supports control over the morphology of the synthesized nanoparticles and is very dependable, quick, and simple (Azharuddin et al., 2019b). This technique relies on the principles of dipole interaction (molecules have a tendency to align and oscillate in step with the oscillating electrical field of the microwaves, causing collision and friction that produces heat) and ionic conduction (The electricfield generates ionic motion as the molecules try to orient themselves to the rapidly changing field, causing instantaneous super heating), producing a heating effect that causes the reduction of metal ions to nanoparticles.[14] The morphological properties of the nanoparticles are mainly controlled by the microwave irradiation period and the reactant concentration. Recently, the injection of humate-polyanion at different stages of the synthesis was used to adjust physical features, such as monodispersity and particle size of super paramagnetic magnetite nanoparticles generated via microwave-assisted synthesis. In the absence of solvents or surfactants, microwave-induced electric discharge was also employed to create Cu, Ni, and Zn nanoparticles from metal particles.

2.4.5 GREEN SYNTHESIS METHODS

The future of biological uses of NMNPs has almost been compromised by the excessive usage of chemicals in chemical synthesis. This paved the way for other eco-friendly methods with a minimal usage of chemicals. The green synthetic techniques that use microbes, biopolymers, and plant extracts have shown to be strong contenders to replace chemical methods of nanoparticle synthesis.[24] There are further classifications, such as bacteria-based synthesis, fungal-based synthesis, algae-based synthesis, plant-based synthesis. All these processes are widely used based on the utility and applicability of the nanoproducts. However, each methods have its own drawbacks and majority of these processes cannot be expanded for mass manufacturing. As a result, there is still work to be done in developing alternative methods to create nanoparticles with regulated and adjustable features.

2.5 PHYSICAL AND CHEMICAL PROPERTIES

Most of the physical and chemical properties of particles depends on size. The general properties of nanoparticles include large surface area, mechanically strong, and optically active and chemically reactive, making them unique and suitable applicants for various applications.

2.5.1 OPTICAL PROPERTIES

One of the essential characteristics of metal oxides is their optical conductivity, which may be empirically determined through measurements of reflectance and absorption. When the oxide characteristic size (primary/secondary particle size) is in or out of the range of the photon wavelength,[10] scattering can exhibit substantial changes. But reflectivity is obviously size-dependent, absorption features often control the major absorption behavior of solids. Light absorption changes to be both discrete-like and size-dependent as a result of quantum-size confinement. Transitions between discrete or quantized electron and hole electronic levels provide both linear optical properties (one exciton per particle) and nonlinear optical properties (many excitons) for nanocrystalline semiconductors. Although other theories, such as the free exciton collision model (FECM) or those based on the bond length–strength correlation, have been developed to account for several shortcomings of the EMA theory, the effective mass theory (EMA) is the most elegant and general theory to explain the size dependence of the optical properties of nanometer semiconductors.

2.5.2 TRANSPORT PROPERTIES

It is very well known that both ionic and mixed ionic/electronic conductivity in oxide materials can be affected by the solid's nanostructure.[9] According to Boltzmann statistics, the quantity of electronic charge carriers in a metal oxide depends on the band gap energy. Depending on whether electrons or holes are the dominant charge carriers, the electronic conduction is referred to as either n- or p-hopping type. By introducing non-stoichiometry, it is possible to increase the amount of "free" electrons and holes in an oxide, which are then balanced by the much less mobile oxygen and cation vacancies. Ionic conduction occurs when ions can hop across sites within a crystal lattice as a result of thermal activation. Similar to hoping-type conduction, it is often explained on the basis of a modified Fick's second law. Ionic conduction has been seen using the direct interstitial, vacancy, and grotthuss mechanism types. To reduce strain and electrostatic potential contributions to the total energy, charge species (defects; impurities) in polycrystalline oxides typically segregate to particle boundaries. This results in a contribution to the conductivity parallel to the surface, which becomes significant at the nanoscale regime. As a result of the protected electrostatic potential depletion at surface layers of nanosized materials,[21] the charge carrier (defect) distribution is also severely modified by bulk materials since charge carriers are present across the entire material.

2.5.3 MECHANICAL PROPERTIES

The primary mechanical characteristics relate to observables at low (yield stress and hardness) and high (superplasticity) temperatures.[9] The majority of information on oxide nanoparticles is focused on sinterability, ductibility, and superpasticity analysis. In conventional materials, the yield stress (σ) and harness (H) follow the Hall-Petch (H-P) equation:

$$\sigma \backslash H = \sigma_0 \backslash H_0 + kd^{-1/2}, (2.1)$$

where the initial constants describe friction stress and hardness, d is the primary particle/grain size and k is the corresponding slope.

2.5.4 CHEMICAL PROPERTIES

Absorption and catalysis make use of metal oxides for both their redox and acid/base characteristics. The coordination environment of the surface atoms, the redox capabilities, and the oxidation state at the surface layers are the three main characteristics that are necessary for their use as absorbents or catalysts. There have been several attempts to establish links between the redox and acid/base properties in the literature because they are both connected.[4] According to a straightforward classification, oxides with d or f outer electrons have a larger range of applications than those with only s or p electrons in their valence orbitals. Redox catalysts are solids that under specific reaction conditions concurrently undergo reduction and reoxidation by releasing surface lattice oxygen anions and absorbing oxygen from the gas phase.[9] This procedure necessitates dynamic operation and microscopic reversibility.

2.6 APPLICATIONS OF METAL OXIDE NANOPARTICLES

Metal oxides are considered as an important class of materials from both technological and scientific points of view owing to their unique properties. Metal oxide semiconductors have been identified as the most useful nanomaterials in industries due to their prominent applications in diverse areas of electronics, sensing, optoelectronics, biosensing, and catalysis. The schematic illustration of various applications of metal oxide in different industries is given in Figure 2.2.

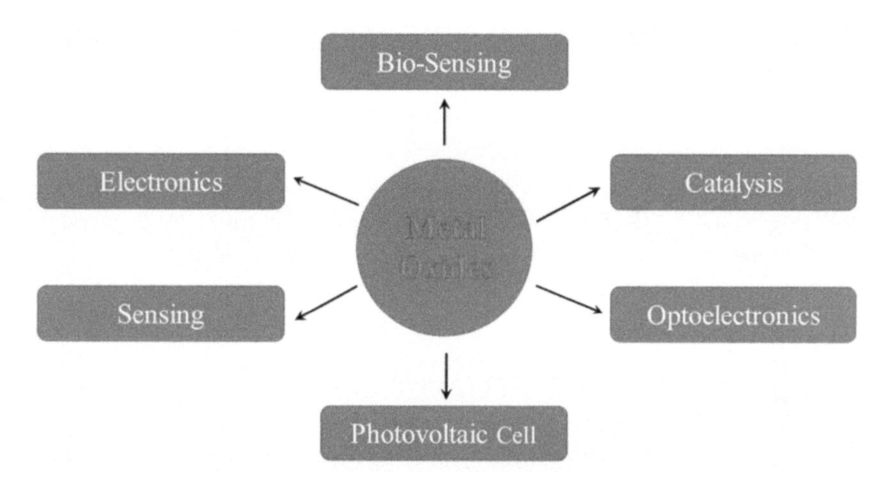

FIGURE 2.2 Applications of metal oxide nanomaterials in diverse fields of industry.

Some of the most commonly used metal oxides nanomaterials include zinc oxide (ZnO), titanium oxide (TiO_2), iron oxide (Fe_2O_3, Fe_3O_4, etc.), tungsten oxide (WO_3), tin oxide (SnO_2), cobalt oxide (Co_3O_4), cupric oxide (CuO), cuprous oxide (Cu_2O), and cadmium oxide (CdO). Among these metal oxides, iron oxide, cobalt oxide (Co_3O_4), cupric oxide (CuO), and zinc oxide (ZnO) have been widely studied recently. Metal oxide nanoparticles are widely employed in the fabrication of devices, such as fuel cells, sensors, piezoelectric devices, and microelectronic circuits.[20]

Making nanostructures or nanoarrays with unique features in comparison to bulk or single particle species is an aim in the developing field of nanotechnology. Due to their small size and a high density of corner or edge surface sites, metal oxide nanoparticles can display distinct physical and chemical characteristics. Three significant classes of fundamental properties in any material are anticipated to be influenced by particle size. The first one includes the structural properties, specifically the cell parameters and lattice symmetry. Typically, bulk oxides are reliable, stable, and have crystal structures that are clearly characterized. Although changes in thermodynamic stability are associated with the size that can lead to modifications in cell parameters and/or structural transformations, it is important to take into account the growing importance of surface free energy and stress with decreasing particle size. In extreme cases, metal oxide nanoparticles can vanish due to interactions with their surrounding environment and a high surface free energy. But in order to display mechanical or structural stability, metal oxide nanoparticles must have a low surface free energy.

The thin film metal oxide nanomaterials have an amorphous or crystalline structure that is equally found useful in several industrial applications, such as the fabrication of microelectronic circuits, fuel cells, catalysts, sensors, photovoltaic cells, and piezoelectric cells. The structural, electrical, and optical properties of the thin films, such as conductivity, bandgap can be tuned depending on their industrial requirements by modifying the synthesis conditions and the deposition process. In this chapter, the possibilities of metal oxides for efficient energy applications are elaborated.

2.7 ADVANCED ENERGY APPLICATIONS

Metal oxide nanostructures are promising candidates for a number of advanced industrial applications, including catalysis, gas sensors, fuel cell membranes, electrochromic (EC) windows, and lithium–ion batteries.[11,16] These applications are made possible by the accompanying changes in electrical and optical properties during ion insertion/removal processes. Recently, they have also been found useful in the design of energy-saving and harvesting devices, such as photovoltaics, lithium-ion batteries, fuel cells, supercapacitors, and even in hydrogen production byphotolysis.

Metal and metal oxide nanomaterials are offering true elucidations and chances to ensure supportable energy for the upcoming energy requirements. Owing to their unique physical and chemical properties, structure, and configuration, they are considered as most promising nanomaterials in energy devices.

Moreover, nanostructures such as one-dimensional nanorods, nanofibers, and nanotubes have gained more attention recently because of their remarkable structural, physical, mechanical, and optical properties and confinement effects. They find remarkable applications in the field of energy storage devices and energy conversion applications due to their high surface-to-volume ratio, efficient charge transport properties, and porosities. The schematic representation of advanced energy applications of metal oxide nanomaterials is given in Figure 2.3. The photovoltaic devices and lithium–ion batteries developed using metal oxide nanofibers have remarkably good electrochemical stability, cycle life, power conversion efficiency, better reversible ability, etc.

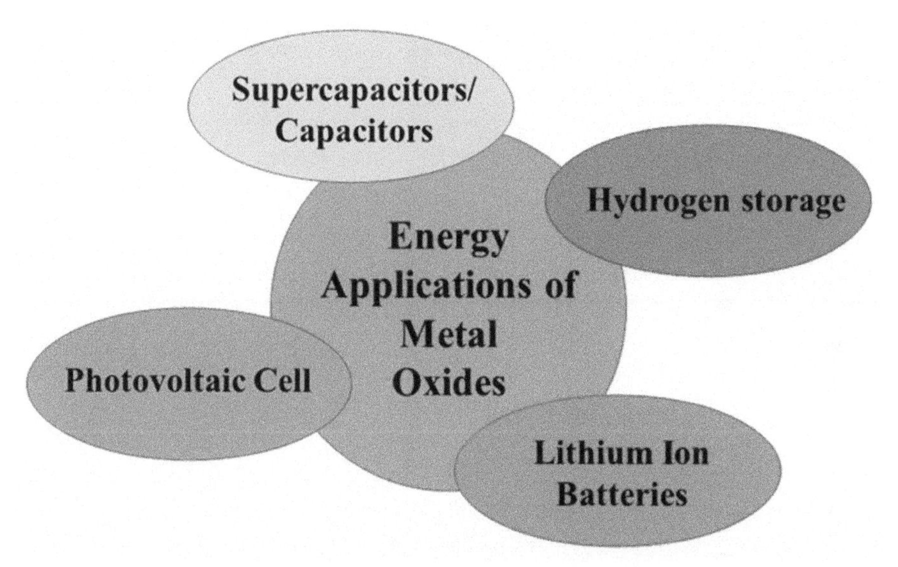

FIGURE 2.3 Advanced energy applications of metal oxide nanomaterials in various industries.

Numerous transitional metal oxides, such as those made up of tungsten, molybdenum, vanadium, manganese, cobalt, and nickel have been thoroughly studied as ion insertion hosts for Li^+, Na^+, and H^+ ions. The solid-state diffusion of the ions frequently places a restriction on the insertion reaction's kinetics. The time constant of this process is determined by the chemical diffusion coefficient and length of the diffusion path. The length of the diffusion path is determined by the microstructure, whereas the former depends on the chemical and crystal structure of the metal oxide. The smallest dimension in the case of nanoparticles corresponds to the diffusion path length. Therefore, the key to a material with quick insertion kinetics and superior overall device performance is designing a nanoparticle with a small radius while maintaining the proper crystalline phase.

2.7.1 SUPERCAPACITORS/CAPACITORS

Metal oxide nanomaterials have been a focus of intense scientific research for industrial applications in the development of energy storage devices. They offer distinguished physical and chemical properties for the fabrication of various supercapacitor devices. However, compared with single metal oxide nanomaterials, bimetallic oxides are more desirable rather than single metal oxides to overcome some of their major constraints such as feeble electric conductivity, low capacitance, and low energy density at this capacitor-level power.

Bimetallic oxide materials offer numerous opportunities to improve asymmetric supercapacitors and reach battery-level energy density, confirming numerous advancement reports about aqueous asymmetric and hybrid supercapacitors. Bimetallic oxides have shown a continuous improvement in conductivity, surface specific area, and the abundance of electrochemically activesites compared with their bulk counterparts, further resulting in a tremendous boost in their electrochemical properties.[1] This improvement has been achieved by appropriately adjusting the composition and nanostructure of the bimetallic oxides and composing different carbons.

Furthermore, in a wide working voltage window, innovative bimetallic oxides with various features as cathodes have demonstrated interesting flexibility and commendable electrochemical capabilities for developing and producing high voltage devices. To balance the relationship between electric conductivity, porosity, theoretical specific capacity, and electrochemically active sites, it is necessary to reasonably modify the species and contents of the bimetallic oxides based on the composition and concentration of the electrolytes.

Bimetallic oxide/carbon composites with high conductivity and specific capacitance, in particular, expertly assist in reducing mass discrepancies between the cathode and anode, and consequently boost the energy density while maintaining the power density.[26] The use of organic electrolytes increases this energy density and aids in the development of hybrid systems.

It has been determined that cutting-edge bimetallic oxide/C composites can increase capacitive characteristics by maximizing the benefits of each constituent. Despite the fact that hybrid systems have good electrochemical performances, research into their physicochemical properties and key reaction processes is still in its early stages. For the steerable combination of two metal elements, low consumption, green, environmental-friendly, and moderate preparation technologies based on molecular assemblies are urgently needed, particularly by implementing the controllable doping of metal types and counter-content. Furthermore, it is important to note that bimetallic oxides readily accept oxygen vacancies produced by a variety of methods. As a result of enough active sites and larger carrier/donor density, this would relatively improve the oxides' capacitive properties. The aforesaid strategies should be judiciously chosen and properly matched in accordance with the specific requirements of diverse supercapacitor systems.

Transitional metal oxide materials have been identified as suitable candidates to be used as electrodes of supercapacitors due to their availability, their diverse constituents, different morphologies, large surface area, and high theoretical specific capacitance. Moreover, they offer a conspicuous capacitance improvement by tuning defects and surface/interfaces on the nanoscale. However, the main demerits include low electrical conductivity, volume expansion, and sluggish ions diffusion affects their practical applications. Hence, several research interest has been devoted in this direction to overcome the major demerits and improve their electrochemical properties. The design of metal oxide materials by tuning their composition, morphology, and electroconductivity has boosted the physical and chemical performances of metal oxides to a great extent for practical applications. Moreover, metal doping with metal oxide materials reduces the charge transfer impedance of supercapacitors, with a comparatively high specific capacitance. In addition, nanostructured metal oxides with porous morphology provide a large surface-to-volume ratio, ensure better contact between the active materials and electrolytes, ionic transportation in the electrolytes, and more efficient utilization of the active materials during the charging-discharging process. Moreover, the availability of more electro-active sites and good stabilities of the metal oxide materials ensure high pseudocapacitance and cyclic stability. Also, incorporation of metal oxide/C composites including carbon nanotubes, carbon nanofibers, graphene, and amorphous carbon materials, are reported to have dramatically high electroconductivity, specific capacitance, and rate performances. The reports indicate that the introduction of oxygen vacancies into metal oxides produces a larger interlayer spacing for enhanced charge storage kinetics. These unique features marked the breakthroughs of metal oxides for pseudosupercapacitors that may ameliorate the energy density of supercapacitors at the battery level in order to bridge the gap between batteries and supercapacitors.

Additionally, a number of metal–ion hybrid supercapacitor devices have significantly improved, suggesting that these devices may hold the key to bridging the gap between supercapacitors and ordinary batteries. But since the development of metal–ion hybrid supercapacitors is still in its early stages and the energy storage mechanisms are still unknown, more research is necessary to fully understand the interactions between the electrode and electrolyte (a schematic representation of supercapacitor electrodes and electrolyte solution is given in Figure 2.4).

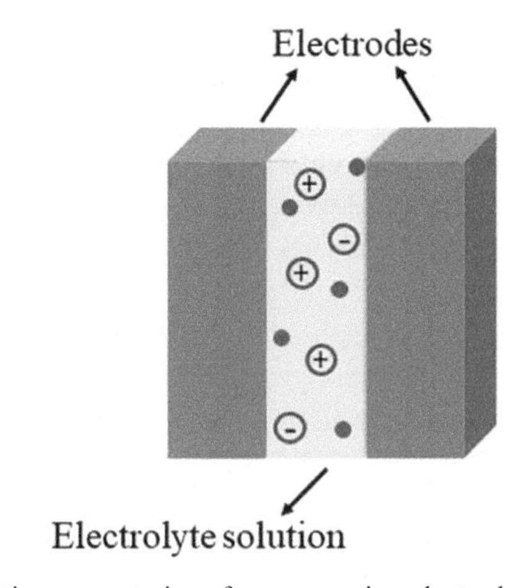

FIGURE 2.4 Schematic representation of supercapacitor electrodes and electrolyte solution.

To reveal potential electrochemical pathways at scale, theoretical calculations and simulations are crucial. In situ technologies, such as spectroscopy and microscopy, have also undergone further advancements and are essential for providing clear experimental results. Therefore, in near-future studies, electrochemical mechanisms, device designs, and synthetic method developments all call for collaborative efforts to promote bimetallic oxide-based supercapacitor devices.

2.7.2 HYDROGEN STORAGE

Given the current worries about climate change, energy security, and pollution, alternative energy sources are highly sought for. One such substitute is hydrogen gas, which may be generated from domestic sources and used to power fuel cells, which are emission-free energy producers. Despite the benefits of using hydrogen as a fuel, there remain questions about secure high-pressure storage procedures, particularly for automotive applications.

Several methods exist to address this problem, one of which is the storage of hydrogen inside porous frameworks. While transition metals can adsorb hydrogen through metallic bonding and dissociation processes, most porous materials do so through weak van derWaals interactions.[8]

Metal/metal oxide @MOF composites appear to be attractive materials for hydrogen storage due to their high surface areas and improved adsorption enthalpy at a metal surface (Figure 2.5).

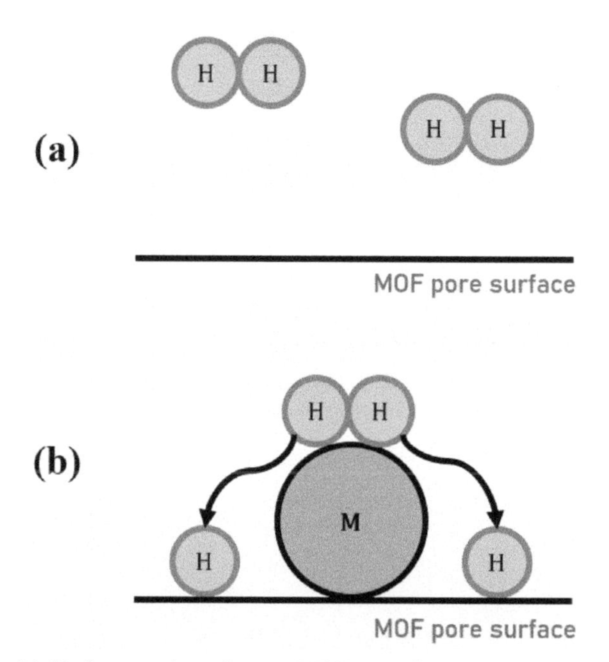

FIGURE 2.5 (a) Hydrogen adsorption on MOFs usually occurs by weak van derWaals interactions. The introduction of metal nanoparticles allows for the stronger mechanism (b) where dissociation and subsequent spillover may occur.

2.7.3 BATTERIES

Energy storage devices such as batteries and supercapacitors with high energy density and low cost are highly desired nowadays. The batteries have become an important and most commonly used energy storage device. A battery consists of one or more voltaic cells and each voltaic cell consists of two half cells connected in series. The major parts of a battery include a positive electrode (anode), a negative electrode (cathode), and an electrolyte. The schematic representation of a battery is shown in Figure 2.6. During redox reactions, the cations are reduced at one electrode while the anions are oxidized at the other electrode. The electrode materials should possess certain properties such as excellent electrochemical reversible reactivity, chemical and thermal stabilities, and sustainability for practical applications. Metal oxide nanomaterials are highly promising candidates in this regard, and thus a large number of studies have been devoted to obtaining electrochemically efficient anode nanomaterials.

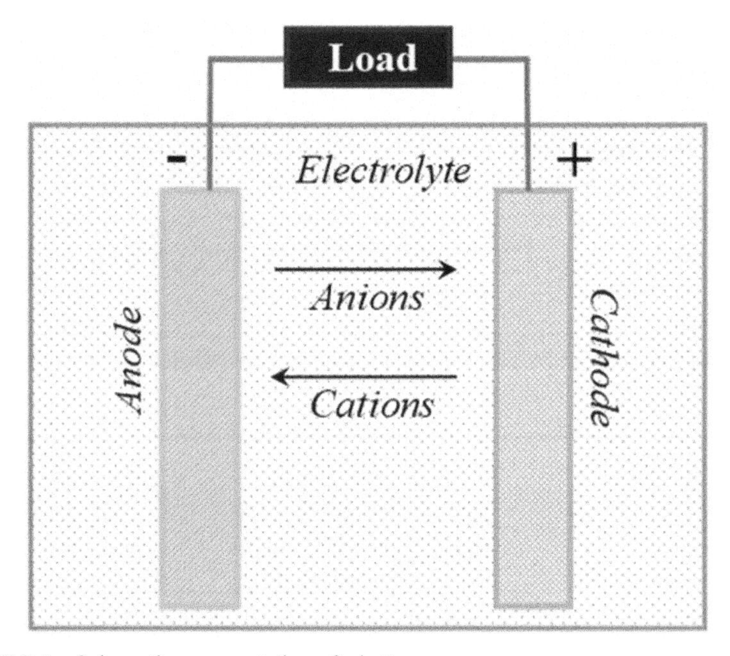

FIGURE 2.6 Schematic representation of a battery.

The efficient properties of metal oxide nanomaterials have provided solutions to many technical problems in achieving high-performance anode materials. This is because the performance indexes of the anode material depend on the material properties, such as crystal structure, electronic conductivity, and ionic diffusivity.[29] Some drawbacks of conventional batteries are the poor electronic and ionic transport, agglomeration of active nanomaterial, change in the volume of electrode material during lithiation, pulverization of the active materials after long cycles, poor contact with the current collector, etc.[31] These lead to some issues, such as capacity fade, poor cycle life, etc.,[30] which can be tackled by employing porous metal oxide nanomaterials because these materials can endure stress and prevent pulverization. The functional metal oxide nanomaterials with different porous morphologies can incorporate volume expansion and helps to preserve the structural integrity of the electrode materials. Especially, the hollow metal oxide nanostructures with very thin shells offer good structural stability and diffusion kinetics against volume change during lithiation due to the fact that the hollow nanostructures accommodate more volume expansion, which results in enhanced electrochemical performance.[30]

2.7.4 PHOTOVOLTAIC CELLS

The photovoltaic cell also known as a solar cell is an important source of sustainable energy that converts light energy into electrical energy. It mainly consists of a p-n junction diode which when illuminated with electromagnetic radiation with an energy greater than the bandgap energy of the nanomaterial, produces an electric current. The light photons absorbed by a semiconductor diodeproduce electron–hole pairs or excitons. They are separated by an internal potential and the subsequent flow of electrons and holes produces an electric current.

Metal oxides have been used in the development of photovoltaic cells for many years due to their remarkable properties and the feasibility to be fabricated by simple, low-cost fabrication methods. The performance efficiency of a photovoltaic cell depends on various factors, such as materials used, radiation falling upon the module, etc. The steady growth of photovoltaics is expected to be continuing as the main source of energy generation. In a perovskite solar cell, a perovskite layer is sandwiched between an electron transfer layer and a hole transfer layer which absorbs photons and generates electron–hole pairs. This is followed by a subsequent photoexcitation, as a result, the electron transfer layer accepts electrons from the perovskite layer and conducts them to the electrode. However, it is crucial to separate photogenerated carriers and pass them to the carrier transfer layer before recombination. Thus, it is evident that the electron transfer layer determines the photovoltaic performance by inducing the charge carrier extraction, charge transport, and recombination process. Inorganic n-type semiconducting metal oxides, such as TiO_2, SnO_2, and ZnO are promising electron transfer layers for being highly efficient and stable perovskite solar cells.

For an ideal electron transfer layer consisting of metal oxide nanomaterials such as TiO_2, SnO_2, and so on, the conduction band minimum should be well-matched with perovskite layers to ease electron extraction and transport. However, the valence band maximum should be lower than that of perovskites for blocking holes, which suppresses carrier recombination. High carrier mobility is favored for the electron transfer layer which influences the balanced flux of carriers transported to the electron transfer layer and hole transfer layer, respectively. Good electron mobility reduces charge accumulation at the interface and prevents carrier recombination. The electron transfer layer possesses sufficient light transmittance and ensures the stability of perovskite in sunlight. The morphology of the constituent nanomaterials plays a key role in facilitating the crystallization of perovskite layers and reduces the traps at the interface.

The production of solar cells would heavily depend on the dimension and electronic property of the electron transfer layer. Moreover, large-scale fabrication is a major parameter to be dealt with to realize commercial applications of solar cells. The development of suitable deposition methods to fabricate uniform and pinhole-free electron transfer layers is also a major factor to prepare large-area solar cells.

2.8 CONCLUSIONS

Metals and metal oxide nanomaterials are playing a pivotal role in emerging electronic technologies, such as batteries, photovoltaic cells, hydrogen storage devices, and so on. These nanomaterials generally have good catalytic efficiencies, and large flexibilities with promising applications for environmental purification, sensing, material-air barriers, and fuel cells. Despite many of these advantages, the main problem regarding the material sustainability and availability for electrochemical and optical transduction requires intense multidisciplinary research efforts. Moreover, the exact growth mechanisms involved in producing oxide nanostructures are often a matter of speculation due to the fact that several factors, such as ionic strength, temperature, solvent viscosity, and the presence of organic ligands play a significant role in developing the morphology of the products.

KEYWORDS

- **metal oxide**
- **energy applications**
- **battery**
- **supercapacitors**
- **photovoltaic**

REFERENCES

1. An, C.; Zhang, Y.; Guo, H.; Wang, Y. Metal Oxide-Based Supercapacitors: Progress and Prospectives. *Nanoscale Adv.* **2019,** *1* (12), 4644–4658.
2. Azharuddin, M.; Zhu, G. H.; Das, D.; Ozgur, E.; Uzun, L.; Turner, A. P.; Patra, H. K. A Repertoire of Biomedical Applications of Noble Metal Nanoparticles. *Chem. Commun.* **2019a,** *55* (49), 6964–6996.
3. Bhattacharyya, S.; Kudgus, R. A.; Bhattacharya, R.; Mukherjee, P. Inorganic Nanoparticles in Cancer Therapy. *Pharm. Res.* **2011,** *28* (2), 237–259.
4. Busca, G. *The Surface Acidity and Basicity of Solidoxides and Zeolites*. Chemical Industries: Marcel Dekker: New York, **2006;** Vol 108; p 247.
5. Ealia, S. A. M.; Saravanakumar, M. P. In *A Review on The Classification, Characterisation, Synthesis of Nanoparticles and their Application*, IOP Conference Series: Materials Science and Engineering, IOP Publishing. Vol. 263, 2017; p 032019
6. Ealia, S. A. M.; Saravanakumar, M. In *A Review on the Classification, Characterisation, Synthesis of Nanoparticles and their Application*, IOP Conference Series: Materials Science and Engineering, IOP Publishing; Vol. 263, 2017; p 032019.
7. Emerich, D. F.; Thanos, C. G. Nanotechnology and Medicine. *Expert Opin. Biol. Ther.* **2003,** *3* (4), 655–663.
8. Falcaro, P.; Ricco, R.; Yazdi, A.; Imaz, I.; Furukawa, S.; Maspoch, D.; Ameloot, R.; Evans, J. D.; Doonan, C. J. Application of Metal and Metal Oxide Nanoparticles@mofs. *Coord. Chem. Rev.* **2016,** *307,* 237–254.
9. Fern´andez-Garcia, M.; Rodriguez, J. Metal Oxide Nanoparticles. Technical Report. Brookhaven National Lab. (BNL): Upton, NY (United States), 2007.
10. Ganachari, S. V.; Hublikar, L.; Yaradoddi, J. S.; Math, S. S. Metal Oxide Nanomaterials for Environmental Applications. In *Handbook of Ecomaterials*, Vol. 4; 2019; pp 2357–2368.
11. Granqvist, C. G. *Handbook of Inorganic Electrochromic Materials.* Elsevier, 1995.
12. Habibullah, G.; Viktorova, J.; Ruml, T. Current Strategies for Noble Metal Nanoparticle Synthesis. *Nanoscale Res. Lett.* **2021,** *16* (1), 1–12.
13. Hatakeyama, Y.; Onishi, K.; Nishikawa, K. Effects of Sputtering Conditions on Formation of Gold Nanoparticles in Sputter Deposition Technique. *RSC Adv.* **2011,** *1* (9), 1815–1821.

14. Kokel, A.; Schäfer, C.; Török, B. Microwave Assisted Reactions in Green Chemistry. In *Encyclopedia of Sustainability Science and Technology*; Springer: New York, 2018; pp 1–40.
15. Korshed, P.; Li, L.; Ngo, D.-T.; Wang, T. Effect of Storage Conditions on the Long-Term Stability of Bactericidal Effects for Laser Generated Silver Nanoparticles. *Nanomaterials* **2018,** *8* (4), 218.
16. Lee, S.-H.; Tracy, C. E.; Yan, Y.; Pitts, J. R.; Deb, S. K. Solid-State Nanocomposite Electrochromic Pseudocapacitors. *Electrochem. Solid-State Lett.* **2005,** *8* (4), A188.
17. Malik, M. A.; Wani, M. Y.; Hashim, M. A. Microemulsion Method: A Novel Route to Synthesize Organic and Inorganic Nanomaterials: 1st Nano Update. *Arab. J. Chem.* **2012,** 5 (4), 397–417.
18. Murugesan, K.; Sivakumar, P.; Palanisamy, P. An Overview on Synthesis of Metal Oxide Nanoparticles. **2016,** *2,* 58–66.
19. Nguyen, M. T.; Yonezawa, T. Sputtering Onto a Liquid: Interesting Physical Preparation Method Formulti-Metallic Nanoparticles. *Sci. Technol. Adv. Mater.* **2018,** *19* (1), 883–898.
20. Oskam, G. Metal Oxide Nanoparticles: Synthesis, Characterization and Application. *J. Sol-gel Sci. Technol.* **2006,** *37* (3), 161–164.
21. Rodriguez, J. A.; Fernández-García, M. *Synthesis, Properties, and Applications of Oxide Nanomaterials*; John Wiley & Sons, 2007.
22. Rotello, V. *Nanoparticles: Building Blocks for Nanotechnology*; Springer Science & Business Media, 2004.
23. Schreyer, H.; Eckert, R.; Immohr, S.; de Bellis, J.; Felderhoff, M.; Schüth, F. Milling Down to Nanometers: Ageneral Process for the Direct Dry Synthesis of Supported Metal Catalysts. *Angew. Chem. Int. Ed.* **2019,** *58* (33), 11262–11265.
24. Siddiqi, K. S.; Husen, A.; Rao, R. A. Areview on Biosynthesis of Silver Nanoparticles and their Biocidal Properties. *J. Nanobiotechnol.* **2018,** *16* (1), 1–28.
25. Sportelli, M. C.; Izzi, M.; Volpe, A.; Clemente, M.; Picca, R. A.; Ancona, A.; Lugar`a, P. M.; Palazzo, G.; Cioffi, N. The Pros and Cons of the Use of Laser Ablation Synthesis for the Production of Silver Nano-Antimicrobials. *Antibiotics* **2018,** *7* (3), 67.
26. Wang, G.; Zhang, L.; Zhang, J. A Review of Electrode Materials for Electrochemical Supercapacitors. *Chem. Soc. Rev.* **2012,** *41* (2):797–828.
27. Wu, S.; Li, M.; Sun, Y. In Situ Synchrotronx-ray Characterization Shines Light on the Nucleation and Growth Kinetics of Colloidal Nanoparticles. *Angew. Chem. Int. Ed.* **2019,** *58* (27), 8987–8995.
28. Zou, C.; Yang, B.; Bin, D.; Wang, J.; Li, S.; Yang, P.; Wang, C.; Shiraishi, Y.; Du, Y. Electrochemical Synthesisof Gold Nanoparticles Decorated Flower-like Graphene for High Sensitivity Detection of Nitrite. *J. Coll. Interface Sci.* **2017,** *488,* 135–141.
29. Wang, Y.; Xie, K.; Guo, X.; Zhou, W.; Song, G.; Cheng, S. Mesoporous Silica Nanoparticles as High-performance Anode Materials for Lithium-ion Batteries. *New J. Chem.* **2016,** *40,* 8202.
30. Yang, Z.; Wu, H.; Zheng, Z.; Cheng, Y.; Li, P.; Zhang, Q.; Wang, M. Tin Nanoparticles Encapsulated Carbon Nanoboxes as High-Performance Anode for Lithium-Ion Batteries. *Front. Chem.* **2018,** *6,* 533.
31. Sun, J.; Cui, B.; Chu, F.; Yun, C.; He, M.; Li, L.; Song, Y. Printable Nanomaterials for the Fabrication of High-Performance Supercapacitors. *Nanomaterials* **2018,** *8* (7), 528.

CHAPTER 3

Metal Oxide Nanoparticles and Their Importance in Energy Devices

G. SANTHOSH[1] and G. P. NAYAKA[2]

[1]*Department of Mechanical Engineering, NMAM Institute of Technology, NITTE (Deemed to be University), Nitte-off-campus center, NITTE, India*

[2]*Physical and Materials Chemistry Division, CSIR-National Chemistry Laboratory, Pune, Maharashtra, India*

ABSTRACT

In the past decades, energy demand has raised to the maximum possible extent due to the consumption and use of the energy available. The increase in population, and the use of smart devices have led to the invention of various energy storage devices. Metal oxide nanoparticles are among great materials and play crucial role in the development and performance of energy storage devices, such as batteries, supercapacitors, and solar and fuels cells. Hence, a better understanding of the physical, chemical, and structural properties of metal oxides will enable us to design and develop advanced techniques to enhance the performance of energy storage devices. Bimetallic metal oxides and multi-metal oxide materials are highly preferred in energy device applications due to their high capacitance, high energy density, and power density compared with single metal oxide materials. In this chapter, we will discuss about the metal oxides strategies to improve the ionic conductivity of the energy devices by incorporating the nanoparticles with high surface area and high aspect ratio.

Technological Advancement in Clean Energy Production: Constraints and Solutions for Energy and Electricity Development. Amritanshu Shukla, Kian Hariri Asli, Neha Kanwar Rawat, Ann Rose Abraham, & A. K. Haghi (Eds.)

3.1 INTRODUCTION

The past few decades have witnessed the growing demand for energy; global energy consumption is currently drastically increasing. As a result, several studies are being conducted worldwide on energy storage technologies including supercapacitors and rechargeable batteries.[1,2] Even though energy storage technologies have advanced significantly over the past several decades, it is still exceedingly difficult to fulfil the demand for designing and manufacturing effective storage systems. Greater efforts are being made to produce and construct appropriate electrodes to fit the intended performance as the need for flexible, lightweight, and high energy density storage devices grows. In contrast to its three-dimensional (3D) bulk graphite, graphene-based materials for energy storage applications have been created over the past 10 years.[3,4] The Figure 3.1 shows the graphene and graphene oxide obtained using conventional synthesis methods.

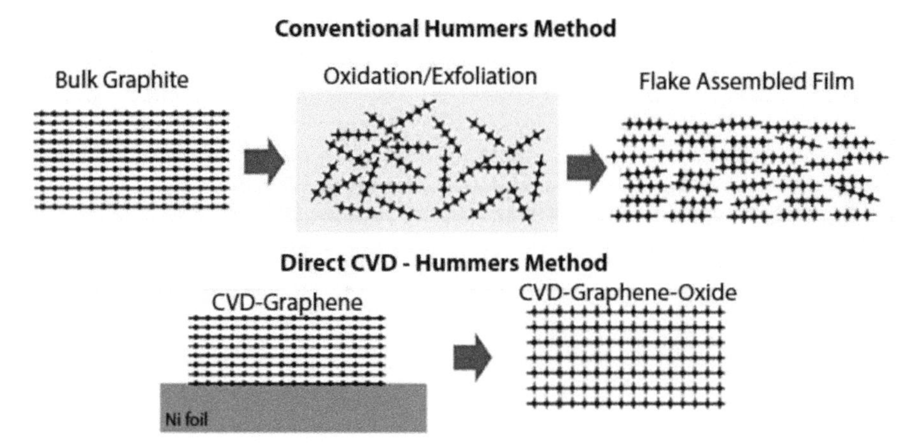

FIGURE 3.1 Graphene and graphene oxide derived from graphite from conventional Hummer's and CVD methods.

The application of graphene-based composites in energy storage technologies has been explored by researchers. However, using a standard method to assemble these composites results in unfair characteristics that limit the effectiveness of devices.[6–9] Because of its wide variety of characteristics, including surface area, mechanical strength, chemical, electrical, and thermal stability, graphene is regarded as one of the most appropriate materials for energy device applications.[10] Numerous morphological advantages of graphene, including its high accessible surface with exposed active sites and quick response kinetics, aid in the electrochemical energy storage process and lead to high power and energy densities.[11]

3.1.1 ANODE MATERIALS FOR ELECTRODES

Numerous graphene-based anodes with good electrochemical characteristics are being developed via unending research efforts. However, the energy devices' energy discharge due to charge and discharge cycles makes them challenging to commercialize. Graphene-based materials are now a strong contender for energy devices due to the difficulties in commercializing other anode materials.

For lithium-ion batteries (LIBs), Bharwaj and colleagues have created four distinct nanostructures, including graphite, multilayer carbon nanotubes (MlCNTs), and oxidized and non-oxidized graphene nanoribbons (OGNr and NoGNr). Reduced graphene ribbons demonstrated the maximum potential with an efficiency of 53%, according to the electrochemical tests of the time.[12]

Reduced graphene oxide has been synthesized by Uthaisar et al.[13] utilizing a variety of synthetic techniques. It is discovered that the reduced graphene oxide made using Hummers' process has a high oxygen content (33 mol%) and 1.2 carbon/oxygen ratio (C/O ratio). Furthermore, it is discovered that adding 7–17 mol% of oxygen causes the C/O ratio to rise from 4.2 to 11.6. According to electrochemical research, reduced graphene oxide has a capacity of 500–1000 mAh/g for charge discharge, whereas unreduced graphene oxide has an enhanced capacity of 1000 and 500 mAh/g. The behavior is linked to the defective oxygen groups' increased ability to absorb lithium. The charge capacity of GO produced with various levels of oxidation was investigated by Lee et al.[14] The authors demonstrated the excellent stability of synthetic GO by using a material with a maximum capacity of 2000 mAh/g.

Chen and colleagues[15] provided a straightforward method to create porous graphene nanosheets (PGNs). According to the electrochemical data, PGNs are poorly stacked to give the maximum storage capacity at a high discharge rate. After 100 cycles, the supplied capacity reaches 480 mAh/g, and after 300 cycles at 2 A/g, it approaches 320 mAh/g. By using MgO templates and the CVD technique, Fan et al.[16] also created PGNs. The findings point to nanoporous materials with a good cycle stability and a reversible capacity of 1723 mAg/h.

Investigations into the graphene paper produced by Hu et al.[17] show that paper thickness has a significant impact on storage capacity. The electrochemical experiments are performed on graphene papers that are 1.5 and 3 m thick. The reversible capacity of the 1.5 m thick paper is 200 mAh/g, while the 3 m thick material is 140 mAh/g. The findings indicate that due to GNs being restacked, capacity declines as paper thickness increases.

The behavior of graphene sheets with carbon coatings was researched by Yang et al.[18] These results validate the rate capability of composite anodes and other anodes based on graphene. The thin layer of carbon enhances the rate capacity and cycle life at various current rates.

3.1.2 CATHODE MATERIALS FOR ELECTRODES

Graphene-integrated materials used as cathodes in energy devices researched predominantly to enrich the overall performance of the devices.

He and colleagues[19] created $Li[Li_{0.2}Mn_{0.5}4Ni_{0.13}Co_{0.13}]O_2$ with graphene coating using a spray-drying technique. The scientists' findings confirm that adding a graphene coating to a device might improve its rate and efficiency. By using a hydrothermal technique, Sun et al.[20] created rGO-based composite films containing free-standing $VO_{2.07}$. High reversible capacity of 160 mAhg1 and strong capacity retention of 83% after 200 cycles are shown in the results. By using the coprecipitation approach, Ding et al.[21] created nano-$LiMn_1/3Ni_{1/3}$ $Co1/3O_2$ particles based on graphene. Graphene was evenly deposited on prepared nanostructures that were around 50 nm in size. The coated nanostructures show improved discharge in the electrochemical experiments, with a capacity retention of 95.5%.

Sandwiched composite structures based on graphene were created by Prabakar et al.[22] The substance showed an effective conducting network that prevented LiNi0.5Mn1.5O4 from forming a solid–electrolyte contact. The produced composite structure containing 2.5% graphene demonstrated higher electrochemical results in terms of rate capability. $LiNi_{0.5}Mn_{1.5}O_4$ showed good capacity retention, which is ascribed to graphene's ability to create enough conducting channels by preventing electrolyte breakdown.

Wang and colleagues[23] demonstrated a carbon-coated $LiFePO_4$ modified with rGO. With only 5% rGO, the cathode material demonstrated extremely high specific capacity. The cathode material's rate performance is thought to be dominant, with discharge capacities of 160.4 mAhg1 and 115.0 mAhg1 at 0.2°C and 20°C, respectively. The materials have also shown higher cycling stability, with 10% degradation at 10C after 1000 cycles.

3.2 GRAPHENE-INTEGRATED ELECTRODES FOR LITHIUM-ION BATTERIES (LIBS)

The world has been engulfed by battery technology. Batteries have had a significant influence on our modern lives by allowing technologies to be portable and long lasting. The LIB is now the most dominant and commonly used battery (LIB). LIBs are preferred over other batteries due to their unique performance characteristics, which include (i) high gravimetric and volumetric energy density, (ii) high coulombic efficiency and energy efficiency, (iii) extended cycle life, (iv) no memory effect, and (v) design flexibility. No surprise, LIBs have become an essential component of many modern products and appliances, including smart phones, laptop computers, and other portable electronic devices. Because of the high expectations placed on LIBs, there is a growing trend of research into producing next-generation LIBs for electric cars (EVs).

3.2.1 GRAPHENE-INTEGRATED CATHODES

The three most frequently used metal oxide structures for cathode materials are (i) the olivine LiFePO4 structure, which is regarded as a secure alternative with outstanding thermal stability and practical capacity (170 $mAhg^{-1}$) but has the significant drawback of having a low ionic and lithium diffusion rate. (ii) The spinel-structured $LiMn_2O_4$ (LMO) is unable to give a high practical capacity (120 $mAhg^{-1}$), although having the advantages of a good ion diffusion rate and structural stability. (iii) The $LiMO_2$ [M = Co, Mn, Ni] layered structure, which essentially generates the largest capacity (180 $mAhg^{-1}$) but lacks in chemical and/or structural stability with modification in lithium concentration during cycling. The synergistic impact of graphene can improve the practical capacity, electrical conductivity, chemical, and structural stability of metal oxides.

3.2.1.1 GRAPHENE/$LiFePO_4$ AS CATHODE

$LiFePO_4$ (LFP) is the primary focus of current research on flexible cathode materials. The strong P-O bonds in the LFP structure give it a high level of stability. Tetrahedral sites are occupied by phosphorus, whereas octahedral sites are occupied by Li^+ and Fe^{2+}, and it is noted that one-dimensional diffusion channel is created by this type of configuration, and Li^+ diffusion is only possible in the edge-shared LiO_6 units. Impurities or defects that are present along the 1D diffusion channel length further obstruct the diffusion as well. The low electrical conductivity (10^{-9} Scm^{-1}) and low Li^+ diffusion rate (~10–14 cm^2s^{-1}) prevent the capacity from increasing. The use of graphene, which offers a conductive network, can overcome these two shortcomings.

LFP/graphite electrode, developed by Ding in 2010, has been fully utilized as a flexible LIB cathode material. For the LFP/rGO electrode, Ding's group was able to attain a capacity of 160 $mAhg^{-1}$ at 0.2°C, whereas the greatest capacity recorded by bare LFP is only 113 $mAhg^{-1}$.[24] Encouraged by the outcomes, Zhou et al.[25] thought about the advantages of using nanochemistry to create composites of LFP and graphene. The team was able to properly spread LFP nanoparticles on the 3D graphene nanostructures by using a straightforward spray-drying technique. As a result, Li^+ with a high capacity of 70 $mAhg^{-1}$ at 60°C diffused readily. A similar method was used to create LFP/rGO with a carbon coating, which was an in situ synthesis of LFP nanoparticles.[26]

3.2.1.2 GRAPHENE/$LiMn_2O_4$ AS CATHODE

With Li^+ occupying the tetrahedral sites and Mn ions occupying the octahedral sites, LMO is an oxygen-containing compound with a cubic close-packed arrangement of oxygen. A 3D channel is used for lithium diffusion. The Jahn-Teller effect of the high spin Mn^{3+}:$3d^4(t^3_{2g}e^1_g)$ causes capacity fading in the LMO at the operational voltage, causing a phase transition (cubic to tetragonal). While cycling, Mn^{2+} dissolves into the electrolyte, reducing the metal oxide's reversible capacity to a meager 100 $mAhg^{-1}$. There have been a variety of creative ways to address these real-world issues, but one that stands out is the usage of graphene. During the intercalation and de-intercalation of Li^+ from and into the electrode, the conductive network of graphene is essential in enhancing the lithium diffusion.

A sandwiched graphene and Ni doped LMO structure was created by Prabakar and his colleagues.[27] The high coulombic efficiency of the graphene/$LiNi_{0.5}Mn_{1.5}O_4$ structure was attained by graphene's ability to resist electrolytic breakdown on the electrode surface. In comparison to the undoped electrode, the electrode's capacity and cyclic retention increased (107 $mAhg^{-1}$ at 50°C).[28] A graphene/LMO cathode with a conformal coating of graphene on LMO NPs has been created by Chen and colleagues. Due to the effective charge transfer that results from such a coating, high cycle efficiency may be maintained even at high rates and low temperatures. The investigated electrode was able to maintain 96% of its capacity at 0.2°C at −20°C.[29]

Recently, composites containing many types of metal oxides have come to light as a possible replacement for the materials used in LIB cathodes. A capacity of 129.3 $mAhg^{-1}$ (at 1°C) and a good capacity retention of 89.3% after 500 cycles have been observed in $Y_2O_3/LiMn_2O_4$ wrapped graphene.[30] Similar results were drawn for graphene/ZnO/LMO, which demonstrated good electrochemical performance and cyclic stability.[31]

3.2.1.3 GRAPHENE-LAYERED CATHODE MATERIAL

Transition metals, such as Co, Mn, or Ni are typically present in the layered lithium metal oxides ($LiMO_2$) that are frequently used as cathode materials in LIBs. Although the layered structures offer great capacity, these oxides' structural and chemical stability remain under doubt. The metal ions from the metal octahedral sites switch positions with the lithium octahedral sites during cycling via a nearby tetrahedral site.[32] If the octahedral site stabilization energy (OSSE) is low, this type of structural instability will be more prominent. Metal ions' OSSE values are in the following order: Co > Ni > Mn. This is the cause of Mn^{3+} simple migration and conversion from layered to spinel structure. The most stable metal oxide structure is provided by Co^{3+}, but when more than 50% of the lithium in $LiCoO_2$ is extracted, Co^{3+} becomes chemically unstable. When compared with Mn, Co is also more costly and poisonous; the toxicity of Ni sits between Co and Mn. However, $LiNiO_2$ is not a good option for LIBs since Ni^{3+} also experiences chemical instability when the concentration of lithium varies. Furthermore, since Ni^{3+} decomposes at temperatures above 250°C, creating stoichiometric $LiNiO_2$ is difficult. The failure of layered architectures with a single transition metal oxide to produce a commercially viable electrode material can be attributed to these serious drawbacks. Instead, within the last 10 years, scientists have become interested in layered structures made of $Li(Ni_xMn_yCo_{1-x-y})O_2$ ($0 \leq x, y \leq 1$ and $0 \leq x+y \leq 1$), commonly known as NMC electrodes, which include mixed transition metal oxides. The NMC cathode has a high charge capacity like that of $LiNiO_2$, structural stability from Mn^{3+}, and a sizable rate capability similar to that of $LiCoO_2$.[33] In addition, NMC cathodes are less expensive and hazardous than $LiCoO_2$.

In order to generate cathode material with good electrochemical performance, a variety of synthetic techniques have been used to halt the synergism of graphene and NMC. As an illustration, spray-drying and heat treatment were used to create NMC particles that were wrapped in rGO. With the addition of rGO, the capacity of NMC increased from 6 $mAhg^{-1}$

to 88.5 $mAhg^{-1}$ (at 5°C).[34] By replacing the carbon black slurry with rGO during the production of the electrode, it is simple to introduce rGO as an active material. The composite electrode was able to produce 110 $mAhg^{-1}$ and 55 $mAhg^{-1}$ capacities at 6°C and 20°C, respectively, when rGO was used to replace half of the slurry. The pure NCM cathode material unexpectedly showed zero capacity up to 6°C in the absence of rGO.[35]

3.2.2 GRAPHENE-INTEGRATED ANODES

For a high-performance LIB, the cathode should have a high voltage while the voltage should be low for anode. At the same time, both the electrodes must produce high capacity and high energy density. Mechanical stability, a large surface area, strong electrical conductivity, and good charge carrier mobility are only a few of the benefits of graphene. Since Li and C have a relatively low binding energy, the LiC_6 phase is unstable. Lithium storage capacity is also low because to the coulombic repulsion between lithium atoms positioned in the graphene layer's opposing orientations. The formation of the solid–electrolyte interface (SEI) layer, which prohibits the contact between the electrode and the electrolyte, and the dendrite growth are the two main obstacles to create anodes with the necessary properties.

The electrode surface area is highly associated with SEI development. Due to Li's high reactivity, some of the Li ions that are released during cycling are instead trapped in the SEI layer during SEI creation. Higher surface area of the electrodes use more Li during SEI development. The initial irreversible capacity rises as a result of the irreversible Li insertion in the SEI layer. Hence, an ideal SEI layer has to be thin and dense is ion conducting yet electrically insulating. Since SEI coatings are fragile in reality, they may ultimately shatter through repeated charge/discharge cycles, which would worsen dendritic formation. The main anode material for LIBs is graphene, and graphene has these aforementioned two issues. As an alternative, the organic framework can be combined with the host materials through (i) insertion (such as lithium titanate), (ii) alloying (such as Si, Sn, or Ge), and (iii) conversion.

3.2.2.1 GRAPHENE/$Li_4Ti_5O_{12}$ AS ANODE

$Li_4Ti_5O_{12}$ (LTO) is considered as one of the prominent anode for LIBs as it has zero strain insertion reaction of Li without any structural change, which provides good cycle life for batteries. It also has a relatively high operating voltage (1.55 V vs Li/Li$^+$), which prevents the production of lithium dendrites and makes it more stable and safer than other electrode materials. However, it has certain inherent limitations, including low electrical conductivity (<10^{-13} Scm^{-1}) and low Li$^+$ diffusion. In general, these difficulties are solved by mixing LTO with carbon compounds such as graphene.

The sol–gel technique is a well-known synthetic method for producing LTO particles with uniform dispersion. Graphene influences the size and aggregation of LTO particles. For example, using the sol–gel process, LTO particles with sizes ranging from 200 to 500 nm have been synthesized to produce graphene/LTO nanocomposite with increased rate capability. During the 1000 charge–discharge cycles, the electrode's reversible capacity grew from 44.1 mAhg^{-1} to 108.4 mAhg^{-1} at 10°C.[56] Many attempts have been documented on creating graphene/LTO composites with better rate capability and high capacity.[37] To advance the research, Jeong et al. employed N-doped rGO to scatter LTO particles. The importance of nitrogen is to promote electrical conductivity and prevent graphene layer restacking. LTO/N-doped rGO achieved a maximum electrical conductivity of 1.60 Scm^{-1} at 2.3 wt% nitrogen.[38]

3.2.2.1 GRAPHENE/METAL OXIDES AS ANODES

3.2.2.1.1 Transition Metal Oxides

Transition metal oxides are one of the promising materials in battery technology, the transition metals oxides interact with Li by conversion mechanism. In this mechanism, the active material undergoes complete reduction from metal. Not only metal oxides, other transition elements, such as sulfides, phosphides, and nitrides follow the same mechanism. Co_3O_4 nanoparticles exhibit exceedingly high reversible electrochemical reactions, making them a likely candidate for use as an active electrode material. The porous characteristic of graphene, on the other hand, inhibits any significant volume change in Co_3O_4 during conversion and helps to improve conductivity.[39] Considering the advantages of graphene, 3D sandwich-structured Co_3O_4/graphene and grapheme-coated hybrid Co_3O_4 having excellent electrochemical performance have been studied as anode electrodes for LIBs..[40–42] Zhu et al. and group have fabricated nanostructured Co_3O_4 electrodes for LIBs. The group followed two-step calcination technique to prepare hollow Co_3O_4 nanoparticles in nitrogen, sulfur and

Co-doped rGO framework. The porous structure of metal oxides is responsible for the high performance of the material as electrode, the fabricated electrode exhibits a capacity of 1590 $mAhg^{-1}$ at 1 mAg^{-1} after 600 cycles.[43]

The core-void-shell structure is considered to be one of the best alternative structures to provide higher electrochemical behaviors. The shell is the hollow structure of graphene which provides or act as booster to the electrochemical performance of the metal oxides occupying the core, further it also reduces agglomeration of the nanoparticles. The presence of void in the core-void-shell structure provides large volume changes, the change in volume offers electrode stability during the charge–discharge cycling. The increased surface area also increases the number of reactive sites on the electrode surface, which improves rate capability. Fe_3O_4 nanoparticles were enclosed in a hollow graphene shell and the resulting composite electrode had a high capacity of 1236.6 $mAhg^{-1}$, outperforming pure Fe_3O_4 or a simple physical mixture of graphene and Fe_3O_4.[44] The magnetic nanoparticles have also used with batteries as a supporting material, such combination provides improved electronic and ionic diffusion resulting in better and improved capacity.[45]

3.2.2.1.2 Tin Oxides (SnO_2)

SnO_2 is one of the widely appreciated metal oxide in battery applications, SnO_2 can be used as an efficient anode due to its high theoretical capacity, that is, 782 $mAhg^{-1}$. However, when subjected to high volume change, SnO_2 shows decreased electrochemical behaviors. The interaction of SnO_2- lithium is obtained by two-step "conversion-alloying" mechanism. Further, as loke transition metal oxides SnO_2 also converted to metal form after interacting with lithium in the first step and in the second step, stable alloy of Li_4Sn is formed. The formation of Li_4Sn is shown in the below equation.

$$SnO_2 + 4Li^+ + 4e^- \rightarrow Sn + 2Li_2O \tag{3.1}$$

$$Sn + xLi^+ + xe^- \rightarrow Li_xSn\ (0 \le x \le 4) \tag{3.2}$$

The advantages of SnO_2 in batteries created a road map to functionalize and coat them with other electrochemically active materials. Graphene-based carbon-coated SnO_2 have been prepared using glucose as a soft template, which controls the growth of SnO_2. The prepared electrode delivers a specific capacitance of 2238.2 $mAhg^{-1}$ at 0.1°C and exhibit a capacity retention of 1467.8 $mAhg^{-1}$ after 150 cycles.[46]

Arnaiza and group fabricated SnO_2/rGO electrodes, these electrodes when combined with activated carbon produced exceptionally good energy density of 60 $Whkg^{-1}$.[47] Many other relevant studies on the durability of SnO_2 as electrode was studied by many researchers, SnO_2/S-doped rGO in the presence of sulfur delivers a capacity of 897 $mAhg^{-1}$ with good capacity retention of 88% at the end of 500 cycles.[48] Further, flexible "graphene"-based electrodes with fluoride doped and combined with metal oxides also been reported with exceptionally good electrochemical behavior and charge–discharge capacity.[49,50]

3.2.2.1.3 Titanium Oxides (TiO_2)

TiO_2 metal oxides also have considered one of the best materials for battery applications, TiO_2, like LTO, are said to follow the same insertion and extraction mechanism Li-ions without disturbing any lattice structures. However, TiO_2 also suffers from low conductivity like other metal oxides. This behavior of TiO_2 can be improved by incorporating graphene into electrodes.

The ordered structure of TiO_2/N-doped graphene have been fabricated as anode electrodes. The fabricated electrodes have large surface area with porous structure. The doped TiO_2 anodic electrode exhibits higher ion and electron transport in the batteries. The anodic electrode display increased conductivity and electrochemical behavior with 165 $mAhg^{-1}$ at 1°C at 200 cycles.[51]

Wang et al. have studied the influence of graphene content as primary battery electrode material; graphene is found to be the effective material to improve the electrochemical performance as electrode material. It is found that graphene-based composites improve the capacity to 264 $mAhg^{-1}$ at 1°C with the capacity retention of 171 $mAhg^{-1}$.[52]

3.3 CONCLUSIONS

The number of papers on innovative cathodic and anodic materials for LIBs has increased dramatically during the past 10 years. As a scaffolding material, graphene has made a significant contribution to improving the active material's capacity, rate capability, and cycle efficiency. Because of its high conductivity (for quicker electron/ion transport), defect sites (for active lithiation/delithiation), and porous structure, graphene works well as an electrode material (for maximizing Li storage capacity and diminishing large volume and structural changes during cycling). In addition to these intrinsic qualities, graphene has the capacity to disperse the active ingredient uniformly and is resistant to particle aggregation during repeated cycling. Additionally, graphene offers the mechanical resilience that is crucial for real-world uses of a free-standing electrode.

The versatility of graphene is found in the different structural configurations it can easily take on, with sandwich-like, core-shell, and hierarchical structures drawing the most interest. There is still room for improvement even if graphene-based electrodes appear promising for the next-generation LIB technology. High surface area and conductivity must be achieved in the design of the composite electrodes. In reality, by concentrating on improving the active materials' ability to convert energy into electrical energy, a significant amount of the discharge capacity and coulombic efficiency waste may be reduced. The lack of understanding of electrode–electrolyte interactions is another significant gap in the study on electrodes made of graphene. A significant portion of the reported composite electrodes experienced early capacity loss.

The choice of active substances and electrolytes for high-performance batteries will undoubtedly benefit from a fuller understanding of the mechanisms behind interactions that take place at these interfaces. Future graphene-based electrode perspectives should focus on creating an easy and economical technology to create composite electrodes for large-scale manufacturing. Next-generation batteries must be created with a suitable electrolyte, a high voltage cathode, and a low-voltage anode. Flame retardants, which can reduce the risk of fire and so increase battery safety, will also be an intriguing addition.

KEYWORDS

- **batteries**
- **supercapacitors**
- **solar**
- **fuels cells**
- **metal oxides**
- **nanoparticles.**

REFERENCES

1. Cong, H. P.; Chen, J. F.; Yu, S. H. Graphene-Based Macroscopic Assemblies and Architectures: An Emerging Material System. *Chem. Soc. Rev.* **2014,** *43*, 7295–7325.
2. Han, C.; Zhang, N.; Xu, Y. J. Structural Diversity of Graphene Materials and their Multifarious Roles in Heterogeneous Photocatalysis. *Nano Today* **2016,** *11,* 351–372.
3. Goodenough, J. B. Energy Storage Materials: A Perspective. *Energy Storage Mater.* **2015,** *1,* 158–161.
4. Geim, A. K.; Novoselov, K. S. The Rise of Graphene. *Nat. Mater.* **2007,** *6,* 183–191.
5. Lockett, M.; Sarmiento, V.; Balingit, M.; Oropeza-Guzman, M. T.; Vazquez-Mena, O. Direct Chemical Conversion of Continuous CVD Graphene/Graphite Films to Graphene Oxide Without Exfoliation. *Carbon* **2019,** *158* (1), 202–209.
6. Wang, X.; Shi, G. Flexible Graphene Devices Related to Energy Conversion and Storage. *Energy Environ. Sci.* **2015,** *8,* 790.
7. Novoselov, K. S.; Falko, V. I.; Colombo, L.; Gellert, P. R.; Schwab, M. G.; Kim, K. A Roadmap for Graphene. *Nature* **2012,** *490,* 192–200.
8. Mao, M.; Hu, J.; Liu, H. Graphene-based Materials for Flexible Electrochemical Energy Storage. *Int. J. Energy Res.* **2015,** *39,* 727–740.
9. Hu, K.; Kulkarni, D. D.; Choi, I.; Tsukruk, V. V. Graphene-Polymer Nanocomposites for Structural and Functional Applications. *Prog. Polym. Sci.* **2014,** *39,* 1934–1972.
10. Bonaccorso, F.; Colombo, L.; Yu, G. H.; Stoller, M.; Tozzini, V.; Ferrari, A. C.; Ruoff, R. S.; Pellegrini, V. Graphene, Related Two-dimensional Crystals, and Hybrid Systems for Energy Conversion and Storage. *Science* **2015,** *347,* 124651.
11. Raccichini, R.; Varzi, A.; Passerini, S.; Scrosati, B. The Role of Graphene for Electrochemical Energy Storage. *Nat. Mater.* **2015,** *14,* 271–279.
12. Bharwaj, T.; Antic, A.; Paven, B.; Barone, V.; Fahlman, B. D. Enhanced Electrochemical Lithium Storage by Graphene Nanoribbons. *J. Am. Chem. Soc.* **2010,** *132,* 12556–12558.
13. Uthaisar, C.; Barone, V.; Fahlman, B. D. On the Chemical Nature of Thermally Reduced Graphene Oxide and Its Electrochemical Li Intake Capacity. *Carbon* **2013,** *61,* 558–567.
14. Lee, W.; Suzuki, S.; Miyayama, M. Lithium Storage Properties of Graphene Sheets Derived from Graphite Oxides with Different Oxidation Degree. *Ceram. Int.* **2013,** *39,* S753–S756.
15. Kuo, S. L.; Liu, W. R.; Wu, H. C. Lithium Storage Behavior of Graphene Nanosheets-based Materials. *J. Chin. Chem. Soc.* **2012,** *59* (10), 1220–1225.
16. Liu, Y.; Fan, Q.; Tang, N.; Wan, X.; Liu, L.; Lv, L.; Du, Y. Study of Electronic and Magnetic Properties of Nitrogen Doped Graphene Oxide. *Carbon* **2013,** *60,* 538–561.
17. Hu, Y.; Li, X.; Geng, D.; Cai, M.; Li, R.; Sun, X. Influence of Paper Thickness on the Electrochemical Performances of Graphene Papers as an Anode for Lithium Ion Batteries. *Electrochem. Acta* **2013,** *91,* 227–233.
18. Yang, Y. Q.; Wu, K.; Pang, R. Q.; Zhou, X. J.; Zhang, Y.; Wu, X. C.; Wu, C. G.; Wu, H. X.; Guo, S. W. Graphene Sheets Coated with a Thin Layer of Nitrogen-enriched Carbon as a High-performance Anode for Lithium-ion Batteries. *RSC Adv.* **2013,** *3,* 14016–14020.
19. He, Z.; Wang, Z.; Guo, H.; Li, X.; Xianwen, W.; Yue, P.; Wang, J. (2013). A Simple Method of Preparing Graphene coated $Li[Li_{0.2}Mn_{0.5}4Ni_{0.13}Co_{0.13}]O_2$ for Lithium-Ion Batteries. *Mater. Lett.* **2013,** *91,* 261–264.
20. Sun, Y.; Yang, S. B.; Lv, L. P.; Lieberwirth, I.; Zhang, L. C.; Ding, C. X.; Chen, C. H. A Composite Film of Reduced Graphene Oxide Modified Vanadium Oxide Nanoribbons

as a Free Standing Cathode Material for Rechargeable Lithium Batteries. *J. Power Sour.* **2013,** *241,* 168–172.

21. Ding, Y. H.; Ren, H. M.; Huang, Y. Y.; Chang, F. H.; He, X.; Fen, J. Q.; Zhang, P. (2013). Co-precipitation Synthesis and Electrochemical Properties of Graphene Supported $LiMn_1/3Ni_1/3Co_1/3O_2$ Cathode Materials for Lithium-ion Batteries. *Nanotechnology* **2013,** *24,* 375401–375408.
22. Prabakar, S. J. R.; Hwang, Y. H.; Lee, B.; Sohn, K. S.; Pyo, M. Graphene-Sandwiched $LiNi_{0.5}Mn_{1.5}O_4$ Cathode Composites for Enhanced High Voltage Performance in Li Ion Batteries. *J. Electrochem. Soc.* **2013,** *160* (6), A832–A837.
23. Wang, B.; Wang, D.; Wang, Q.; Liu, T.; Guo, C.; Zhao, X. Improvement of the Electrochemical Performance of Carbon-coated $LiFePO_4$ Modified with Reduced Graphene Oxide. *J. Mater. Chem. A* **2013,** *1,* 135–144.
24. Ding, Y.; Jiang, Y.; Xu, F.; et al. Preparation of Nano-structured $LiFePO_4$/Graphene Composites by Co-precipitation Method. *Electrochem. Commun.* **2010,** *12,* 10–13.
25. Zhou, X.; Wang, F.; Zhu, Y.; et al. Graphene Modified $LiFePO_4$ Cathode Materials for High Power Lithium Ion Batteries. *J. Mater. Chem.* **2011,** *21,* 3353–3358.
26. Ha, S. H.; Lee, Y. J. Core-shell $LiFePO_4$/Carbon-Coated Reduced Graphene Oxide Hybrids High-power Lithium-ion Battery Cathodes. *Chemistry* **2015,** *21,* 2132–2138.
27. Prabakar, S. J. R.; Hwang, Y. H.; Lee, B.; et al. Graphene-sandwiched $LiNi_{0.5}Mn_{1.5}O_4$ Cathode Composites for Enhanced High Voltage Performance in Li Ion Batteries. *J. Electrochem. Soc.* **2013,** *160,* A832–A837.
28. Sreelakshmi, K. V.; Sasi, S.; Balakrishnan, A.; et al. Hybrid Composites of $LiMn_2O_4$–Graphene as Rechargeable Electrodes in Energy Storage Devices. *Energy Technol.* **2014,** *2,* 257–262.
29. Chen, K. S.; Xu, R.; Luu, N. S.; et al. Comprehensive Enhancement of Nanostructured Lithium-ion Battery Cathode Materials via Conformal Graphene Dispersion. *Nano Lett.* **2017,** *17,* 2539–2546.
30. Ju, B.; Wang, X.; Wu, C.; et al. Electrochemical Performance of the Graphene/Y_2O_3/$LiMn_2O_4$ Hybrid as Cathode for Lithium-ion Battery. *J. Alloys Comp.* **2014,** *584,* 454–460.
31. Aziz, S.; Zhao, J.; Cain, C.; et al. Nanostructured $LiMn_2O_4$/Graphene/ZnO Composites as Electrodes for Lithium Ion Batteries. *J. Mater. Sci. Technol.* **2014,** *30,* 427–433.
32. Chebiam, R. V.; Prado, F.; Manthiram, A. Structural Instability of Delithiated $Li_{1-x}Ni_{1-y}Co_yO_2$ Cathodes. *J. Electrochem. Soc.* **2001,** *148,* A49−A53.
33. Choi, J.; Manthiram, A. Role of Chemical and Structural Stabilities on the Electrochemical Properties of Layered $LiNi_{1/3}Mn_{1/3}Co_{1/3}O_2$ Cathodes. *J. Electrochem. Soc.* **2005,** *152,* A1714.
34. He, J. R.; Chen, Y. F.; Li, P. J.; et al. Synthesis and Electrochemical Properties of Graphene Modified $LiCo_{1/3}Ni_{1/3}Mn_{1/3}O_2$ Cathodes for Lithium Ion Batteries. *RSC Adv.* **2014,** *4,* 2568–2572.
35. Jiang, K. C.; Xin, S.; Lee, J. S.; et al. Improved Kinetics of $LiNi_{1/3}Mn_{1/3}Co_{1/3}O_2$ Cathode Material through Reduced Graphene Oxide Networks. *Phys. Chem. Chem. Phys.* **2012,** *14,* 2934–2939.
36. Liu, H. P.; Wen, G. W.; Bi, S. F.; et al. High Rate Cycling Performance of Nanosized $Li_4Ti_5O_{12}$/Graphene Composites for Lithium Ion Batteries. *Electrochimica Acta* **2016,** *192,* 38–44.
37. Cao, N.; Wen, L. N.; Song, Z. H., et al. $Li_4Ti_5O_{12}$/Reduced Graphene Oxide Composite as a High-rate Anode Material for Lithium Ion Batteries. *Electrochimica Acta* **2016,** *209,* 235–243.

38. Jeong, J. H.; Kim, M. S.; Kim, Y. H.; et al. High-rate $Li_4Ti_5O_{12}$/N-doped Reduced Graphene Oxide Composite Using Cyanamide both as Nanospacer and a Nitrogen Doping Source. *J. Power Sour.* **2016,** *336,* 376–384.
39. Qu, G. L.; Geng, H. B.; Ge, D. H.; et al. Graphene-coated Mesoporous Co_3O_4 Fibres as an Efficient Anode Material for Li-ion Batteries. *RSC Adv.* **2016,** *6,* 71006–71011.
40. Yang, Q.; Wu, J.; Huang, K.; et al. Layer-by Layer Self-assembly of Graphene-like Co_3O_4 Nanosheet/Graphene Hybrids: Towards High-performance Anode Materials for Lithium-ion Batteries. *J. Alloys Comp.* **2016,** *667,* 29–35.
41. Yang, Y.; Huang, J.; Zeng, J.; et al. Direct Electrophoretic Deposition of Binder-free Co_3O_4/Graphene Sandwich-like Hybrid Electrode as Remarkable Lithium Ion Battery Anode. *ACS Appl. Mater. Interfaces* **2017,** *9,* 32801–32811.
42. Geng, H. B.; Guo, Y. Y.; Ding, X. G.; et al. Porous Cubes Constructed by Cobalt Oxide Nanocrystals with Graphene Sheet Coatings for Enhanced Lithium Storage Properties. *Nanoscale* **2016,** 8, 7688–7694.
43. Zhu, J.; Tu, W.; Pan, H.; et al. Self-templating Synthesis of Hollow Co_3O_4 Nanoparticles Embedded in N,S-dual Reduced Graphene Oxide for Lithium Ion Batteries. *ACS Nano* **2020**. DOI: 10.1021/acsnano.0c00712.
44. Jiang, Y.; Jiang, Z. J.; Yang, L. A High-Performance Anode for Lithium Ion Batteries: Fe_3O_4 Microspheres Encapsulated in Hollow Graphene Shells. *J. Mater. Chem. A* **2015,** *3,* 11847–11856.
45. Zhang, N.; Yan, X.; Huang, Y.; et al. Electrostatically Assembled Magnetite Nanoparticles/ Graphene Foam as a Binder-free Anode for Lithium Ion Battery. *Langmuir* **2017,** *33,* 8899–8905.
46. Zhang, Q.; Gao, Q.; Qian, W.; et al. Graphene-based Carbon Coated Tin Oxide as a Lithium Ion Battery Anode Material with High Performance. *J. Mater. Chem. A* **2017,** *5,* 19136–19142.
47. Arnaiza, M.; Botasa, C.; Carriazoa, D.; et al. Reduced Graphene Oxide Decorated with SnO_2 Nanoparticles as Negative Electrode for Lithium Ion Capacitors. *Electrochimica Acta* **2018,** *284,* 542–550.
48. Wu, K.; Shi, B.; Qi, L.; et al. SnO_2 Quantum dots@3D Sulfur-doped Reduced Graphene Oxides as Active and Durable Anode for Lithium Ion Batteries. *Electrochimica Acta* **2018,** *291,* 24–30.
49. Xu, H.; Shi, L.; Wang, Z.; et al. Fluorine-doped Tin Oxide Nanocrystal/Reduced Graphene Oxide Composites as Lithium Ion Battery Anode Material with High Capacity and Cycling Stability. *ACS Appl. Mater. Interfaces* **2015,** *7,* 27486–27493.
50. Guo, J. X.; Zhu, H. F.; Sun, Y. F.; et al. Flexible Foams of Graphene Entrapped SnO_2-Co_3O_4 Nanocubes with Remarkably Large and Fast Lithium Storage. *J. Mater. Chem. A* **2016,** *4,* 16101–16107.
51. Jiang, X.; Yang, X.; Zhu, Y.; et al. 3D Nitrogen-doped Graphene Foams Embedded with Ultrafine TiO_2 Nanoparticles for High-performance Lithium-Ion Batteries. *J. Mater. Chem. A* **2014,** *2,* 11124–11133.
52. Wang, J. F.; Zhang, J. J.; He, D. N. Flower-like TiO_2-B Particles Wrapped by Graphene with Different Contents as an Anode Material for Lithium-Ion Batteries. *Nanostruct. Nanoobjects* **2018,** *15,* 216–223.

CHAPTER 4

Role of Nanomaterials in Lithium-Ion Batteries

B. P. SHIVAMURTHY[1,2], G. SANTHOSH[3], and G. P. NAYAKA[1,2]

[1]*Physical and Materials Chemistry Division, CSIR-National Chemical Laboratory, Pune, Maharashtra, India*

[2]*Academy of Scientific and Innovative Research (AcSIR) Ghaziabad, Uttar Pradesh, India*

[3]*Department of Mechanical Engineering, NMAM Institute of Technology, Nitte, Karnataka, India*

ABSTRACT

Nanomaterials for the application of batteries have been studied significantly in past decades. These studies show that more works must be done on this subject in order to be able to better investigate the application of nanomaterials with different morphology and size for flexible printed batteries. In this chapter we will review the impact of nanomaterials on the development of high performing and low-cost electrode materials for lithium-ion batteries.

4.1 INTRODUCTION

Nanomaterials are not new; people unknowingly came across nano-nano-materials-based objects before the arrival of the nanotechnology concept. History has a lot of nanomaterials proofs including the usage of nanomaterials during the early civilization life stage of the ancient Romans, ancient

Technological Advancement in Clean Energy Production: Constraints and Solutions for Energy and Electricity Development. Amritanshu Shukla, Kian Hariri Asli, Neha Kanwar Rawat, Ann Rose Abraham, & A. K. Haghi (Eds.)

Egyptians, ancient Indians, Mesopotamia, and Maya. The most popular examples are the Lycurgus cup (4th century CE) used by ancient Romans and the nanomaterial-based hair dyes used by ancient Egyptians. In the last few decades, the nanotechnology area exploded widely. Across all the fields, a lot of nanomaterials have been prepared and ruling the materials science that can be coined as the era of nanotechnology. The unique mechanical, electrical, and optical properties of nanomaterials made them apply in various fields such as catalysis, energy storage, water storage, medicines, agriculture, etc., here the discussion moves on to the application of nanomaterials in the energy storage sector, especially concerning the batteries.

The global energy demands continuously increase as technology grows and the population increases. Energy production and storage is one of the fundamental challenges of the 21st century and it gained a lot of importance in the present electric vehicles (EVs) era. The production of energy mainly relies on fossil fuels, efficient energy storage devices are the present need to store the energy which is generated from intermittent sources such as wind, solar, tidal, and hydroenergy, the integration of these into grid storage is increased interest in the energy storage sector. An electrochemical device that stores electrical energy in the form of chemical energy in electrodes during the charging process, and releases as during the discharge process, such kind of device is termed the battery.

Nanoscience has opened up a way to explore more possibilities in energy storage research, the application of nanomaterials for energy conversion and storage provides great opportunities to enhance the performance, energy density, and ease of transportation. Nanomaterials could be the first-choice materials in energy storage, especially for rechargeable batteries, which could enhance the material properties in favor of better performance and enable novel battery chemistries for sustainable growth. When compared to conventional batteries, the batteries with nanostructured materials offers greatly improved ionic and electronic conductivities which enables the efficient intercalation process by occupying all the intercalation sites with a faster diffusion of ions between the electrodes, which greatly enhances the specific capacity and energy density of the materials. Such kinds of unique capabilities of nanostructured electrodes are more tolerant toward the high current densities at the high rate charge/discharge process.

The fast-growing market of EVs has a large scope for nanomaterials in the production of components of vehicles. Precisely, rechargeable batteries are the most important part of EVs and are the focused area of research. On a very special note, nanotechnology is the development of electrode materials for batteries with high energy density, power density, low cost, high stability, and larger capacity retention performance.[1] The advancement of battery technology is not possible without the development of novel materials with suitable physicochemical properties of the materials in the nanoscale for high-performing batteries. The advanced cathode and anode materials for lithium (Li)-ion batteries (LIBs) majorly rely on the nanocomposite materials and nanometric thick surface coatings to improve materialistic properties which would better the ionic and electronic conduction pathways to avoid unwanted side reactions could enhance the performance of the battery with improved energy density and power density.[1]

Design and development of novel materials can reduce the cost, enhance performance efficiency, and increase life can bring significant progress in commercializing these technologies. In this point of view, nanostructured materials have great potential because of the unique properties shown by nanomaterials utilizing their smaller dimensions. In this regard, the nanomaterials are appealing as electrode materials to increase energy density and high-rate cyclability.

Here we present an overview of the impact of nanomaterials on the development of high-performing and low-cost electrode materials for batteries.

4.2 ADVANTAGES AND DISADVANTAGES OF NANOMATERIALS FOR RECHARGEABLE BATTERY APPLICATION

4.2.1 ADVANTAGES OF NANOMATERIALS

Nanotechnology can be used to improve the performance of battery materials in two manners one is nanoscale engineering of active substance in the electrode and the other one is utilizing nanotechnology to improve the active material by modifying the active materials by either nanocoating or using composite nanomaterials along with active materials. These techniques are advantageous in many ways to enhance the performance of battery materials. Some of the major advantages are listed as follows:

1. The reversible Li intercalation into mesoporous nanomaterials enables the electrode reactions that could not take place with microstructured materials proving the merit of nanomaterials.[2]
2. Nanosized particles enhance the electronic transfer kinetics within the nanomaterials, which would lead to the improved performance of the electrode materials.[3]

3. The characteristic time constant (t) and the diffusion constant (D) are inversely related to each other as in the equation $t = L^2/D$ and the square of the length of diffusion (L) varies directly with time. This indicates that lesser dimension particles can exhibit significantly high-rate Li-ion insertion and deinsertion reactions, which helps the faster diffusion kinetics to improve the electrode performance. The nanoparticles have lesser dimensions than microparticles, which can enable faster intercalation reactions. The faster transfer of Li^+ ions due to a shorter diffusion path can enhance the energy density and power density of the batteries.[3]
4. Low dimensions of nanomaterials are more advantageous than bulk materials as the faster ion transfer and oxidation reactions occur due to low polarization of the electrodes that makes faster penetration of ions with lesser diffusion pathway and energy density increases.[4]
5. The nanomaterials have a very high surface area and large surface-to-volume ratio which provides a larger contact area with electrolytes to form a better interface for the free flow of Li^+ ions. This increases interfacial Faradaic reactions on the electrode surface which enhances the capacity significantly.[5]
6. The electrode potentials of the reactions can be changed by the modification of the chemical potentials of the metal ions of the nanoparticles which optimizes the performance of the batteries.[6]
7. The material pulverization during the charge/discharge process causes performance decay in batteries, this could be protected by employing nanomaterials as the smaller nanoparticles can accommodate the structural changes during the reversible charge/discharge processes feasibly.[7]
8. The nanoparticles can accommodate the strain better during Li insertion/extraction of metal ions and could help the electrode active materials to increase the cycle life of the batteries.

4.2.2 DISADVANTAGES OF NANOMATERIALS

1. Synthetic methods of nanomaterials are comparatively difficult and control of parameters is quite complicated to get the least dimensions.
2. Nanomaterials with a very high surface area may lead to adverse parasitic reactions while charge/discharge or exposure with electrolyte, which may degrade the performance and it might be difficult to form a stable interparticle interface.

3. The difference in the density of nanoparticles and microparticles can vary the volume electrode density causing lower volumetric density.
4. The strong attractive interactions of very low-dimension materials are prone to continuous cycling provoked aggregation and surface passivation can also lead to capacity degradation.

4.3 CLASSIFICATION OF NANOMATERIALS FOR BATTERY APPLICATIONS

Nanomaterials can broadly be classified into organic, inorganic, and hybrid nanomaterials. The majority of the organic nanomaterials are carbonaceous materials such as graphitic nanomaterials, carbon nanofibers, carbon nanotubes (CNTs), single-walled CNTs, multiwalled CNTs, and fullerenes. In inorganic nanomaterials, the most common metals (Au, Ag, Cu, Zn, etc.), metal oxides (Fe_2O_3, CuO, ZnO, Al_2O_3, etc.), and quantum dots (CdSe, ZnS, etc.), and in hybrid nanomaterials, inorganic and carbonaceous nanomaterials combinations are best. The materials which display at least one of their properties in the nanoscale (< 100 nm) are termed nanomaterials. The nanomaterials can be categorized into four different types based on the dimensionalities of those materials which are represented in Figure 4.1.

4.3.1 ZERO-DIMENSIONAL (0D) NANOMATERIALS

This class of nanomaterials with all three dimensions is at the nanoscale. For example, quantum dots, fullerenes, etc.

The most common and successful method to control the morphology of nanoparticles is the use of additives. Such additives are generally termed surfactants (e.g., cetyl trimethyl ammonium bromide and sodium dodecyl sulfate), ligands (e.g., amines and thiols), polymers (e.g., polyvinylpyrrolidone and polyvinyl alcohol), and foreign ions (e.g., Ag^+, Cu^{2+}, Cl^-, Br^-).[8]

The morphology can be modified by generating defects in materials, such as twinning of particles, stacking, and the twist of grain boundaries. This could be an effective approach to tune up the morphology of nanomaterials and obtain unusual shapes such as octahedrons, stars, decahedrons, bipyramids, and multipods. The defect formation possibilities can be increased by slowing down the nucleation by employing mild reducing agents or by maintaining low-temperature reaction conditions.[12] The tropical morphologies, such as core-shell, heterodimer, and dumbbell, can be prepared by combining multiple components in nanoparticles such as the phase control and composition control of nanomaterials.[13] A lot of zero-dimensional (0D) nanomaterials have been used in LIBs by considering more advantages over bulk structures like faster Li^+ ion diffusion and quick electron transport, availability of high surface area for electrode–electrolyte interaction, and accommodation of volume changes due to charge–discharge cycles.

4.3.2 ONE-DIMENSIONAL (1D) NANOMATERIALS

This class of nanomaterials with any one dimension at the nanoscale is termed 1D nanomaterials. For example, nanotubes, nanofibers, nanorods, etc.

The solid nanofibers are synthesized commonly using a template-assisted route. In this case, the growth mechanism is simple and straight because the template directs the shape evolution. However, the process of development of one-dimensional (1D) nanomaterials gets complex when there is no physical template or in case of unavailability of an inherent crystallographic anisotropy.[9]

4.3.3 TWO-DIMENSIONAL (2D) NANOMATERIALS

This class of nanomaterials has two dimensions at the nanoscale and is termed 2D nanomaterials. For example, round disks, hexagonal/triangular/quadrangular plates, nanosheets, nanofilms, etc.

2D materials are also synthesized by using template-assisted methods, the morphology of these materials can be controlled and guided by the sacrificial templates (either soft or hard templates).[10] Recently, another most advantageous strategy through a direct liquid-phase exfoliation of bulk layered materials to prepare monolayered flakes or sheets on in bulk scale.[11]

4.3.4 THREE-DIMENSIONAL (3D) NANOMATERIALS

These are the bulk nanomaterials, any of the dimensions are not confined to the nanoscale. For example, embedded clusters, equiaxed crystallites, arrays of nanorods, dispersion of nanomaterials, etc.

The most popular and versatile method of preparation of 3D materials is the self-templating approach, which offers a wide variety of mechanisms to introduce voids, such as Ostwald ripening, Kirkendall effect, galvanic replacement, and heterogeneous contraction.[12,13] The 3D nanoparticle fabrication through self-assemblies can be done by controlled deposition of monodispersed nanoparticles over a substrate. Self-assembling of 3D nanoparticles happens due to an equilibrium between intermolecular forces.[14]

4.4 OVERVIEW OF NANOMATERIALS FOR RECHARGEABLE BATTERIES

There have been reported numerous experimental methods for the preparation and modification of nanomaterials to develop electrode materials for batteries. Such as (1) mechanical ball-milling (solid-state), (2) chemical-vapor deposition (CVD), (3) the template method, (4) electrochemical deposition, (5) hydrothermal reaction, (6) dehydration, (7) thermo-sintering, (8) pulsed laser deposition (PLD), (9) ultrasound sonication, (10) sol–gel synthesis, (11) colloidal synthesis, (12) microwave synthesis. The major components of the battery are the cathode, anode, and electrolyte.

4.4.1 NANOSTRUCTURED POSITIVE ELECTRODE MATERIALS

Cathode materials are the class of electrode materials that contribute more to the improvement of battery performance and also the maximum production cost depends on these materials. Mainly in the performance of LIBs governed by the active material used as the positive electrode, there are many different types of cathode materials have been reported so far, but $LiCoO_2$, $LiMn_2O_4$, and $LiFePO_4$ have been identified as prominent candidates for large-scale applications. Still, the practical specific capacity is less than 200 mAh g^{-1}. Therefore, research on the development of more efficient cathode materials is still underway and nanomaterials are the frontline runners in this regard. The latest trending other chemistries like Na-ion batteries, Zn batteries, Mg batteries, and other types of batteries also use a lot of nanomaterials.

The research group of Ref. [15] reported the nanostructured $Li(Ni_{0.86}Co_{0.10}Mn_{0.04})O_2$ and $Li(Ni_{0.70}Co_{0.10}Mn_{0.20})O_2$ materials as cathode materials for Li-ion batteries. Nanofunctional full concentration gradient (FCG) strategy is applied here to compare the effect of nanofunctional gradient cathode over the spherical nanostructures, the FCG materials show a very high capacity of 215 mAh g^{-1} and which is stable cycling over 1000 cycles of charge/discharge. The stable and long calendar life performance of the cathode is achieved by the nanostructural features of the FCG materials.[15]

Figure 4.1 shows scanning electron microscopic (SEM) images and the elemental distribution of Ni, Co, and Mn within a single particle of both the precursor ($(Ni_{0.75}Co_{0.10}Mn_{0.15})(OH)_2$) and the final lithiated product ($LiNi_{0.75}Co_{0.10}Mn_{0.15}O_2$) having a concentration gradient. The images revealed that the cocomposition ratio remains constant and the Ni concentration decreased from the core to the outer layer of the particle whereas the Mn concentration increased from the center to the surface layer of the particle, this indicates the proper concentration gradient material has been formed and which influenced on the performance of the battery.[15]

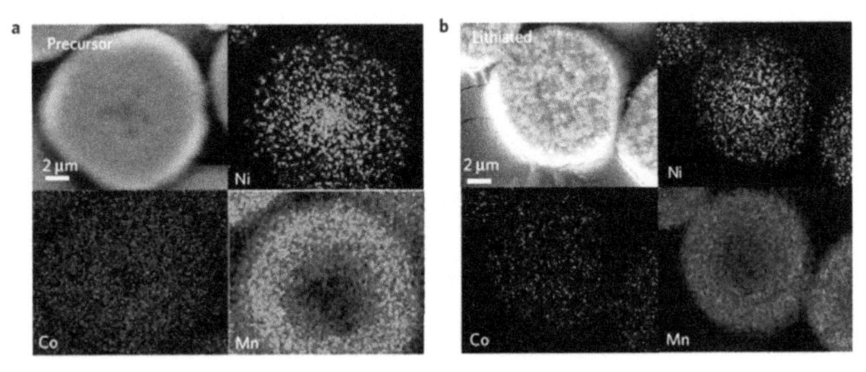

FIGURE 4.1 (a) SEM mapping photograph of Ni, Co, and Mn within a single particle for the precursor and (b) for the lithiated material.[15]

The electrolyte used is 1.2 M $LiPF_6$ in EC/EMC (3:7 by volume) with 1 wt% vinylene carbonate as an electrolyte additive. The cells were characterized between 3.0 and 4.2 V with a constant current of 1C.

Yasuhara et al.[16] reported the cathode material which was modified by interfacial nanodot $BiTiO_3$ treated on $LiCO_2O_3$ cathode for thin film batteries. The modified cathode enhanced the high-rate capability and cyclability of the thin film Li-ion battery.

Discharge capacities as a function of C-rate, which increases from low 1C to ultrahigh 100C in a five-repeated measurement sequence for each C-rate. Black, blue, and red circles correspond to Bare, Planar BTO, and Dot BTO, respectively. Which increases from low 1C to ultrahigh 100C in a five-repeated measurement sequence for each C-rate. Black, blue, and red circles correspond, respectively.[16]

Core-shell structured o-$LiMnO_2$@Li_2CO_3 nanosheet array cathode materials were reported by Guo et al.[17] for high-performance, wide-range temperature tolerance Li-ion batteries.

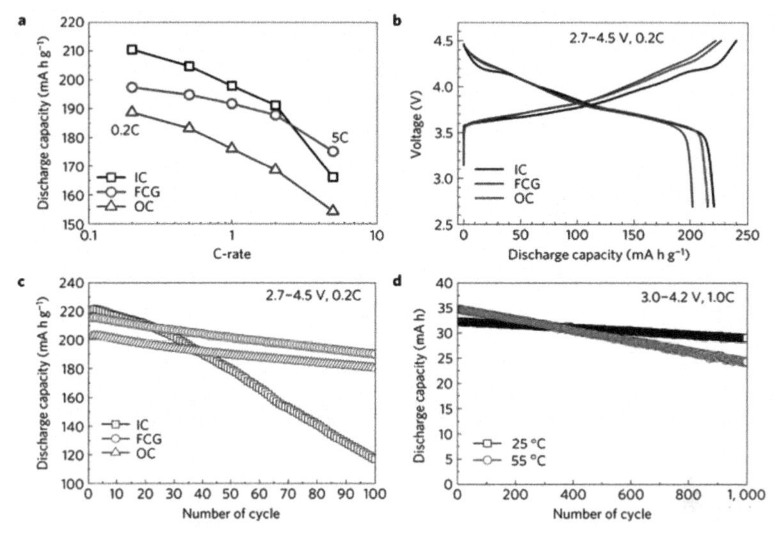

FIGURE 4.2 (a) Comparison of rate capabilities of the FCG with the IC and OC materials (upper cutoff voltage of 4.3 V vs Li^{+}/Li). (b) Initial charge-discharge curves. (c) The cycling performance of half-cells using the FCG, IC, and OC materials cycled between 2.7 and 4.5 V vs Li^{+}/Li using a constant current of C/5 (about 44 mA g^{-1}). (d) The discharge capacity of mesocarbon microbeads (MCMB)/FCG cathode full-cells at room and high temperature.[15]

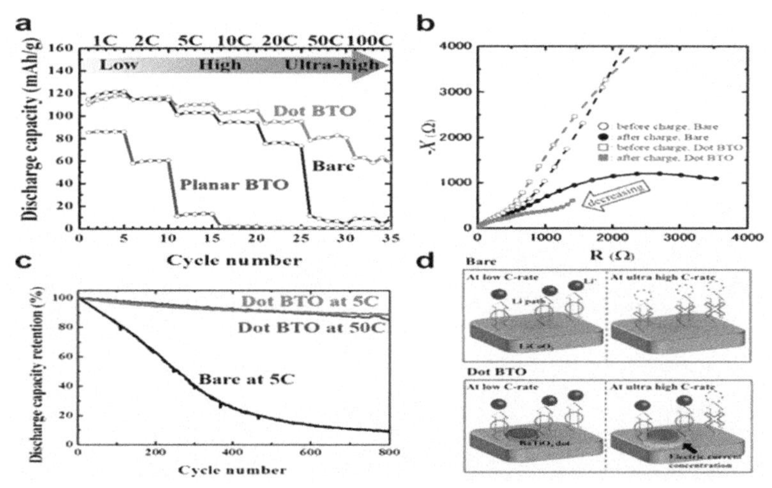

FIGURE 4.3 (a) Discharge capacities Bare, Planar BTO, and Dot BTO as a function of C-rate. (b) Nyquist plots of Bare and Dot BTO before and after charge at 1C. Black open circle, black solid circle, red open circle, and red solid circle correspond to Bare before charge, Bare after charge, Dot BTO before charge, and after charge, respectively. (c) Discharge capacities of Bare at 5C, Dot BTO at 5C, and Dot BTO at 50C repeating 800 times. (d) Schematic of lithium intercalation and deintercalation in Bare and Dot BTO thin films at a low C-rate and ultrahigh C-rate.[16]

The Li_2CO_3 coated $LiMnO_2$ material as a cathode helped in the prevention of Mn dissolution with electrolyte during cycling, and the synergistic effective coating layer with nanostructure effect improved the performance and capacity of the o-LMO cathode material with wide range temperature applicability. They have claimed the capacity of 207 mAh g^{-1} at 0.5 C tested at 60°C temperature, the capacity retention was 79% after 400 cycles. The rate capability at 5C was 128 mAh g^{-1}. The o-LMO cathode performed well in a full cell with lithium titanate nanoarray anode showed a capacity of 200 mAh g^{-1} at 2C and exhibited capacity retention of 67% after 400 cycles.[17]

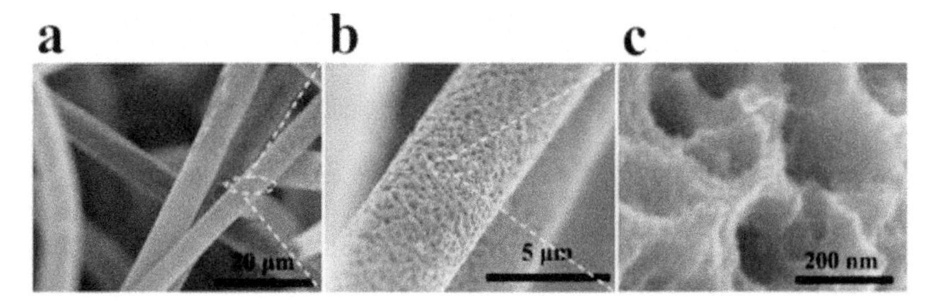

FIGURE 4.4 (a–c) SEM images of the *o*-LMO@ Li_2CO_3 nanosheet array electrode (fabricated using 0.1 M LiOH) at different magnifications.[17]

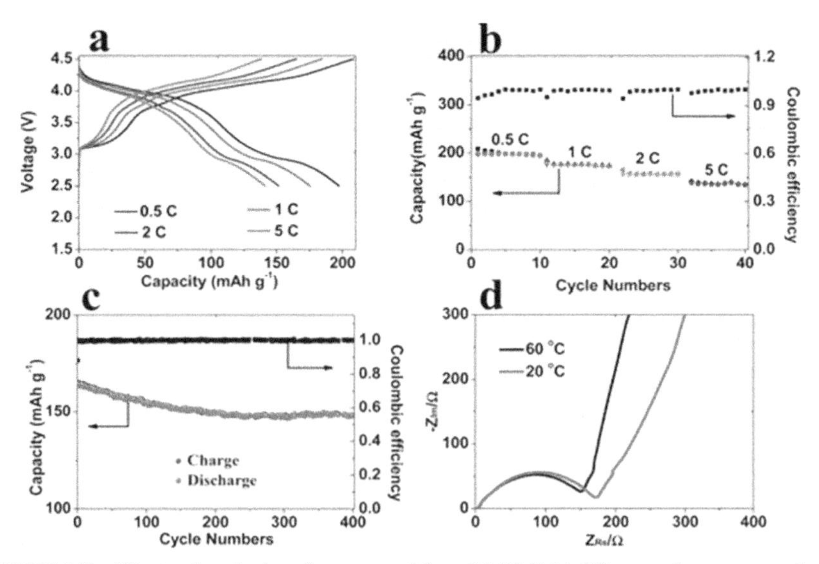

FIGURE 4.5 Electrochemical performance of the *o*-LMO@ Li_2CO_3 nanosheet array cathode (fabricated with 0.1 M LiOH) at 60°C: (a) First charge–discharge curves at different current densities. (b) Rate performance at progressively increased current densities. (c) Cycling performance at 2C. (d) Nyquist plots at different temperatures.[17]

4.4.2 NANOSTRUCTURED NEGATIVE ELECTRODE MATERIALS

In the path leading to the development of next-generation batteries with high energy density and power density, the selection of suitable anode is also a major challenge to overcome. The anode with ease of intercalation of alkali ions into it, along with long cycle life and satisfying safety concerns would be the priority for the selection of anodes. The nanostructured materials can offer good cycle stability and better performance of anode materials and the nanoscale approach offers more options to develop the anode materials for rechargeable batteries.

Li et al.[18] reported an anode material that is applicable for both Li-ion and Na-ion batteries, which is an amorphous red phosphorous embedded mesoporous carbon with a superior capacity of 2250 mAh g^{-1} and excellent capacity of 624 mAh g^{-1} at a high current rate of 12C.[18]

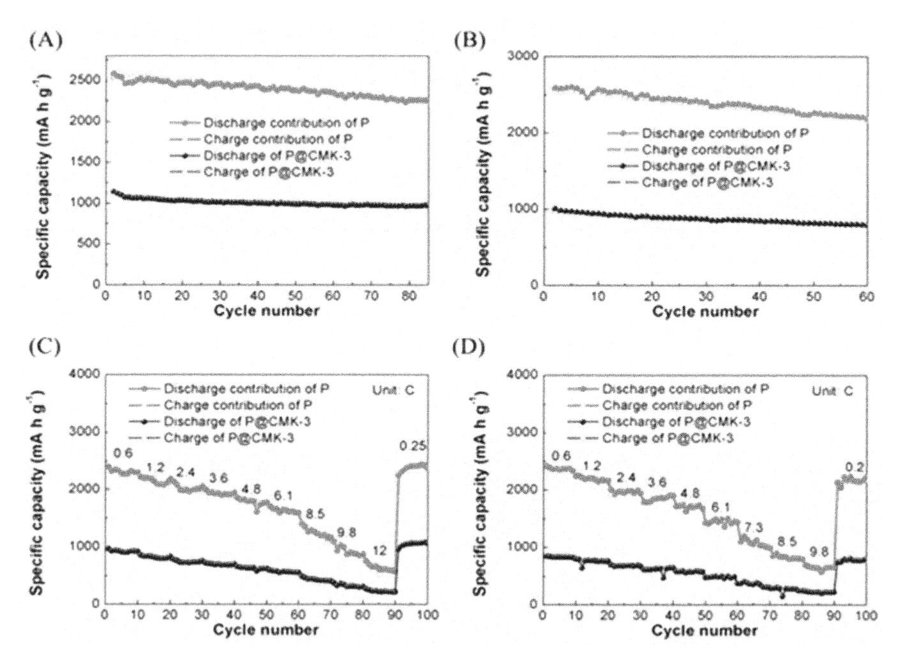

FIGURE 4.6 Electrochemical performance of P@CMK-3 composite for LIBs and NIBs cycled between 0.001 and 2.5 V vs Li^+/Li and 0.001 and 2.0 V vs Na^+/Na. Capacity-cycle number curves of P@CMK-3 electrode at a cycling rate of 0.25 C for LIBs (A) and 0.2 C for NIBs, (B–D) Capacity of P@CMK-3 composite as a function of discharge rate (0.6–12 C for LIBs and 0.6–9.8 C for NIBs). The capacities here are calculated based on the mass of red P and the composite, respectively.[18]

The porous hard carbon materials synthesized from pyrolyzed sucrose as an anode reported by Yang et al.,[19] have investigated the effect of morphology and porous structure on the electrochemical performance and Li^+ ion transport characteristics. They have achieved a capacity 503.5 mAh g^{-1} at 0.2C and it maintains a capacity of 332.8 mAh g^{-1} at a 5C rate. The enhanced performance of porous hard carbon anode is due to the good flexibility and shorter diffusion pathways of nanostructures.[19]

FIGURE 4.7 SEM images of the as-prepared carbons: (a and b) HC, (c) PC-1, (d) PC-2.[19]

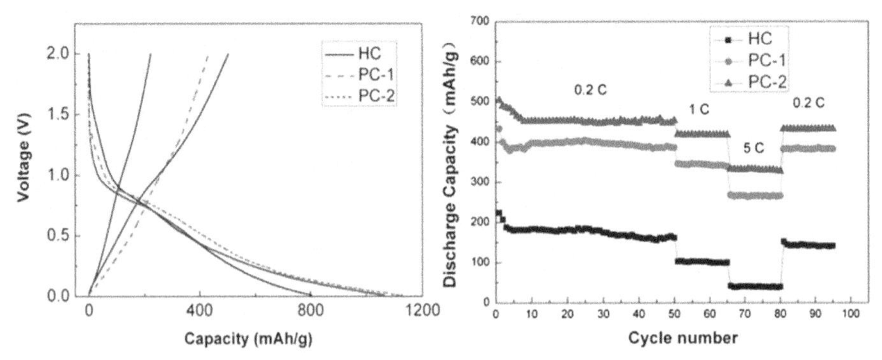

FIGURE 4.8 (a) Charge and discharge curves of HC, PC-1, and PC-2 for the first cycles at a 0.2C current. (b) Rate capacity studies of HC, PC-1, and PC-2.[19]

Two-dimensional edge-based SnS_2 nanosheet arrays on carbon paper is reported by Wang et al.[20] as binder-free anodes for both Li-ion and Na-ion batteries.

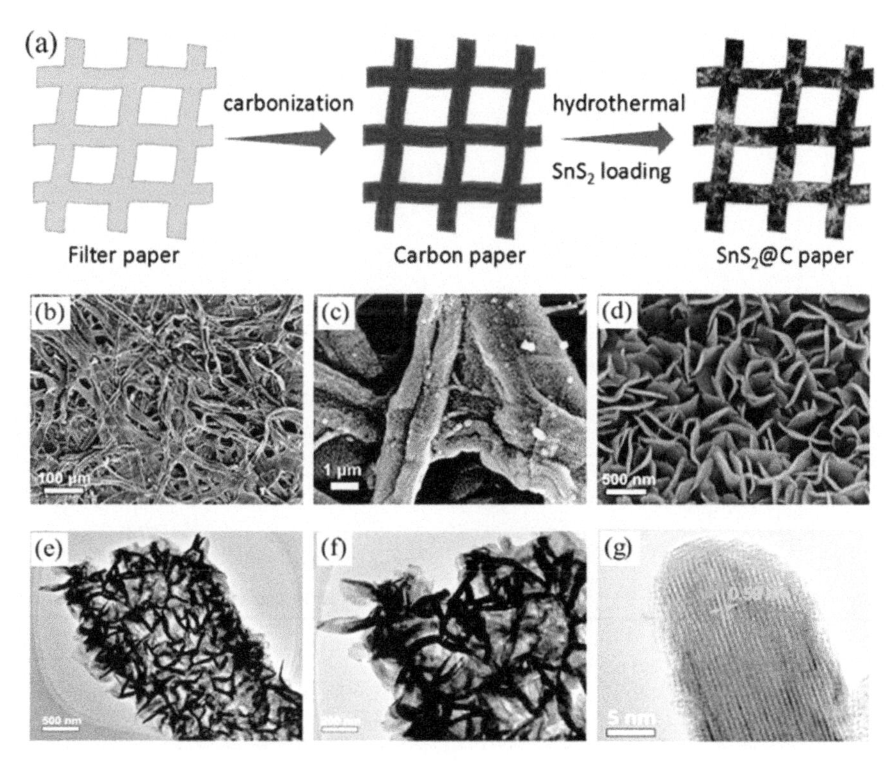

FIGURE 4.9 (a) Schematic illustration of the fabrication procedure of SnS_2@C hybrid paper. (b–d) SEM, (e and f) transmission electron microscopy (TEM), and (g) high-resolution transmission electron microscope (HRTEM) images of SnS_2@C composites.[20]

The rational design and unique hybrid architecture of edge-oriented SnS_2 nanosheet arrays with vertical alignment on carbon paper provided sufficient electrode/electrolyte interaction areas, faster Li^+ ion diffusion, and efficient volume accommodation, which enabled the SnS_2/C nanoarrays to perform well as an efficient anode material for Li-ion and Na-ion batteries. The graphs represented in Figure 4.9 correspond to the SnS_2/C and SnS_2 anode materials, which have shown the initials reversible discharge capacities 1851 mAh g^{-1} and 1304 mAh g^{-1}, respectively. It is evident that nanoarray architecture has boosted the capacity of the SnS_2/C nanoarray anode, which sustained a reversible capacity of 1150 mAh g^{-1}.[20]

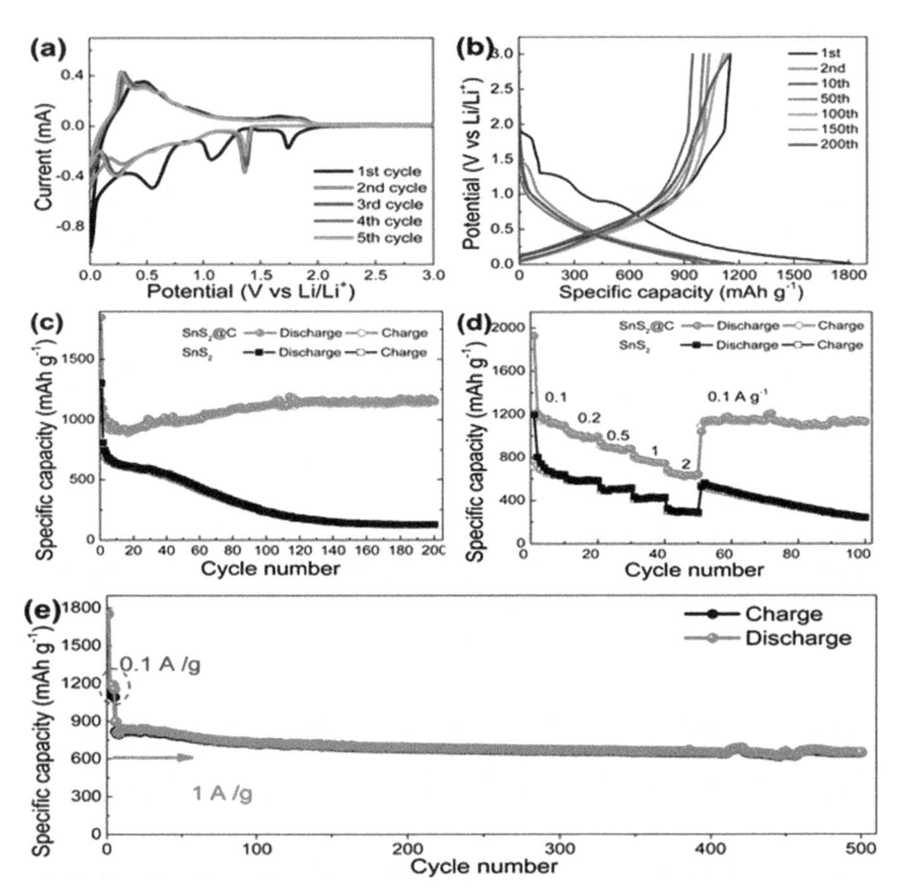

FIGURE 4.10 (a) CV curves, (b) galvanostatic charge/discharge curves, (c) cycling performance (at 100 mA g^{-1} rate), (d) rate capability, and (e) cycling performance (1 A g^{-1}) of SnS_2@C electrodes.[20]

4.4.3 NANOSTRUCTURED ELECTROLYTES FOR RECHARGEABLE BATTERIES

The rechargeable batteries operate on the principle of shuttling alkali ions between the cathode and anode. The electrolyte acts as media for the movement of ions between the anode and cathode during the charge/discharge processes which makes the battery conductive, also, stabilizes the electrode surfaces, and improves the battery performance. Generally, liquid electrolytes are employed in rechargeable batteries, the critical issue of liquid electrolytes is the decomposition of electrolytes occurs at more than ~4.2 V v/s Li/Li^+.

Therefore, the development of electrolytes that can function at a wider voltage window is essential for the realization of next-generation batteries with higher cell voltage with enhanced ionic conductivity. The recent trends follow the development of nanostructured electrolyte designs of liquids as well as solid-state electrolytes for high-performing batteries.

4.4.4 BINDERS FOR RECHARGEABLE BATTERIES

Advanced trends in battery characteristics such as stability and irreversible capacity losses depend not only on the electrode materials but also depends on the binder's properties. Generally, polymers containing carboxy groups, such as polyvinylidene fluoride (PVDF), polyacrylic acid (PAA), and carboxymethyl cellulose (CMC) are used as binders in rechargeable batteries. The weak binder–electrolyte interaction is a critical condition for long-term electrode stability.[21–26]

Kovalenko et al.[27] used the polysaccharide extracted from brown algae as a binder for Si anode processing which yielded a very high capacity as compared to the conventional binder PVDF with a graphitic anode. The study has demonstrated that the performance and cycle stability of the nano-Si anode with the PVDF, CMC, and alginate binders, the alginate binder has shown outstanding performance, and the nano-Si anode processed with alginate binder has delivered a very high reversible discharge capacity 1700–2000 mAh g^{-1} at a current density 4200 mA g^{-1}, which is five times higher than the theoretical capacity of the graphite.[27]

Pieczonka et al.[28] reported lithium polyacrylate (LiPAA) as an advanced binder for the processing of electrodes for Li-ion batteries. They have tested the LiPAA binder with $LiNi_{0.5}Mn_{1.5}O_4$ and graphite prototype batteries in comparison with the PVDF binder. The group has achieved significantly improved electrochemical performance (i.e., cycle life, cell impedance, and rate capability) of the electrodes as compared to conventional PVDF binders. The application of LiPAA binder to high-voltage cathode effectively stabilized the cathode–electrolyte solution interphase (CEI) and provided crucial functionality, demonstrated as an extra Li^+ ion source that could compensate for the Li^+ ion loss during the cycling of full cells, the binder LiPAA provided improved adhesion to the current collector, which helped to overcome the delamination issue which occurs usually in case of lithium nickel manganese oxide (LNMO) and PVDF binders. The group has demonstrated improved electrochemical performance of the battery by using LiPAA as a binder.[28]

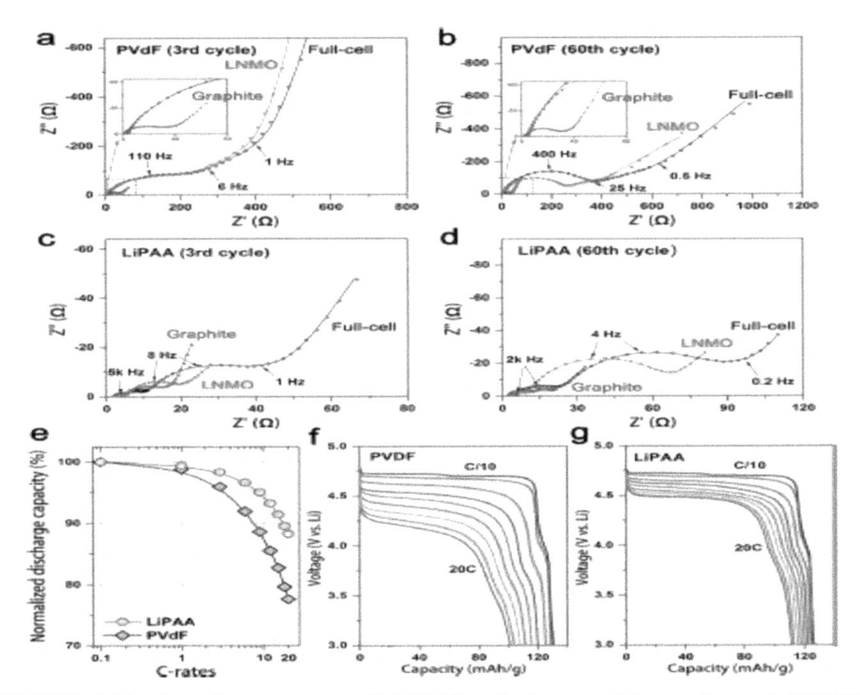

FIGURE 4.11 Impedance spectra of LNMO cathodes, graphite anode, and LNMO/graphite full cells measured in three-electrode cell configuration using Li metal as reference electrodes; Nyquist plots obtained by using PVdF–LNMO cathodes after (a) third and (b) 60th cycles; Nyquist plots obtained by using LiPAA–LNMO cathodes after (c) third and (d) 60th cycles. (e) Comparison of rate capabilities between PVdF–LNMO and LiPAA–LNMO cathodes in a range from C/10 (13 mA g^{-1}) to 20C-rate (2600 mA g^{-1}) at 30°C. The half cells were charged with a constant rate of C/7. The resulting voltage profiles of (f) PVdF–LNMO and (g) LiPAA–LNMO cathodes are plotted.[28]

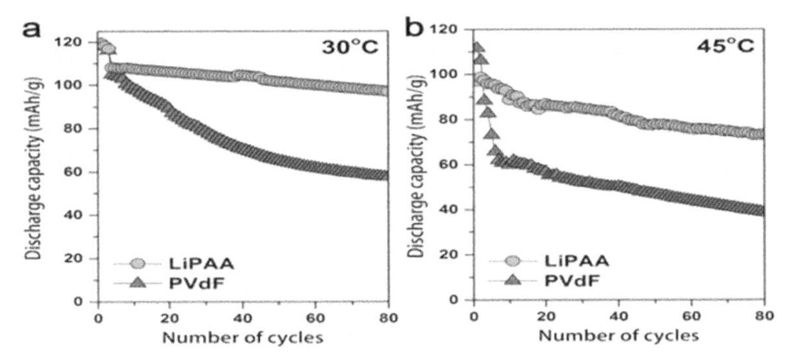

FIGURE 4.12 Capacity retentions of PVdF–LNMO/graphite and LiPAA–LNMO/graphite full cells cycled at 0.5C rate at (a) 30 and (b) 45°C (the initial three cycles were performed at 0.1C rate).[28]

4.5 THREE-DIMENSIONAL FLEXIBLE RECHARGEABLE BATTERIES NANOSTRUCTURED ELECTRODES

Flexible and compact electronic devices are the new trend in the digital era. Flexible/foldable batteries are also in demand to fulfill the needs of this sector. Nanomaterials can offer a lot of options and directions to prepare flexible batteries to develop compact and bendable electronic devices. Nanomaterials have a lot of advantages over all other materials for the development of electrode materials for the flexible batteries which would be part of these flexible electronics.[29–32]

Bin Liu et al. reported the hierarchical three-dimensional $ZnCo_2O_4$ nanowire arrays on carbon cloth as an anode for flexible Li-ion batteries, they have achieved a high reversible capacity of 1300–1400 mAh g^{-1} and stable performance over 160 cycles.

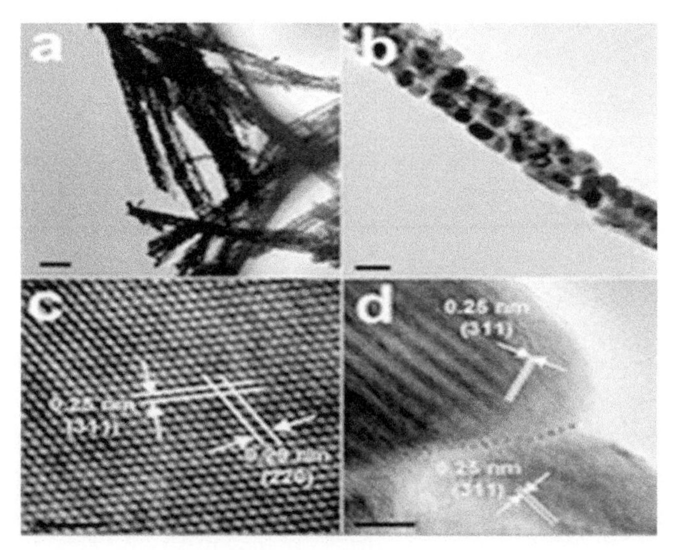

FIGURE 4.13 (a) Low-magnification TEM image, where nanowires with diameters of about 80–100 nm can be seen. A higher-magnification TEM image depicted in (b–d) reveals that a typical $ZnCo_2O_4$ nanowire is a porous nanowire composed of many small nanoparticles instead of the conventional single-crystalline nanowire.[33]

Three-dimensional graphene/$LiFePO_4$ composite nanostructures are reported by Ding et al.[34] for the application of flexible rechargeable Li-ion batteries as cathode materials. The report says the composite nanostructure with the least graphene has shown a high capacity of 163.7 mAh g^{-1} at 0.1

C and 114 mAh g^{-1} at 5 C without incorporation of additional conductivity enhancers.[34]

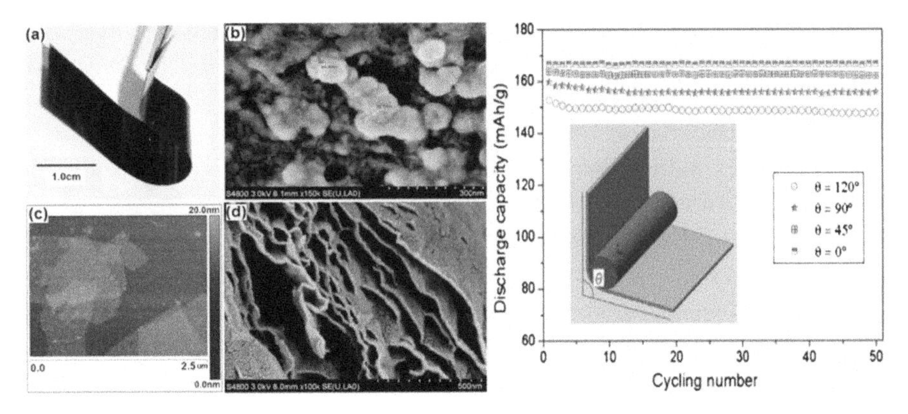

FIGURE 4.14 (a) The flexibility feature of the electrodes without any fractures or cracks. The morphology of GLFP composites and the flexible electrode is shown in (b) SEM image, (c) atomic force microscopy (AFM) image, and (d) cross-section view of a flexible electrode. (e) Correlation of extent of bending angle and discharge capacity.[34]

Graphene/$LiFePO_4$ nanoparticles of ~20 nm thickness are coated on the surface of the graphene sheet. Figure 4.14e represents the correlation between the extent of bending angle and discharge capacity. Is shown that the discharge capacity of the flexible electrodes decreases as the increase in bending angle of an electrode. Here they have achieved a high capacity of 159.7 mAh g^{-1} during the harshest bending conditions of the electrodes at 120° bending angle. The capacity of the electrode was achieved due to the nanoscale features of $LiFePO_4$ cathode active material and the mechanical strength with high electrical conductivity obtained by the nanostructured graphene support.[34]

Another group of scientists Kong et al.[35] reported the free-standing CNTs as an anode for a flexible Li-ion battery. The flexible Li-ion battery with 3D-CNT as an anode delivers a high aerial capacity of 0.25 mAh cm^{-1} at 0.1C rate. They claimed that the 3D CNT anode flexible Li-ion battery showed a high volumetric capacity of 300 mAh cm^{-3}, and a high specific capacity (207 mAh g^{-1}) of the 3D CNTs with consistent open circuit voltage in the range 2.55V–2.75V.[35]

Three-dimensional $CoSe_2$ grains attached to carbon nanofibers (CNFs) binderless flexible films are used as anode material for sodium-ion batteries (SIBs). The prepared nanofibers were well interlinked with the 3D conductive framework which formed the binderless flexible films for flexible Na-ion batteries. The research achieved a high initial capacity of 972 mAh g^{-1} cycled at a current density of 200 mA g^{-1} and maintained 430 mAh g^{-1} after 400 cycles.[36]

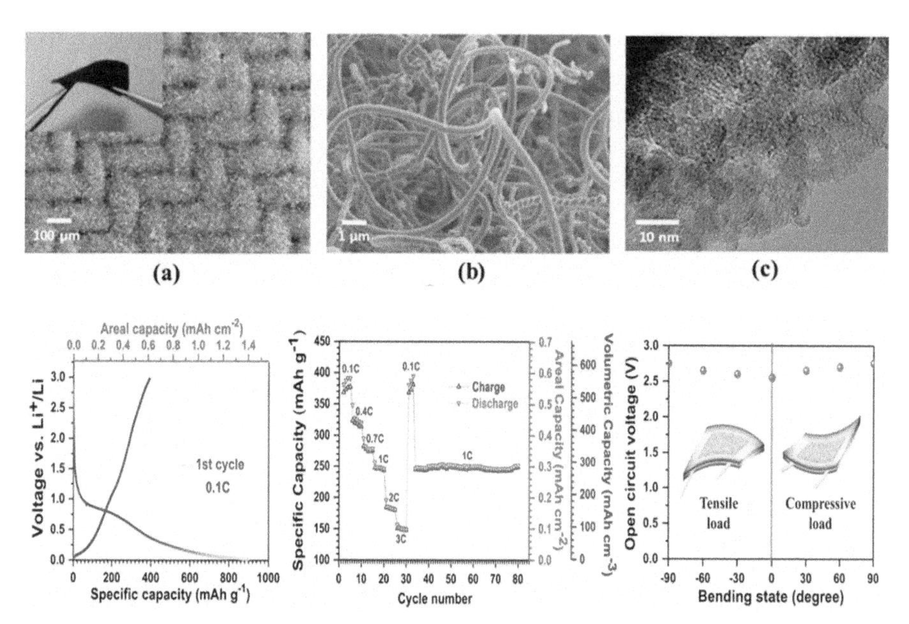

FIGURE 4.15 (a) FE-SEM image of 3D free-standing CNTs. The inset shows a flexible 3D free-standing CNT. (b) Magnified FE-SEM image of entangled CNT structure. (c) An HRTEM image of the defective structure in a CNT. (d) The voltage profiles of 3D free-standing CNTs obtained from the first charge and discharge cycles at 0.1C. (e) The C-rate capability and cycling performance of 3D free-standing CNTs at different C-rates. (f) The plot of the OCV value vs the bending state of the flexible LIB.[35]

FIGURE 4.16 SEM images of (a) Cobalt salt/PAN and (b) Co/CNFs. (c) HR-SEM image of $CoSe_2$/CNFs.[36]

Zhang et al.[37] reported similar material, free-standing $Co_{0.85}Se$ nanosheets on graphene sheets ($Co_{0.85}Se$ NSs/G) composite nanostructures as a high-performance anode for both LIBs and Na-ion batteries (SIBs) which are prepared by simple vacuum filtration and thermal reduction process. Which delivered a high initial reversible capacity of 680 mAh g^{-1}

at 50 mA g^{-1} for a Li-ion battery, whereas the materials used as anode for Na-ion batteries, the reversible capacity reached 388 mAh g^{-1} at 50 mA g^{-1}. In this work, they have achieved a high performance of these sheet-on-sheet nanostructures with strong interfacial interactions and free-standing binder-free electrodes as an anode material for both Li-ion and Na-ion flexible 3D batteries.[37]

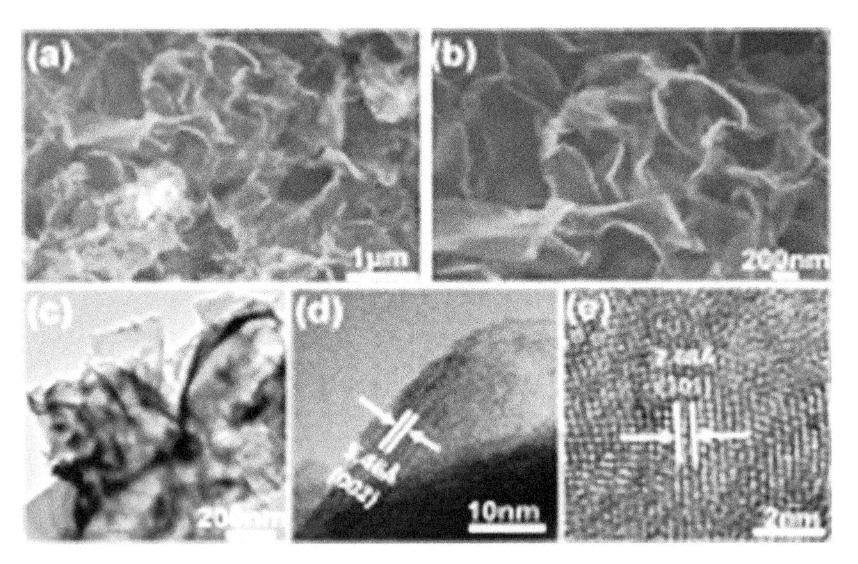

FIGURE 4.17 (a, b) SEM and (c) TEM images of $Co_{0.85}Se$ NSs, HRTEM images of the edge (d) and basal plane (e) of a $Co_{0.85}Se$ NS.[37]

3D flexible Si-composite/CNF electrode consisting of ($Si@Si_3N_4$) NPs and carbon nanofibers prepared and used as the anode which showed the 665 mAh g^{-1} at current density 10 A g^{-1} with the least decrease in capacity with more stability cycled over 2000 cycles. This nanostructured composite material was used as a binder-free flexible electrode which enhanced the electronic conductivity of the electrode induced to perform long cycle stable cycling performance.[38]

Xia et al.[39] reported porous $LiMn_2O_4$ nanowall arrays grown on cloth are used as cathode materials for advanced 3D microbatteries and Li-ion batteries. Which delivered a high capacity ~130 mAh g^{-1} and stable cycling over 200 cycles as compared to the capacity (~110 mAh g^{-1}) of bulk structure cathode active materials. The capacity is enhanced due to the morphological change and also the nanoarray framework of $LiMn_2O_4$ cathode materials.[39]

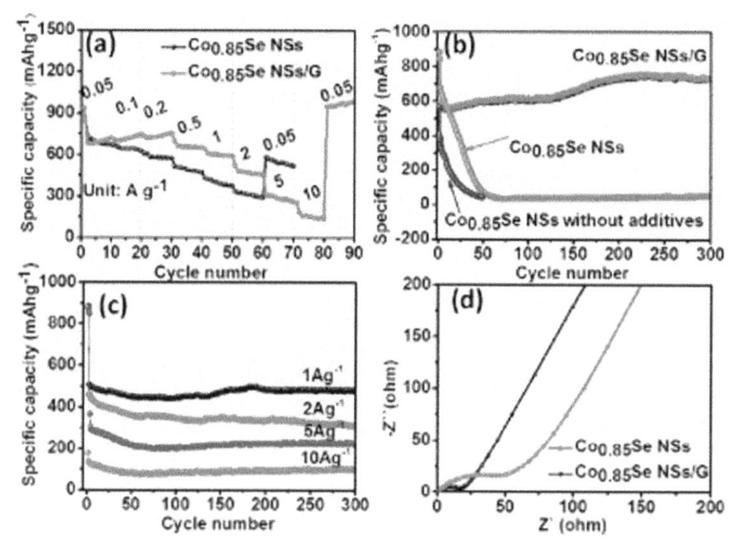

FIGURE 4.18 (a) Li-ion storage rate performance of the $Co_{0.85}Se$ NSs and the $Co_{0.85}Se$ NSs/G film, (b) cyclic properties of the $Co_{0.85}Se$ NSs with and without the addition of a binder and conducting additives and the $Co_{0.85}Se$ NSs/G film at current density of 0.5 Ag^{-1}, (c) cyclic properties of $Co_{0.85}Se$ NSs/G film at high current densities from 1 A g^{-1} to 10 A g^{-1}. (d) EIS of the $Co_{0.85}Se$ NSs and the $Co_{0.85}Se$ NSs/G film.[37]

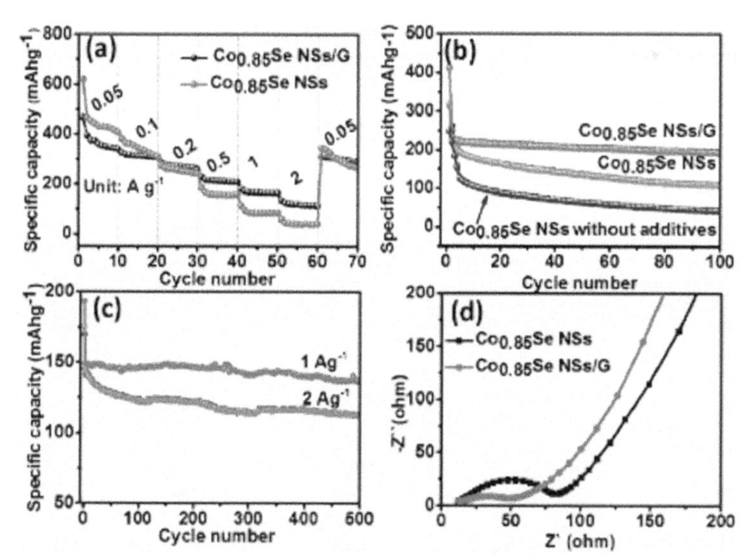

FIGURE 4.19 (a) Na-ion storage rate performance of the $Co_{0.85}Se$ NSs and the $Co_{0.85}Se$ NSs/G film, (b) cyclic properties of the $Co_{0.85}Se$ NSs with and without the addition of a binder and conducting additives and of the Co0.85Se/G film at a current density of 0.5 A g^{-1}, (c) cyclic properties of the $Co_{0.85}Se$ NSs/G film at the current densities of 1 and 2 Ag^{-1}, (d) EIS of the $Co_{0.85}Se$ NSs and the $Co_{0.85}Se$ NSs/G film.[37]

FIGURE 4.20 (a–e) SEM and TEM images of the as-prepared Si-composite/CNF.[38]

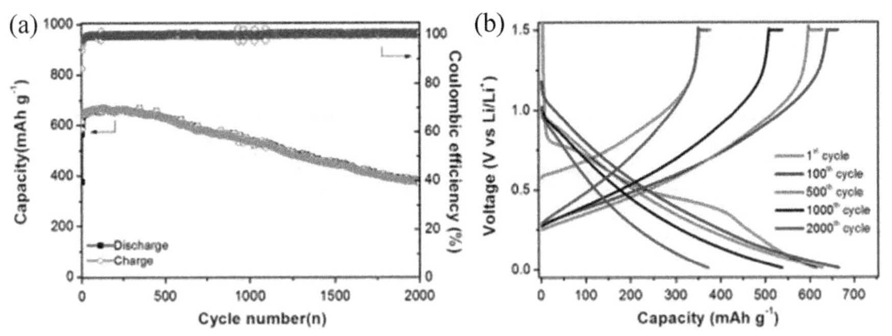

FIGURE 4.21 (a) Charge/discharge capacity and Coulombic efficiency of Si-composite/CNF at a current density of 10 A g^{-1} for 2000 cycles. (b) Charge/discharge profiles of Si-composite/CNF (voltage v/s capacity).[38]

4.6 INFLUENCE OF SIZE AND MORPHOLOGICAL CONTROL OF NANOMATERIALS ON RECHARGEABLE BATTERIES

Nanoscale materials have been extensively explored in the last few decades to improve the performance of Li-ion batteries. Fast fast-growing energy

FIGURE 4.22 (a) FE-SEM image of $LiMn_2O_4$ nanowall arrays grown on the carbon cloth. (b and c) FE-SEM images of Mn_3O_4 nanowall arrays grown on the single carbon fiber with different magnifications. (d and e) FE-SEM images of $LiMn_2O_4$ nanowall arrays grown on the single carbon fiber with different magnifications. TEM images of Mn_3O_4 nanowalls with different magnifications. (f–h) TEM images of $LiMn_2O_4$ nanowalls with different magnifications.[39]

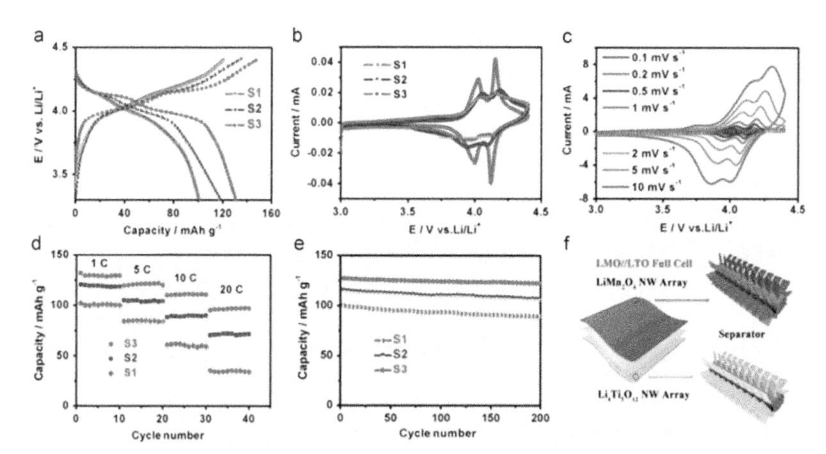

FIGURE 4.23 (a) The first charge/discharge curves of $LiMn_2O_4$ nanowall arrays synthesized at different temperatures. (b) Typical CV curves of $LiMn_2O_4$ nanowall arrays synthesized at different temperatures. (c) CV curves of $LiMn_2O_4$ nanowall arrays synthesized at 240°C at different scan rates. (d) Rate capabilities of $LiMn_2O_4$ nanowall arrays synthesized at different temperatures. (e) Cycle performance of $LiMn_2O_4$ nanowall arrays synthesized at different temperatures. (f) Schematic illustration of the construction of flexible full-cell device combining 3D $LiMn_2O_4$ cathode and 3D $Li_4Ti_5O_{12}$ anode.[39]

storage field allowed to design of more complex morphology-based materials to achieve outstanding performance and address the specific issues related to the performance. In recent years, different kinds of 0D, 1D, 2D, 3D, and hybrid nanomaterials have been used extensively for Li-ion batteries. The role of morphology-dependent nanostructures in Li-ion batteries is not only confined to Li-ion storage but it catalyzes electrochemical reactions and also traps the reaction intermediates. Control of morphology has been a critical parameter that can be modified to achieve the greater capabilities of materials for high-energy storage batteries. Here in this part, we overview and discuss the influence of the morphology of nanomaterials in uplifting the electrode capabilities of the batteries of various battery chemistries. The morphological variation of the nanomaterials could be studied by considering various parameters and depending on various factors. Factors such as (1) the size, structure, and composition of the initial nuclei, (2) the nature, form, and quantities of the precursors, reducing agents, and additives, (3) the selective adsorption of additives to different crystal facets, and (4) the contact with external supports can affect the particle development, and hence, the particle morphology.[40]

Marcus Muller et al. studied the effect of nanostructure and porous morphology on the performance of cathode materials. The research group has prepared the nanostructured and porous $LiNi_{0.33}Co_{0.33}Mn_{0.33}O_2$ (NCM111) cathode materials and examined the influence of nanostructure, and porosity of the material on the performance of the battery. The cathode active with porous nanostructure has shown higher capacity than the bulk structure material, at a 3C rate the difference in capacity is 20 mAh g^{-1}. The trend of the discharge capacity followed over 1100 cycles, the porous nanostructures were affected less by degradation and aging than bulk structure cathode materials. This indicates the clear influence of the morphology and porosity plays a critical role in the performance of the battery.[41]

4.7 CHALLENGES AND POSSIBLE WAYS TO OVERCOME THEM

The application of nanomaterials in batteries may cause plenty of new challenges as well due to their reduced particle size, high surface area, and low packing density. An engineered design of nanomaterials should compensate for the associated disadvantages along with the issues of micrometer-sized materials. The solid electrolyte interface (SEI) layer formation process is a very important stage of battery cycling. This is critical to achieving the longer cycle life of electrodes, and the control over the SEI layer has a critical role in maintaining high initial Coulombic efficiency. This SEI layer formation on the electrode surface consumes the electrolyte and the lithium from the cathode-derived nanoparticle as compared with electrodes composed of microscale materials, this SEI formation process on the surface of nanostructured electrodes takes up a stupendous amount of electrolyte and lithium due to availability of larger area of electrode/electrolyte interface, which may lead to lower the initial charge/discharge efficiency and can potentially lower the overall capacity and energy density of batteries.

4.8 FUTURE SCOPE OF NANOMATERIALS FOR BATTERY APPLICATION

Nanomaterials for the application of batteries have been explored extensively in past decades, more research work needs to be done to study the application of nanomaterials with different morphology/size for flexible/printed batteries. There are different sets of challenges in 2D and 3D flexible battery development; for example, nanostructured ink formulation is a critical process in 2D and 3D printed batteries, where the fine control over morphology, size, rheological, mechanical, and electrical properties is key, all of which can influence by the design of nanomaterials and which results in the performance of the batteries.[42]

KEYWORDS

- **nanomaterials**
- **lithium-ion batteries**
- **low-cost electrode**
- **high performing materials**
- **flexible printed batteries**

REFERENCE

1. Li, B., Meng, Y.; Tang, W. The Role of Nanotechnology in the Design of Materials for Lithium-ion Battery. *E3S Web Conf.* **2021,** *308*, 01009.

2. Jiao, F.; Bruce, P. G. Mesoporous Crystalline β-MnO2-A Reversible Positive Electrode for Rechargeable Lithium Batteries. *Adv. Mater.* **2007,** *19*, 657–660.
3. Aricò, A. S.; Bruce, P.; Scrosati, B.; Tarascon, J. M.; Van Schalkwijk, W. Nanostructured Materials for Advanced Energy Conversion and Storage Devices. *Nat. Mater.* **2005,** *4*, 366–377.
4. Wang, F.; et al. Excess Lithium Storage and Charge Compensation in Nanoscale Li 4+xTi5O12. *Nanotechnology* **2013,** *24*, 424006.
5. Bruce, P. G;, Scrosati, B.; Tarascon, J. M. Nanomaterials for Rechargeable Lithium Batteries. *Angew. Chem. Int. Ed.* **2008,** *47*, 2930–2946.
6. Balaya, P.; et al. Nano-ionics in the context of lithium batteries. *J. Power Sources* **2006,** *159*, 171–178.
7. He, W.; Wen, K.; Niu, Y. Oriented-Attachment Nanocrystals in Lithium Ion Batteries. 2018. DOI: 10.1007/978-3-319-72432-4_3.
8. Tiwari, J. N.; Tiwari, R. N.; Kim, K. S. Zero-Dimensional, One-Dimensional, Two-Dimensional and Three-Dimensional Nanostructured Materials for Advanced Electrochemical Energy Devices. *Prog. Mater. Sci.* **2012,** *57* (4), 724–803.
9. Nanorods and Nanocomposites. In *Nanorods and Nanocomposites*; IntechOpen, 2020. DOI: 10.5772/intechopen.77453.
10. Cao, G.; Liu, D. Template-Based Synthesis of Nanorod, Nanowire, and Nanotube Arrays. *Adv. Coll. Interface Sci.* **2008,** *136*, 45–64.
11. Coleman, J. N.; et al. Two-dimensional nanosheets produced by liquid exfoliation of layered materials. *Science* **2011,** *331*, 568–571.
12. Yu, L.; Hu, H.; Wu, H. B.; Lou, X. W. D. Complex Hollow Nanostructures: Synthesis and Energy-Related Applications. *Adv. Mater.* **2017,** *29*, 1604563.
13. Yu, L.; Yu, X. Y.; Lou, X. W. D. The Design and Synthesis of Hollow Micro-/Nanostructures: Present and Future Trends. *Adv. Mater.* **2018,** *30*, e1800939.
14. Van Gough, D., Juhl, A. T.; Braun, P. V. Programming Structure into 3D Nanomaterials. *Mater. Today* **2009,** *12*, 28–35.
15. Sun, Y. K.; et al. Nanostructured High-Energy Cathode Materials for Advanced Lithium Batteries. *Nat. Mater.* **2012,** *11*, 942–947.
16. Yasuhara, S.; et al. Enhancement of Ultrahigh Rate Chargeability by Interfacial Nanodot BaTiO 3 Treatment on LiCoO 2 Cathode Thin Film Batteries. *Nano Lett.* **2019,** *19*, 1688–1694.
17. Guo, J.; Cai, Y.; Zhang, S.; Chen, S.; Zhang, F. Core-Shell Structured o-LiMnO2@Li2CO3 Nanosheet Array Cathode for High-Performance, Wide-Temperature-Tolerance Lithium-Ion Batteries. *ACS Appl. Mater. Interfaces* **2016,** *8*, 16116–16124.
18. Li, W.; et al. Amorphous Red Phosphorus Embedded in Highly Ordered Mesoporous Carbon with Superior Lithium and Sodium Storage Capacity. *Nano Lett.* **2016,** *16*, 1546–1553.
19. Yang, J.; Zhou, X. Y.; Li, J.; Zou, Y. L.; Tang, J. J. Study of Nano-Porous Hard Carbons as Anode Materials for Lithium Ion Batteries. *Mater. Chem. Phys.* **2012,** *135*, 445–450.
20. Wang, J. G.; et al. Edge-Oriented SnS2 Nanosheet Arrays on Carbon Paper as Advanced Binder-Free Anodes for Li-ion and Na-ion Batteries. *J. Mater. Chem. A* **2017,** *5*.
21. Fransson, L.; Eriksson, T.; Edström, K.; Gustafsson, T.; Thomas, J. O. Influence of Carbon Black and Binder on Li-ion Batteries. *J. Power Sources* **2001,** *101*, 1–9.

22. Bridel, J. S.; Azaïs, T.; Morcrette, M.; Tarascon, J. M.; Larcher, D. Key Parameters Governing the Reversibility of Si/carbon/CMC Electrodes for Li-ion Batteries. *Chem. Mater.* **2010,** *22*, 1229–1241.
23. Zhang, S. S.; Jow, T. R. Study of Poly(acrylonitrile-methyl methacrylate) as Binder for Graphite Anode and LiMn2O4 Cathode of Li-ion Batteries. *J. Power Sources* **2002,** *109*, 422–426.
24. Guy, D.; Lestriez, B.; Guyomard, D. New Composite Electrode Architecture and Improved Battery Performance from the Smart Use of Polymers and Their Properties. *Adv. Mater.* **2004,** *16*, 553–557.
25. Mazouzi, D.; Lestriez, B.; Roué, L.; Guyomard, D. Silicon Composite Electrode with High Capacity and Long Cycle Life. *Electrochem. Solid-State Lett.* **2009,** *12*, A215.
26. Magasinski, A.; et al. Toward Efficient Binders for Li-ion Battery Si-based Anodes: Polyacrylic Acid. *ACS Appl. Mater. Interfaces* **2010,** *2*, 3004–3010.
27. Kovalenko, I.; et al. A Major Constituent of Brown Algae for Use in High-Capacity Li-ion Batteries. *Science* **2011,** *334*, 75–79.
28. Pieczonka, N. P. W.; et al. Lithium Polyacrylate (LiPAA) as an Advanced Binder and a Passivating Agent for High-Voltage Li-Ion Batteries. *Adv. Energy Mater.* **2015,** *5*, 1501008.
29. Gwon, H.; et al. Flexible Energy Storage Devices Based on Graphene Paper. *Energy Environ. Sci.* **2011,** *4*.
30. Liu, J.; Buchholz, D. B.; Chang, R. P. H.; Facchetti, A.; Marks, T. J. High-Performance Flexible Transparent Thin-Film Transistors Using a Hybrid Gate Dielectric and an Amorphous Zinc Indium Tin Oxide Channel. *Adv. Mater.* **2010,** *22*, 8412–8419.
31. Lieber, C. M.; Wang, Z. L. Functional Nanowires. *MRS Bull.* **2007,** *32*, 99–108.
32. Wang, Z.; et al. Transferable and flexible nanorod-assembled TiO2 cloths for dye-sensitized solar cells, photodetectors, and photocatalysts. *ACS Nano* **2011,** *5*, 8412–8419.
33. Liu, B.; et al. Hierarchical Three-Dimensional ZnCo 2O 4 Nanowire Arrays/Carbon Cloth Anodes for a Novel Class of High-Performance Flexible Lithium-ion Batteries. *Nano Lett.* **2012,** *12*, 3005–3011.
34. Ding, Y. H.; Ren, H. M.; Huang, Y. Y.; Chang, F. H.; Zhang, P. Three-Dimensional Graphene/LiFePO4 Nanostructures as Cathode Materials for Flexible Lithium-ion Batteries. *Mater. Res. Bull.* **2013,** *48*, 3713–3716.
35. Kang, C.; Cha, E.; Baskaran, R.; Choi, W. Three-Dimensional Free-Standing Carbon Nanotubes for a Flexible Lithium-ion Battery Anode. *Nanotechnology* **2016,** *27*, 105402.
36. Yin, H.; et al. Long Cycle Life and High Rate Capability of Three Dimensional $CoSe_2$ Grain-Attached Carbon Nanofibers for Flexible Sodium-ion Batteries. *Nano Energy* **2019,** *58*, 715–723.
37. Zhang, G.; Liu, K.; Liu, S.; Song, H.; Zhou, J. Flexible Co0.85Se Nanosheets/Graphene Composite Film as Binder-Free Anode With High Li- and Na-Ion Storage Performance. *J. Alloys Compd.* **2018,** *731*, 714–722.
38. Kim, S. J. *et al.* 3D Flexible Si Based-Composite (Si@Si3N4)/CNF Electrode With Enhanced Cyclability and High Rate Capability for Lithium-ion Batteries. *Nano Energy* **2016,** *27*, 545–553.
39. Xia, H.; et al. Self-Standing Porous LiMn2O4 Nanowall Arrays as Promising Cathodes for Advanced 3D Microbatteries and Flexible Lithium-ion Batteries. *Nano Energy* **2016,** *22*, 475–482.

40. Saito, F., Baron, M. & Dodds, J. In *Morphology Control in Size Reduction Processes*, 2004; pp 3–23. DOI: 10.1007/978-3-662-08863-0_1.
41. Muller, M.; Schneider, L.; Bohn, N.; Binder, J. R.; Bauer, W. Effect of Nanostructured and Open-Porous Particle Morphology on Electrode Processing and Electrochemical Performance of Li-Ion Batteries. *ACS Appl. Energy Mater.* **2021,** *4*, 1993–2003.
42. AbdelHamid, A. A.; Mendoza-Garcia, A.; Ying, J. Y. Advances in and Prospects of Nanomaterials' Morphological Control for Lithium Rechargeable Batteries. *Nano Energy* **2022,** *93*, 106860.

CHAPTER 5

Advanced Nanomaterials for Water Oxidation and Hydrogen Generation

AHUMUZA BENJAMIN[1], N. P. SINGH[2], and MAMATA SINGH[3]

[1]Department of Mechanical Engineering, GITAM School of Technology, Gandhi Institute of Technology and Management (Deemed to be University), Bengaluru, Karnataka, India

[2]Department of Chemistry, GITAM School of Sciences, Gandhi Institute of Technology and Management (Deemed to be University), Bengaluru, Karnataka, India

[3]CeNSE (Center for Nano Science and Engineering), Indian Institute of Science, Bengaluru, Karnataka, India

ABSTRACT

Hydrogen is a green, carbon-free, and renewable fuel capable of addressing today's energy dilemma. Large-scale production of hydrogen is being undertaken via an artificial photosynthetic approach, which mirrors the natural photosynthesis in plants. This development has garnered significant interest in scientific fields such as physics, chemistry, material science, etc. Hydrogen can be produced cheaply and sustainably by splitting H_2O using nanomaterials and semiconductors as photoelectrochemical (PEC), photo, and biomimetic catalysts. However, the entire process may not generate sufficient fuel due to various inherent constraints, including diffusion distance, bandgap, photostability, and carrier lifetime of semiconductors. Thus, it is critical to consider a few remedies and improvements, such as increasing the contact surface and decreasing the minority carrier diffusion distance, which would guarantee increased hydrogen

Technological Advancement in Clean Energy Production: Constraints and Solutions for Energy and Electricity Development. Amritanshu Shukla, Kian Hariri Asli, Neha Kanwar Rawat, Ann Rose Abraham, & A. K. Haghi (Eds.)

fuel production. This chapter briefly introduces water splitting and discusses the various catalysts and mechanisms involved in this process.

5.1 INTRODUCTION TO WATER SPLITTING

The rate of environmental degradation due to pollution from conventional fuels being used daily is increasing exponentially. Hydrogen being eco-friendly, as stated before, stands out as the most suitable fuel for future use. In contrast to gasoline and other fossil fuels, hydrogen fuel has a significantly high energy conversion efficiency.[1] The generation of H_2 employing light energy is an alluring and viable strategy for solving the energy crisis. This procedure does not create damaging byproducts and is safe for the environment.[2–4] Traditionally, carbon dioxide is a byproduct of the H_2 manufacturing process. Therefore, the crucial point is looking for workable alternatives that will proceed without producing greenhouse gases.

Photocatalytic materials have demonstrated the ability to produce hydrogen since 1972, and a wide range of semiconductor materials have since been explored to evaluate their hydrogen production capability.[5] The photocatalysis mechanism proceeds with oxidation and reduction coinciding under simulated solar irradiation. The ideal potential difference for this water redox process is −1.23 V.[6] These materials must have $E_g > 1.6–1.8$ eV for suitability in the practical water-splitting reaction because they need to utilize the excess potential to speed up the process.[7] Water reduction occurs on the conduction band (CB) and oxidation on the valence band (VB) to ensure the effective transfer of electrons and holes. Furthermore, they should possess a more negative CB edge location than the H_2 generation level ($E = -0.43$ eV, H_2/H^+) and a more positive VB edge location than the H_2O oxidation level ($E = 0.8$ eV, O_2/H_2O) to ensure effective O_2 production.[8] In addition to the above parameters, the catalysts should have an optimal band structure that facilitates charge separation, stability against corrosion, and structural stability over prolonged exposure.

5.2 CATALYSTS FOR WATER SPLITTING

5.2.1 PHOTOCATALYSTS

Photocatalysis is a viable approach to generating H_2 efficiently on a large scale. Recent research has shown that photocatalytic nanoparticles are an essential component in the process of transforming solar radiation into H_2 energy.[9–11] These nanoparticles have specific surface properties which determine their charge transfer, light absorption, surface catalysis, etc.[12,13] For every photocatalytic system to be operational, it must meet specific fundamental requirements which are listed as follows:

i. The capacity for greater charge transfer mobility.
ii. The ability to integrate a simple synthesis procedure with an adequate reaction mechanism.
iii. Effective photon penetration of the nanoarchitectures to land in the photoactive region.

In contrast to these requirements, numerous investigations have been done to fine-tune the catalysts at the nanoscale to improve the reaction activity. As discussed in the following, photocatalysts can be classified as transitional metal oxides, sulfides, nitrides, and carbides.

5.2.1.1 METAL CHALCOGENIDES

Chalcogenides are chemicals made up of one or more chalcogen anions and one or more electropositive elements such as tellurium, selenium, and sulfur. Due to their excellent optical and electrical characteristics, metal chalcogenides are widely used as optoelectronics in displays, batteries, and photovoltaics (PVs).[14,15] Sulfur-based compounds are recently the subject of intensive research among metal chalcogenides due to their wide bandgap, favorable PV characteristics, and high charge transport mobility. In addition, intensive research is also being carried out on the novel ternary, quaternary class of chalcogenide catalysts, which offer a greater degree of freedom for band gap tuning without using hazardous elements.[16] However, chalcogenides are photo-corroded when exposed to radiation without sacrificial electron donors. Thus, it is convenient to employ appropriate band engineering to prevent backward reactions and fast recombination of e^-/h^+ pairs. The following are some strategies to alleviate this problem:

- Doping
    - Doping of the metal ion may create traps between the CB and VB of the catalyst. By retaining holes or electrons, dopants serve as recombination-resistant regions and light penetration centers. They may thereby improve the separation of charges needed for photocatalysis.[17,18]

- Doping of the non-metal ions initiates an upward shift of the VB edge since dopants in the impurity phases of non-metal ions are relatively near the VB. This causes the bandgap energy to narrow.[17,18]

- Semiconductor composites

  Merging two different semiconductors with appropriate photocatalysts VB and CB would potentially improve the trapping of photo-generated holes and electrons on respective surfaces of the semiconductor and optimize the simultaneous reduction and oxidation processes of the e^-/h^+ pairs.[19–21]

5.2.1.2 METAL OXIDES

Metal chalcogenides have the potential to photo-corrode into S (or Te, Se) and M^{n+} in case of protracted irradiation. The same applies to metal oxides in case of bandgap excitation, that is,[22]

$$ZnO + hv \Rightarrow e^-_{CB} + h^+_{VB} \tag{5.1}$$

$$ZnO + 2h^+_{VB} \Rightarrow Zn^{2+} + \frac{1}{2}O_2 \tag{5.2}$$

Specific findings have shown that zinc oxide is a potential photocatalyst, provided its susceptibility and photo-corrosion can be enhanced or reduced.[23,24] H_2 production using sulfides of copper/zinc or zinc oxide in a mixed solution of Na_2SO_3 and Na_2S produced better results than when only ZnO was used. This is because the clusters of CuS on the ZnO/ZnS surfaces play a vital role in accelerating the evolution of H_2 and the splitting of e^-/h^+ pairs. Pure ZnS and ZnO structures have lower photocatalytic activity than hollow ZnS shells or core ZnO. Pure g-C_3N_4 structures have low H_2 evolution compared to oxygen-deficient ZnO/Pt-loaded g-C_3N_4 heterostructures, which attain an optimal H_2 evolution.

The highest rate of H2 evolution is 768 mmol/h which is 14.8 times greater than pure zinc oxide registered by 1 wt.% MOS_2/ZnO nanocomposite while using sacrificial reagents of Na_2SO_3 and Na_2S.[26] The addition of Ru exponentially enhances the water-splitting reaction of ZnO/CdO/CdS nanocomposites since Ru decreases the likelihood of the recombination of e^-/h^+ pairs, hence boosting the amount of H_2 produced. However, more excellent results of H_2 production can be obtained from different metal oxides apart from ZnO, for example, BiOI,[27,28] Bi_2O_3,[29,30] MoO_3,[31–33] $BiVO_4$,[34] a-Fe_2O_3,[35] $MgTiO_3$,[36] and TiO_2,[37–39] where the nanocomposite of platinum and dibismuth trioxide added with ruthenium (IV) oxide stands out with 14.5 and 11.6 mmol/g h in 0.03M $H_2C_2O_4$ and 0.3M Na_2SO_3, respectively. This is due to the formation of a donor state in the energy band by the Ru^{3+}, which lowers the rate of recombination of e^-/h^+ and electron trapping behavior of the Pt nanoparticles on the surface of the semiconductor acting as accumulation sites of electrons, thus improving the separation of holes and electrons. Furthermore, adding Cr_2O_3 as a cocatalyst to the above nanocomposite, thus forming Cr_2O_3/Pt/RuO_2:Bi_2O_3, would increase the rate of H_2 evolution since Cr_2O_3 is capable of activating the hydrogen atom as well as adsorbing the hydrogen ion.[29]

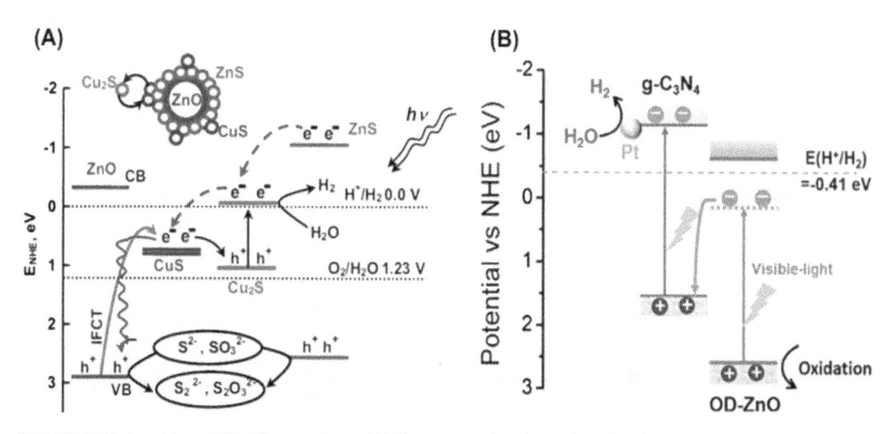

FIGURE 5.1 (A and B) Illustration of different mechanisms for the photocatalytic H_2 generation with semiconductor composites of ZnO.[25]

5.2.2 *PHOTOELECTROCHEMICAL CATALYSTS*

Despite the outstanding performance of photocatalysts in the H_2 evolution reaction (HER), a significant drop in the rate of production is still registered due to the recombination of H_2 and O_2 which is why the system of PEC catalysis comes in handy.[40,41] In this, separate half-reactions of hydrogen and oxygen evolution proceed independently. A PEC cell having an H-type reactor, an ion-exchange membrane, and two photoelectrodes is used.[42,43] The photoelectrodes are formed by depositing photocatalysts on F-based SnO_2 glasses and these electrodes are used to implement the water-splitting reaction in the solution. This kind of setup gives a basic PEC system however, there are many more varieties of PEC systems for water splitting.[43–45] These systems have shown extraordinary potential to attain even greater solar-to-hydrogen efficiency. Their arrangement does not necessitate positioning of the CB medium of the cathode above the H_2O reduction potential since an external PV cell can provide the extra bias. Therefore, to split the water, two voltages are added up, that is, one at the photoanode and the other from the PV cell to come up with sufficient voltage for the water-splitting reaction. Furthermore, the gain in voltage must be greater than the loss in current generated by the PEC cell.

The semiconductors used in solar energy harvesting can be classified as[46–49]:

- n-type: this is used in the photoanode to facilitate direct O_2 evolution.
- p-type: this resides at the photocathode and facilitates the production of holes for direct H_2 evolution.

However, the photoelectrode's electrocatalytic and photoactivity are directly affected by nanomaterials.

PEC water splitting is one of the most convincing methods for converting energy because it has demonstrated effectiveness in its practical performance. The two key concepts vital to it are as follows:

i. The contact reaction at the surface of the electrode may actually be more significant than photoabsorption by the semiconductors since the electrocatalytic process facilitating the H_2O splitting serves as the rate-determining step of the entire process.
ii. Electrochemical catalysis and photon absorption are directly governed by the nanomaterials' architecture. The way nanomaterials acting as PEC catalysts are designed or structured greatly influences how they behave at the photoelectrodes and not how ably they harvest light. However, this point is constantly neglected.

Constant laboratory tests and research have clearly pointed out that a kin architecture design for nanomaterials could be a tuning process to achieve equilibrium between various processes which compete in a complicated system to enhance PEC efficiency.[44,50,51] The main goal is to minimize the density of charge recombination sites while increasing the amount of electro-catalytic centers and creating a more efficient electrolyte/electrode interface.

5.2.2.1 QUANTUM DOTS

Metal oxides with a bandgap greater than 3.0 V are ample options for photoelectrochemical anodes since they can retain minimal sunlight. Band reduction is an interesting aspect of photoelectrochemical studies and research. Small bandgap semiconductor QDs are used to successfully couple standard wide bandgap metal oxides.[52–57] However, the band architecture of semiconductors can be altered by quantum confinement.[58,59] Apart from having particular optical characteristics, quantum dots (QDs) also share common chemical characteristics. Although the chemical science of QDs is still unknown, their charge transfer is very different. Therefore, semiconductor decoration with QDs can modulate the solar absorption band and mediate electrocatalytic reactions.

5.2.2.2 NANOCOMPOSITES

As earlier mentioned, apart from collecting the energy of photons and pushing electrons from the VB to the CB, it is also vital to direct electrons to the current receiver. This can be achieved by adding conductive additives to enhance the material's conductivity. It can also contain other functions. When these additives are added to the material, they affect the growth and nucleation mechanisms and change the PEC nano materials' surface configuration. Furthermore, PEC yields can be enhanced without necessarily affecting the metal oxides' internal lattice configuration by adding noble metal salts to synthetic precursors.[60] This structure enables charge separation and reduces electron-hole coupling since electron mobility through the second nanocomposite component is faster. Some reports claim improved PEC performance using graphics, especially functionalized graphics. The process of separating charges is sophisticated; however, adding graphene to some nanocomposites can improve or degrade their PEC performance. This is because of a different charge transfer mechanism per the graphene's structural and chemical disorder. Since graphene acts as a hole collector, it enhances the PEC performance of metal oxides when coated together and does not alter their corresponding mechanism. Polymers that are good conductors can also be employed for the same role since their surrounding soft matter can significantly enhance the mechanical stability of the active material.[61–63] Conducting polymers can also have bandgaps that span the solar spectrum and thus assist the entire system. However, different methods can be used to synthesize nanostructured conducting polymers to attain the nanoarchitecture of PEC electrodes.[64] The second component of these nanocomposites typically serves as a nucleation site for developing semiconductor materials.

5.2.2.3 SHELL/CORE

Photoelectrochemical materials, for example, metal oxides, have poor conductivity. This is because electrons have to flow via the compound to the current receiver. A shell/core architecture solves the problem by developing a good conductive shell and placing a photoresist on the core. CdS-C-shell-core nanostructures exhibit efficient transfer of electrons and higher light absorbing capacity, thus significantly improving imaging activity. Furthermore, an increase in light scattering leads to a significant improvement in photoelectrochemical activity since it absorbs light along the whole span of the spectral wavelength and efficiently transfers electrons to the carbon layer from the CdS. The external shell creates connections between particles; its electrical conductivity is essential. PEC performance can be degraded due to the plasmonic effect when a metallic core is used to make the core/shell structure. This also affects the photocurrent. Photoactive heterojunctions can be used to develop multilayer shell/core designs since they can change the passband as the outer shell performs electrocatalytic reactions. Core/shell usually describes core-like spherical particles ejected from a shell; however, the same principle can refer to coatings in one-dimensional architectures. The 1D nanostructure of the ZnO-TiO_2 core/shell has a dual function of enhancing the optical activity of pure zincoOxide nanowires and the photolytic stability since the TiO_2 shell plays a protective role in the design. Air treatment decreases recombination and improves the separation of the charges. The electron–hole pairs are formed within the ZnO nanowires on the absorption of incident photons. The TiO_2 shell enhances the separation of charge. In contrast, charge recombination is limited at the contact interface of the semiconductor and electrolyte by the potential barrier in the core/shell nanodesign. Hence the high transfer of electrons of the dimensional zinc oxide nanoparticles in the TiO_2-ZnO core/shell nanoarchitecture is linked to the efficient charge carrier separation between the zinc oxide core and the titanium oxide shell. The SnO_2-CdS core/shell structure can successfully increase the rate of separation of the e^-/h^+ pair, increase the photocatalytic activity in the presence of visible light irradiation, and increase the electron lifetime. The shell and core may be made of the same material from doped and raw forms for this condition.

5.2.2.4 ONE DIMENSIONAL ARCHITECTURE

The vertical growth of the active material on the 1D design-like substrate is favorable for the PEC applications[65–74] since it lowers the recombination of the charge carriers and facilitates charge transport within the proper direction of the current receiver.[75–81] However, the nanomaterial's specific surface area is less than the required for the carrier transfer at the electrode/electrolyte contact. In addition, low light scattering ability reduces the capability of harvesting visible light. Owing to the technological advancement and the novel preparation techniques of 1D materials available today, the development of multilayer materials with 1D architecture for PEC applications is possible. Since the significant consideration is the outer surface of the 1D object, nanotubes are preferred to nanofibers for the PEC applications. Similarly, a sleek outer surface is not of many benefits; hence inducing branches to increase roughness and adding holes to make the character more porous is required.[82–88] Branching is the second step and it may be achieved by employing a different material. This presents a good chance for the multimaterial design for PEC optoelectronics. The benefits of branched nanostructures are not restricted to increasing the contact area for light scattering but also facilitating the rapid transfer of electrons. In addition, the compact size of the ZnO nanodisc aids the separation of electron–hole pairs to forestall recombination loss. The term ID is somewhat controversial, as transport is not strictly one-way. Furthermore, the classical branching design is not 1D. Another design for making a porous structure in an exceedingly 1D direction is the inverse opal,[89–96] where skeletons are made of honeycomb-like ordered walls. This shifts the term from 1D to 3D. Some TiO_2 nanorods can give multiple scattering centers and exhibit increased light-gathering capabilities. In addition, the large extent of the stratified arrays improves electrolyte exposure, whereas the nanorods offer direct pathways for the rapid transfer of electrons. These distinct structures improve charge separation at junctions, facilitating electron transfer.

5.3 MECHANISM OF HYDROGEN EVOLUTION

Oxygen and hydrogen evolution reactions, OER and HER, respectively, are the two essential reactions involved in the water-splitting process.

$$2H^+ + 2e^- \rightarrow H_2\; E^0 = 0.00V \tag{5.3}$$

$$2H_2O \rightarrow O_2 + 4H^+ + 4e^-\; E^0 = +1.23V \tag{5.4}$$

$$H_2O \rightarrow H_2 + \frac{1}{2}O_2$$

$$\Delta E^0 = 1.23V, \Delta G^0 = +237.2\,kJ/mole \tag{5.5}$$

Hence, it's conclusive that for the water-splitting process to proceed successfully, photocatalytic surfaces with the ability to absorb and produce light and electron–hole pairs are vital. These e^-/h^+ pairs can then take part in the HER and OER on the surface of the catalyst. But to achieve this goal, it is necessary to have a more negative CB edge and a more positive VB edge of the catalyst. The most suitable bandgap for the semiconductor should be 1.5–2.5 eV after considering the thermal losses involved in the reaction.

Hydrogen Evolution Reaction: Currently, H_2 production is based on steam reforming. Steam reforming refers to the reaction of CH_4 with H_2O (g) at a raised temperature and high pressure where a suitable catalyst is used to produce CO and H_2 gas. The carbon monoxide produced in the initial step is further exposed to excess steam to generate carbon dioxide and more hydrogen gas.

$$CH_4 + H_2O \rightarrow CO + 3H_2 \tag{5.6}$$

$$CO + H_2O \rightarrow CO_2 + H_2 \tag{5.7}$$

The reaction mechanism leading to hydrogen evolution from water splitting can be summarized in four distinct approaches as follows:

- Photobiological water splitting
  Microorganisms are used to generate H_2 from H_2O in the presence of sunlight. These organisms use photosystems I and II pigments to trap light later transferred to the chlorophyll reaction sites where different oxidants and reductants are formed after charge separation occurs.[97–99] Further oxidants and reductants react in the photosystems to form holes and electrons, which eventually participate in the water-splitting process. Some of the setbacks associated with this process are low H_2 production levels and low solar to H_2 efficiency, which is why photobiological water splitting is unsuitable for large-scale H_2 production.
- Thermochemical water splitting
  The high temperature generated from solar energy or other thermal sources triggers the chemical reaction on the catalyst's surface to produce H_2 and O_2. These temperatures are drawn with the help of massive concentrators or advanced nuclear reactors. Thermochemical water splitting can be direct or hybrid. The direct manner makes use of a two-step course. In contrast, the hybrid procedure is more complex with multiple steps but low operating temperature compared to the direct method and minimal radioactive emission.[100–102]
- Photoelectrochemical (PEC) water splitting
  This process takes place inside a photoelectrochemical cell in the presence of solar radiation

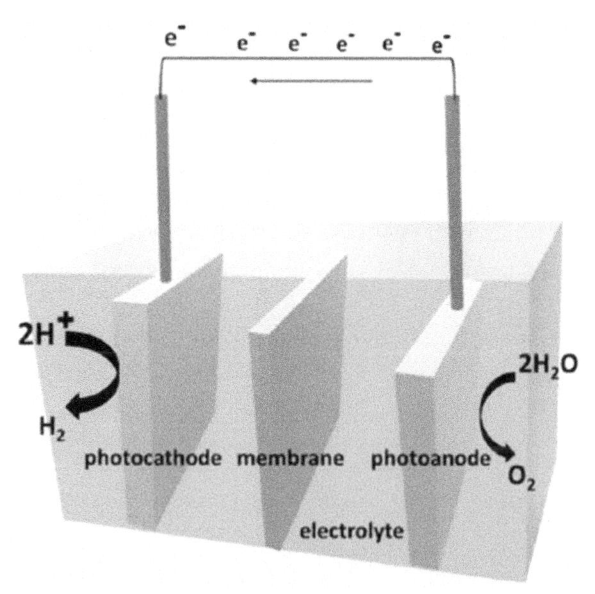

FIGURE 5.2 Illustration of photoelectrochemical cell.[5]

The PEC cell comprises a photoanode where O_2 evolution occurs and a photocathode where H_2 evolution occurs. The electrodes reside in an electrolyte and are connected by a wire which guides the current flow, thus forming a complete circuit. The semiconductors used in the photocathode for H_2 generation should have a bandgap above 1.23 V to supply an external bias sufficient to trigger the reaction. Partial electron depletion in the photoactive surface of the semiconductor may be observed if additional bias is used. This depletion leads to band bending due to surface space layer formation. This process increases the charge carriers' lifespan and lowers the rate at which charges recombine.

5.3.1 MECHANISM OF PHOTOCATALYST FOR HYDROGEN EVOLUTION

Photocatalysis refers to the speeding up of a photoreaction in the presence of a photocatalyst. An adsorbed substrate is used to absorb light. The photocatalytic activity is determined by the ability of the catalyst to form e^-/h^+ pairs that are responsible for generating free radicals which later take part in the secondary reactions. Photocatalytic H_2 production is an example of a photochemical reaction. The figure below shows a representation of a single-cell reaction of photocatalytic water splitting. Photocatalytic water splitting

Ref. [5]. Note: https://pubs.acs.org/doi/10.1021/acsenergylett.9b00940. Further permissions related to the excerpted material should be directed to ACS depends on factors such as the molar concentration of the solution, pH value, type of solution, suspensions in the solution, and their light absorbency. These factors have been studied and investigated extensively in order to come up with an optimized photocatalytic reaction system. However, sacrificial reagents have also been employed so as to fully analyze the behavior of the photocatalytic system under these conditions. These sacrificial reagents are oxidized by the holes irreversibly hence increasing the number of electrons and enhancing the evolution of H_2. It is important to note that this is just an intermediate approach since the holes oxidized the sacrificial reagents which acted as scavengers instead of the photocatalyst. Following is an example of a photocatalytic water-splitting reaction using Cu-doped zinc sulfide as a photocatalyst and sodium sulfide as a sacrificial reagent.

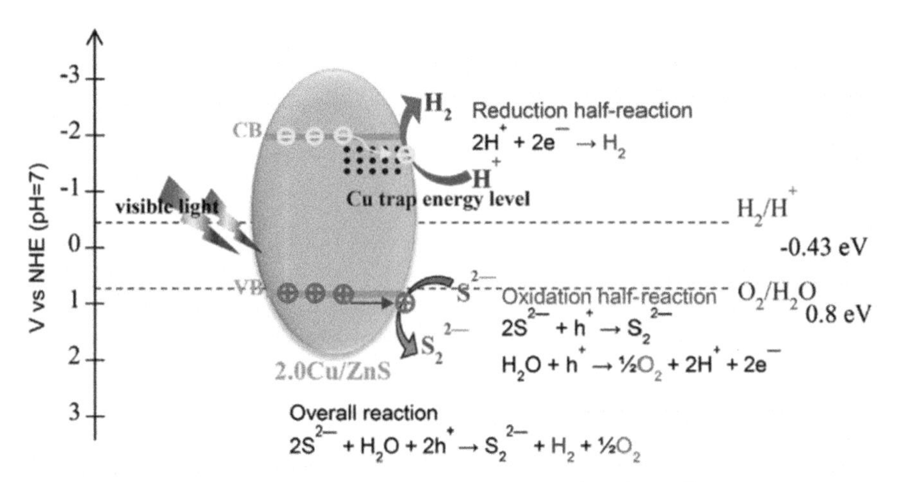

FIGURE 5.3 Illustration of photocatalytic HER using Cu-doped ZnS.[25]

The hydrogen production rate is relatively higher in the presence of a sacrificial reagent since Cu is responsible for the formation of deep trap energy levels between the VB and CB of ZnS which results in efficient transfer of charges.

5.3.2 MECHANISM OF PHOTOELECTROCHEMICAL (PEC) CATALYST FOR HYDROGEN EVOLUTION

The working principle of the PEC H_2 evolution is simple, however, the actual underlying processes are sophisticated and still unclear to scientists at their molecular level. The fundamental setup of a PEC cell consists of a photo-cathode and anode dipped in an electrolyte. The role of the anode is to absorb light and generate e^-/h^+ pairs whereas the cathode is where H_2 generation takes place when protons are reduced by electrons. An external bias from an electrical power source or solar cell is required to transport the electrons to the metal cathode. The excess holes at the anode oxidize H_2O molecules in O_2 and protons. These protons a transmitted to the cathode where they are reduced by electrons to produce H_2. As discussed earlier, there are a number of setups for the PEC system, of which some are basic while others are complicated however they all share the same fundamental working principle.

$$\text{Photoanode: } H_2O + 2H^+ \rightarrow 2H^+ + \frac{1}{2} O_2 \; E^{\circ}_{\text{oxidation}} = -1.23 \text{ V}$$

$$\text{Cathode reaction: } 2H^+ + 2e^- \rightarrow H_2 \; E^{\circ}_{\text{reduction}} = 0 \text{ V}$$

The PEC device is operated at steady-state where the photocurrent induced is directly proportional to H_2 generation activity and the photon to hydrogen generation efficiency (η) is given by the ratio of the energy required to split the water to the power input by the light. Taking an assumption that all e^-/h^+ are used up in the redox reaction, efficiency is given by

$$\eta = \frac{I(1.23 - V_{bias})}{J_{light} \times A} \tag{5.8}$$

where I (in A) is the photocurrent,

1.23 V is the theoretical H_2O splitting potential,

V (in volts) is the external bias potential,

J (W/m^2) is the flux density of light,

A (m^2) is the area of the irradiated surface.

However, efficiency can still be calculated from IPCE (incident photon-to-current efficiency) which is the ratio of photocurrent to the rate of irradiated photons as a function of wavelength of the incident light.

$$IPCE = \frac{I(1240 \times I_{PH})}{\lambda \times J_{light}} \tag{5.9}$$

where I_{PH} (A/m^2) is the density of photocurrent,

λ (nm) is the wavelength of light,

P (W/m^2) is the flux density of photons,

1240 is the unit correction factor.

PEC water splitting can be summarized in four basic steps as follows:

i. Absorption of light and generation of charge carriers in the semiconductor.
ii. Separation of charges and transfer of charges from the electrodes and the electrode/electrolyte interface.
iii. Extraction of charge carriers and formation of PEC product at the semiconductor/electrolyte interface.
iv. Utilization of the PEC products in the solution.

A threshold voltage of at least 1.23 V is sufficient to split water upon photoirradiance of wavelength approximately 1000 nm, hence water splitting by irradiation with visible light is possible if solar energy is effectively utilized in the PEC system. However, photons with energy (>1.23 V) are required in practical applications since some energy losses such as e^-/h^+ recombination, potential drops at the semiconductor/electrolyte interface, and voltage loss because of the contact and electrode resistance may occur during the PEC water splitting process.

KEYWORDS

- **water splitting**
- **photocatalysts**
- **photoelectrochemical catalysts**
- **hydrogen generation**
- **alternative fuel**

REFERENCES

1. Chamousis, R. Hydrogen: Fuel of the Future, n.d. [Online]. www.esrl.noaa.gov/gmd/ccgg/trends/co2_data_mlo.html.
2. Kageshima, Y.; Shiga, S.; Ode, T.; Takagi, F.; Shiiba, H.; Htay, M.T.; Hashimoto, Y.; Teshima, K.; Domen, K.; Nishikiori, H. Photocatalytic and Photoelectrochemical Hydrogen Evolution from Water Over Cu2SnxGe1- xS3Particles. *J. Am. Chem. Soc.* **2021,** *143*, 5698–5708. https://doi.org/10.1021/JACS.0C12140/SUPPL_FILE/JA0C12140_SI_001.PDF.
3. Chakraborty, S.; Edwards, E. H.; Kandemir, B.; Bren, K. L. Photochemical Hydrogen Evolution from Neutral Water with a Cobalt Metallopeptide Catalyst. *Inorg. Chem.* **2019,** *58*, 16402–16410. https://doi.org/10.1021/ACS.INORGCHEM.9B02067.

4. Lee, G. J.; Anandan, S.; Masten, S. J.; Wu, J. J. Photocatalytic Hydrogen Evolution from Water Splitting Using Cu Doped ZnS Microspheres Under Visible Light Irradiation. *Renew. Energy* **2016,** *89*, 18–26. https://doi.org/10.1016/j.renene.2015.11.083.
5. Ganguly, P.; Harb, M.; Cao, Z.; Cavallo, L.; Breen, A.; Dervin, S.; Dionysiou, D. D.; Pillai, S. C. 2D Nanomaterials for Photocatalytic Hydrogen Production. *ACS Energy Lett.* **2019,** *4* 1687–1709. https://doi.org/10.1021/acsenergylett.9b00940.
6. Jang, J. S.; Kim, H. G.; Lee, J. S. Heterojunction Semiconductors: A Strategy to Develop Efficient Photocatalytic Materials for Visible Light Water Splitting. *Catal. Today* **2012,** *185* (1), 270–277. https://doi.org/10.1016/j.cattod.2011.07.008.
7. Fujishima, A.; Honda, K. Electrochemical Photolysis of Water at a Semiconductor Electrode. *Nature* **1972,** *238*, 37–38. https://doi.org/10.1038/238037a0.
8. Chen, S.; Thind, S. S.; Chen, A. Nanostructured Materials for Water Splitting - State of the Art and Future Needs: A Mini-Review. *Electrochem. Commun.* **2016,** *63*, 10–17. https://doi.org/10.1016/j.elecom.2015.12.003.
9. Li, R.; Li, C. Photocatalytic Water Splitting on Semiconductor-Based Photocatalysts. In *Advances in Catalysis*; Academic Press Inc., 2017; pp 1–57. https://doi.org/10.1016/bs.acat.2017.09.001.
10. Ahmed, N.; Morikawa, M.; Izumi, Y. Photocatalytic Conversion of Carbon Dioxide into Methanol Using Optimized Layered Double Hydroxide Catalysts. *Catal. Today* **2012,** *185*, 263–269. https://doi.org/10.1016/j.cattod.2011.08.010.
11. Jang, J. S.; Kim, H. G.; Lee, J. S. Heterojunction Semiconductors: A Strategy to Develop Efficient Photocatalytic Materials for Visible Light Water Splitting. *Catal. Today* **2012,** *185*, 270–277. https://doi.org/10.1016/J.CATTOD.2011.07.008.
12. Aguilera González, E. N.; Estrada Flores, S.; Martínez Luévanos, A. Nanomaterials: Recent Advances for Hydrogen Production. In *Handbook of Nanomaterials and Nanocomposites for Energy and Environmental Applications*, 2021; pp 1–27. https://doi.org/10.1007/978-3-030-11155-7_33-1.
13. Shenoy, K. H.; Moses, V. Nanomaterials for Hydrogen Generation: A Review. n.d.
14. Fan, K.; Zou, H.; Lu, Y.; Chen, H.; Li, F.; Liu, J.; Sun, L.; Tong, L.; Toney, M. F.; Sui, M.; Yu, J. Direct Observation of Structural Evolution of Metal Chalcogenide in Electrocatalytic Water Oxidation. *ACS Nano* **2018,** *12*, 12369–12379. https://doi.org/ 10.1021/acsnano.8b06312.
15. Cho, D. H.; Lee, W. J.; Park, S. W.; Wi, J. H.; Han, W. S.; Kim, J.; Cho, M. H.; Kim, D.; Chung, Y. D. Non-toxically Enhanced Sulfur Reaction for Formation of Chalcogenide Thin Films Using a Thermal Cracker. *J. Mater. Chem. A Mater.* **2014,** *2*, 14593–14599. https://doi.org/10.1039/C4TA02507E.
16. Aldakov, D.; Lefrançois, A.; Reiss, P. Ternary and Quaternary Metal Chalcogenide Nanocrystals: Synthesis, Properties and Applications. *J. Mater. Chem. C Mater.* **2013,** *1*, 3756–3776. https://doi.org/10.1039/c3tc30273c.
17. Paul, A. K.; Prabu, M.; Madras, G.; Natarajan, S. Effect of Metal Ion Doping on the Photocatalytic Activity of Aluminophosphates. *J. Chem. Sci.* **2010,** *122*, 771–785. https://doi.org/10.1007/S12039-010-0065-0.
18. Marschall, R.;Wang, L. Non-metal Doping of Transition Metal Oxides for Visible-Light Photocatalysis. *Catal. Today* **2014,** *225*, 111–135. https://doi.org/10.1016/J.CATTOD.2013.10.088.
19. Habisreutinger, S. N.; Schmidt-Mende, L.; Stolarczyk, J. K. Photocatalytic Reduction of CO2 on TiO2 and Other Semiconductors. *Angew. Chem. Int. Ed.* **2013,** *52*, 7372–7408. https://doi.org/10.1002/ANIE.201207199.

20. Marschall, R.; Marschall, R. Semiconductor Composites: Strategies for Enhancing Charge Carrier Separation to Improve Photocatalytic Activity. *Adv. Funct. Mater.* **2014,** *24*, 2421–2440. https://doi.org/10.1002/ADFM.201303214.
21. Marschall, R. Photocatalysis: Semiconductor Composites: Strategies for Enhancing Charge Carrier Separation to Improve Photocatalytic Activity. *Adv. Funct. Mater.* **2014,** *24*, 2420–2420. https://doi.org/10.1002/ADFM.201470108.
22. Kudo, A.; Miseki, Y. Heterogeneous Photocatalyst Materials for Water Splitting. *Chem. Soc. Rev.* **2009,** *38*, 253–278. https://doi.org/10.1039/b800489g.
23. Agarwal, H.; Venkat Kumar, S.; Rajeshkumar, S. A Review on Green Synthesis of Zinc Oxide Nanoparticles – An Eco-friendly Approach. *Res. Eff. Technol.* **2017,** *3*, 406–413. https://doi.org/10.1016/j.reffit.2017.03.002.
24. Ong, C. B.; Ng, L. Y.; Mohammad, A. W. A Review of ZnO Nanoparticles as Solar Photocatalysts: Synthesis, Mechanisms and Applications. *Renew. Sustain. Energy Rev.* **2018,** *81*, 536–551. https://doi.org/10.1016/J.RSER.2017.08.020.
25. Wu, J. J.; Lee, G. J. Advanced Nanomaterials for Water Splitting and Hydrogen Generation. In *Nanomaterials for Green Energy*; Elsevier, 2018; pp 145–167. https://doi.org/10.1016/B978-0-12-813731-4.00005-9.
26. Yuan, Y. J.; Wang, F.; Hu, B.; Lu, H. W.; Yu, Z. T.; Zou, Z. G. Significant Enhancement in Photocatalytic Hydrogen Evolution from Water Using a MoS2 Nanosheet-Coated ZnO Heterostructure Photocatalyst. *Dalton Trans.* **2015,** *44,* 10997–11003. https://doi.org/10.1039/C5DT00906E.
27. Lee, G. J.; Zheng, Y. C.; Wu, J. J. Fabrication of Hierarchical Bismuth Oxyhalides (BiOX, X = Cl, Br, I) Materials and Application of Photocatalytic Hydrogen Production from Water Splitting. *Catal. Today* **2018,** *307,* 197–204. https://doi.org/10.1016/J.CATTOD.2017.04.044.
28. Vinoth, S.; Ong, W. J.; Pandikumar, A. Defect Engineering of BiOX (X = Cl, Br, I) Based Photocatalysts for Energy and Environmental Applications: Current Progress and Future Perspectives. *Coord. Chem. Rev.* **2022,** *464*.. https://doi.org/10.1016/j.ccr.2022.214541.
29. Hsieh, S.; Lee, G.; Davies, S.; Masten, S.; Wu, J. J. Synthesis of Cr2O3 and Pt Doped RuO2/Bi2O3 Photocatalysts for Hydrogen Production from Water Splitting, 2013.
30. Hsieh, S. H.; Lee, G. J.; Chen, C. Y.; Chen, J.H.; Ma, S. H.; Horng, T. L.; Chen, K. H.; Wu, J. J. Synthesis of Pt Doped Bi2O3/RuO2 Photocatalysts for Hydrogen Production from Water Splitting Using Visible Light. *J. Nanosci. Nanotechnol.* **2012,** *12,* 5930–5936. https://doi.org/10.1166/JNN.2012.6396.
31. Ma, C.; Zhou, J.; Cui, Z.; Wang, Y.; Zou, Z. In Situ Growth MoO3 Nanoflake on Conjugated Polymer: An Advanced Photocatalyst for Hydrogen Evolution from Water Solution Under Solar Light. *Solar Energy Mater. Solar Cells* **2016,** *150*, 102–111. https://doi.org/10.1016/J.SOLMAT.2016.02.010.
32. Chen, X.; Shen, S.; Guo, L.; Mao, S. S. Semiconductor-Based Photocatalytic Hydrogen Generation. *Chem. Rev.* **2010,** *110*, 6503–6570. https://doi.org/10.1021/CR1001645.
33. Lou, S. N.; Ng, Y. H.; Ng, C.; Scott, J.; Amal, R. Harvesting, Storing and Utilising Solar Energy Using MoO3: Modulating Structural Distortion Through pH Adjustment. *ChemSusChem.* **2014,** *7*, 1934–1941. https://doi.org/10.1002/CSSC.201400047.
34. Zhu, R.;Tian, F.; Cao, G.; Ouyang, F. Construction of Z Scheme System of ZnIn2S4/RGO/BiVO4 and its Performance for Hydrogen Generation Under Visible Light. *Int. J. Hydrogen Energy* **2017,** *42*, 17350–17361. https://doi.org/10.1016/J.IJHYDENE.2017.02.091.

35. Rai, S.; Ikram, A.; Sahai, S.; Dass, S.; Shrivastav, R.; Satsangi, V. R. Photoactivity of MWCNTs Modified α-Fe2O3 Photoelectrode Towards Efficient Solar Water Splitting. *Renew. Energy* **2015,** *83*, 447–454. https://doi.org/10.1016/J.RENENE.2015.04.053.
36. Zhu, W.; Han, D.; Niu, L.; Wu, T.; Guan, H. Z-scheme Si/MgTiO3 Porous Heterostructures: Noble Metal and Sacrificial Agent Free Photocatalytic Hydrogen Evolution. *Int. J. Hydrogen Energy* **2016,** *41,* 14713–14720. https://doi.org/10.1016/J.IJHYDENE.2016.06.118.
37. Esrafili, A.; Salimi, M.; Jonidi Jafari, A.; Reza Sobhi, H.; Gholami, M.; Rezaei Kalantary, R. Pt-based TiO2 Photocatalytic Systems: A Systematic Review. *J. Mol. Liq.* **2022,** *352*. https://doi.org/10.1016/j.molliq.2022.118685.
38. Hwang, Y. J.; Yang, S.; Lee, H. Surface Analysis of N-doped TiO2 Nanorods and Their Enhanced Photocatalytic Oxidation Activity. *Appl. Catal. B* **2017,** *204*, 209–215. https://doi.org/10.1016/j.apcatb.2016.11.038.
39. Yu, L.; Shao, Y.; Li, D. Direct Combination of Hydrogen Evolution from Water and Methane Conversion in a Photocatalytic System Over Pt/TiO2. *Appl. Catal. B* **2017,** *204*, 216–223. https://doi.org/10.1016/J.APCATB.2016.11.039.
40. Zhao, E.; Du, K.; Yin, P. F.; Ran, J.; Mao, J.; Ling, T.; Qiao, S. Z. Advancing Photoelectrochemical Energy Conversion through Atomic Design of Catalysts. *Adv. Sci.* **2022,** *9*. https://doi.org/10.1002/advs.202104363.
41. Ahmad, H.; Kamarudin, S. K.; Minggu, L. J.; Kassim, M. Hydrogen from Photo-Catalytic Water Splitting Process: A Review. *Renew. Sustain. Energy Rev.* **2015,** *43*, 599–610. https://doi.org/10.1016/j.rser.2014.10.101.
42. Eftekhari, A.; Babu, V. J.; Ramakrishna, S. Photoelectrode Nanomaterials for Photoelectrochemical Water Splitting. *Int. J. Hydrogen Energy* **2017,** *42*, 11078–11109. https://doi.org/10.1016/j.ijhydene.2017.03.029.
43. Mali, M. G.; An, S.; Liou, M.; Al-Deyab, S. S.; Yoon, S. S. Photoelectrochemical Solar Water Splitting Using Electrospun TiO 2 Nanofibers. *Appl. Surf. Sci.* **2015,** *328*, 109–114. https://doi.org/10.1016/j.apsusc.2014.12.022.
44. Fujishima, A.; Honda, K. Electrochemical Photolysis of Water at a Semiconductor Electrode. *Nature* **1972,** *238,* 37–38. https://doi.org/10.1038/238037A0.
45. Schley, N. D.; Blakemore, J. D.; Subbaiyan, N. K.; Incarvito, C. D.; Dsouza, F.; Crabtree, R. H.; Brudvig, G. W. Distinguishing Homogeneous from Heterogeneous Catalysis in Electrode-Driven Water Oxidation with Molecular Iridium Complexes. *J. Am. Chem. Soc.* **2011,** *133,* 10473–10481. https://doi.org/10.1021/ja2004522.
46. Chen, X.; Shen, S.; Guo, L.; Mao, S. S. Semiconductor-Based Photocatalytic Hydrogen Generation. *Chem. Rev.* **2010,** *110*, 6503–6570. https://doi.org/10.1021/CR1001645.
47. Sakai, Y.; Sugahara, S.; Matsumura, M.; Nakato, Y.; Tsubomura, H. Photoelectrochemical Water Splitting by Tandem Type and Heterojunction Amorphous Silicon Electrodes. *Can. J. Chem.* **1988,** *66*, 1853–1856. https://doi.org/10.1139/V88-299.
48. Maeda, K. Photocatalytic Water Splitting Using Semiconductor Particles: History and Recent Developments. *J. Photochem. Photobiol. C Photochem. Rev.* **2011,** *12*, 237–268. https://doi.org/10.1016/J.JPHOTOCHEMREV.2011.07.001.
49. Maeda, K.; Domen, K. Photocatalytic Water Splitting: Recent Progress and Future Challenges. *J. Phys. Chem. Lett.* **2010,** *1*, 2655–2661. https://doi.org/10.1021/JZ1007966.
50. Yu, S. Q.; Ling, Y. H.; Zhang, J.; Qin, F.; Zhang, Z. J. Efficient Photoelectrochemical Water Splitting and Impedance Analysis of WO3−x Nanoflake Electrodes. *Int. J. Hydrogen Energy* **2017,** *42*, 20879–20887. https://doi.org/10.1016/j.ijhydene.2017.01.177.

51. Jia, J.; Seitz, L. C.; Benck, J. D.; Huo, Y.; Chen, Y.; Ng, J. W. D.; Bilir, T.; Harris, J. S.; Jaramillo, T. F. Solar Water Splitting by Photovoltaic-Electrolysis with a Solar-to-Hydrogen Efficiency Over 30%. *Nat. Commun.* **2016,** *7*. https://doi.org/10.1038/ncomms13237.
52. Sudhagar, P.; Herraiz-Cardona, I.; Park, H.; Song, T.; Noh, S. H.; Gimenez, S.; Sero, I. M.; Fabregat-Santiago, F.; Bisquert, J.; Terashima, C.; Paik, U.; Kang, Y. S.; Fujishima, A.; Han, T. H. Exploring Graphene Quantum Dots/TiO2 Interface in Photoelectrochemical Reactions: Solar to Fuel Conversion. *Electrochim. Acta* **2016,** *187*, 249–255. https://doi.org/10.1016/J.ELECTACTA.2015.11.048.
53. Xu, Z.; Yin, M.; Sun, J.; Ding, G.; Lu, L.; Chang, P.; Chen, X.; Li, D. 3D Periodic Multiscale TiO2 Architecture: A Platform Decorated With Graphene Quantum Dots for Enhanced Photoelectrochemical Water Splitting. *Nanotechnology* **2016,** *27*, 115401. https://doi.org/10.1088/0957-4484/27/11/115401.
54. Li, J.; Gao, X.; Liu, B.; Feng, Q.; Li, X. B.; Huang, M. Y.; Liu, Z.; Zhang, J.; Tung, C. H.; Wu, L. Z. Graphdiyne: A Metal-Free Material as Hole Transfer Layer to Fabricate Quantum Dot-Sensitized Photocathodes for Hydrogen Production. *J. Am. Chem. Soc.* **2016,** *138,* 3954–3957. https://doi.org/10.1021/JACS.5B12758/SUPPL_FILE/JA5B12758_SI_001.PDF.
55. Zhang, L.; Baumanis, C.; Robben, L.; Kandiel, T.; Bahnemann, D. Bi2WO6 Inverse Opals: Facile Fabrication and Efficient Visible-Light-Driven Photocatalytic and Photoelectrochemical Water-Splitting Activity. *Small* **2011,** *7*, 2714–2720. https://doi.org/10.1002/SMLL.201101152.
56. Zhang, L.; Baumanis, C.; Robben, L.; Kandiel, T.; Bahnemann, D. Bi2WO6 Inverse Opals: Facile Fabrication and Efficient Visible-Light-Driven Photocatalytic and Photoelectrochemical Water-Splitting Activity. *Small* **2011,** *7*, 2714–2720. https://doi.org/10.1002/SMLL.201101152.
57. Lee, J.; Cho, C. Y.; Lee, D. C.; Moon, J. H. Bilayer Quantum Dot-Decorated Mesoscopic Inverse Opals for High Volumetric Photoelectrochemical Water Splitting Efficiency. *RSC Adv.* **2016,** *6*, 8756–8762. https://doi.org/10.1039/C5RA27049A.
58. Wang, G.; Li, Y. Nickel Catalyst Boosts Solar Hydrogen Generation of CdSe Nanocrystals. *ChemCatChem.* **2013,** *5*, 1294–1295. https://doi.org/10.1002/CCTC.201300034.
59. Liu, L.; Peng, Q.; Li, Y. Preparation of CdSe Quantum Dots with Full Color Emission Based on a Room Temperature Injection Technique. *Inorg. Chem.* **2008,** *47*, 5022–5028. https://doi.org/10.1021/IC800368U/ASSET/IMAGES/MEDIUM/IC-2008-00368U_0001.GIF.
60. Shen, S.; Li, M.; Guo, L.; Jiang, J.; Mao, S. S. Surface Passivation of Undoped Hematite Nanorod Arrays via Aqueous Solution Growth for Improved Photoelectrochemical Water Splitting. *J. Colloid Interface Sci.* **2014,** *427*, 20–24. https://doi.org/10.1016/J.JCIS.2013.10.063.
61. Lattach, Y.; Fortage, J.; Deronzier, A.; Moutet, J. C. Polypyrrole-Ru(2,2′-bipyridine)32+/MoSx Structured Composite Film as a Photocathode for the Hydrogen Evolution Reaction. *ACS Appl. Mater. Interfaces* **2015,** *7*, 4476–4480. https://doi.org/10.1021/ACSAMI.5B00401/SUPPL_FILE/AM5B00401_SI_001.PDF.
62. Lattach, Y.; Fortage, J.; Deronzier, A.; Moutet, J. C. Polypyrrole-Ru(2,2′-bipyridine)32+/MoSx Structured Composite Film As a Photocathode for the Hydrogen Evolution Reaction. *ACS Appl. Mater. Interfaces* **2015,** *7*, 4476–4480. https://doi.org/10.1021/ACSAMI.5B00401/SUPPL_FILE/AM5B00401_SI_001.PDF.

63. Wang, Z.; Xiao, P.; Qiao, L.; Meng, X.; Zhang, Y.; Li, X.; Yang, F. Polypyrrole Sensitized ZnO Nanorod Arrays for Efficient Photo-Electrochemical Splitting of Water. *Phys. B Condens. Matter.* **2013,** *419*, 51–56. https://doi.org/10.1016/J.PHYSB.2013.03.021.
64. Eftekhari, A. In *Nanostructured Conductive Polymers*, 2011. https://books.google.com/books?hl=en&lr=&id=Us_uzGsblfkC&oi=fnd&pg=PT7&ots=G2rTad8xci&sig=GTyWH7jslO8GPeKnl3PAMz4a_e4 (accessed September 2, 2022).
65. Li, Y.; Feng, J.; Li, H.; Wei, X.; Wang, R.; Zhou, A. Photoelectrochemical Splitting of Natural Seawater with α-Fe2O3/WO3 Nanorod Arrays. *Int. J. Hydrogen Energy* **2016,** *41,* 4096–4105. https://doi.org/10.1016/J.IJHYDENE.2016.01.027.
66. Li, Q.; Zheng, M.; Ma, L.; Zhong, M.; Zhu, C.; Zhang, B.; Wang, F.; Song, J.; Ma, L.; Shen, W. Unique Three-Dimensional InP Nanopore Arrays for Improved Photoelectrochemical Hydrogen Production. *ACS Appl. Mater. Interfaces* **2016,** *8*, 22493–22500. https://doi.org/10.1021/ACSAMI.6B06200/SUPPL_FILE/AM6B06200_SI_001.PDF.
67. Zhang, Z.; Tan, X.; Yu, T.; Jia, L.; Huang, X. Time-Dependent Formation of Oxygen Vacancies in Black TiO2 Nanotube Arrays and the Effect on Photoelectrocatalytic and Photoelectrochemical Properties. *Int. J. Hydrogen Energy* **2016,** *41*, 11634–11643. https://doi.org/10.1016/J.IJHYDENE.2015.12.200.
68. Stoll, T.; Zafeiropoulos, G.; Tsampas, M. N. Solar Fuel Production in a Novel Polymeric Electrolyte Membrane Photoelectrochemical (PEM-PEC) Cell with a Web of Titania Nanotube Arrays as Photoanode and Gaseous Reactants. *Int. J. Hydrogen Energy* **2016,** *41*, 17807–17817. https://doi.org/10.1016/J.IJHYDENE.2016.07.230.
69. Yan, J.; Wu, S.; Zhai, X.; Gao, X.; Li, X. Si Microwire Array Photoelectrochemical Cells: Stabilized and Improved Performances with Surface Modification of Pt Nanoparticles and TiO2 Ultrathin Film. *J. Power Sources* **2017,** *342*, 460–466. https://doi.org/10.1016/J.JPOWSOUR.2016.12.086.
70. Huang, X.; Yang, L.; Hao, S.; Zheng, B.; Yan, L.; Qu, F.; Asiri, A. M.; Sun, X. N-Doped Carbon Dots: A Metal-Free Co-catalyst on Hematite Nanorod Arrays Toward Efficient Photoelectrochemical Water Oxidation. *Inorg. Chem. Front.* **2017,** *4*, 537–540. https://doi.org/10.1039/C6QI00517A.
71. Iqbal, N.; Khan, I.; Yamani, Z. H. A.; Qurashi, A. A Facile One-Step Strategy for In-situ Fabrication of WO3-BiVO4 Nanoarrays for Solar-Driven Photoelectrochemical Water Splitting Applications. *Solar Energy* **2017,** *144*, 604–611. https://doi.org/10.1016/J.SOLENER.2017.01.057.
72. Baek, M.; Kim, D.; Yong, K. Simple but Effective Way to Enhance Photoelectrochemical Solar-Water-Splitting Performance of ZnO Nanorod Arrays: Charge-Trapping zn(oh)2 Annihilation and Oxygen Vacancy Generation by Vacuum Annealing. *ACS Appl. Mater. Interfaces* **2017,** *9*, 2317–2325. https://doi.org/10.1021/ACSAMI.6B12555/SUPPL_FILE/AM6B12555_SI_001.PDF.
73. Li, Y.; Wang, R.; Li, H.; Wei, X.; Feng, J.; Liu, K.; Dang, Y.; Zhou, A. Efficient and Stable Photoelectrochemical Seawater Splitting with TiO_2@g-C3N4 Nanorod Arrays Decorated by Co-Pi. *J. Phys. Chem. C* **2015,** *119*, 20283–20292. https://doi.org/10.1021/ACS.JPCC.5B05427/SUPPL_FILE/JP5B05427_SI_001.PDF.
74. Gao, L.; Cui, Y.; Wang, J.; Cavalli, A.; Standing, A.; Vu, T. T. T.; Verheijen, M. A.; Haverkort, J. E. M.; Bakkers, E. P. A. M.; Notten, P. H. L. Photoelectrochemical Hydrogen Production on InP Nanowire Arrays with Molybdenum Sulfide Electrocatalysts. *Nano Lett.* **2014,** *14*, 3715–3719. https://doi.org/10.1021/NL404540F/SUPPL_FILE/NL404540F_SI_001.PDF.

75. Thiyagarajan, P.; Ahn, H. J.; Lee, J. S.; Yoon, J. C.; Jang, J. H. Hierarchical Metal/Semiconductor Nanostructure for Efficient Water Splitting. *Small* **2013,** *9*, 2341–2347. https://doi.org/10.1002/SMLL.201202756.
76. Xie, S.; Lu, X.; Zhai, T.; Li, W.; Yu, M.; Liang, C.; Tong, Y. Enhanced Photoactivity and Stability of Carbon and Nitrogen Co-treated ZnO Nanorod Arrays for Photoelectrochemical Water Splitting. *J. Mater. Chem.* **2012,** *22*, 14272–14275. https://doi.org/10.1039/C2JM32605A.
77. Lu, X.; Wang, G.; Xie, S.; Shi, J.; Li, W.; Tong, Y.; Li, Y. Efficient Photocatalytic Hydrogen Evolution Over Hydrogenated ZnO Nanorod Arrays. *Chem. Commun.* **2012,** *48*, 7717–7719. https://doi.org/10.1039/C2CC31773G.
78. Feng, X.; Shankar, K.; Varghese, O. K.; Paulose, M.; Latempa, T. J.; Grimes, C. A. Vertically Aligned Single Crystal TiO_2 Nanowire arrays Grown Directly on Transparent Conducting Oxide Coated Glass: Synthesis Details and Applications. *Nano Lett.* **2008,** *8*, 3781–3786. https://doi.org/10.1021/NL802096A/SUPPL_FILE/NL802096A_SI_001.PDF.
79. Wolcott, A.; Smith, W. A.; Kuykendall, T. R.; Zhao, Y.; Zhang, J. Z. Photoelectrochemical Water Splitting Using Dense and Aligned TiO_2 Nanorod Arrays. *Small* **2009,** *5*, 104–111. https://doi.org/10.1002/SMLL.200800902.
80. Li, Y.; Zhang, J. Z. Hydrogen Generation from Photoelectrochemical Water Splitting Based on Nanomaterials. *Laser Photon Rev.* **2010,** *4*, 517–528. https://doi.org/10.1002/LPOR.200910025.
81. Cho, I. S.; Logar, M.; Lee, C. H.; Cai, L.; Prinz, F. B.; Zheng, X. Rapid and Controllable Flame Reduction of TiO_2 Nanowires for Enhanced Solar Water-Splitting. *Nano Lett.* **2014,** *14*, 24–31. https://doi.org/10.1021/NL4026902/SUPPL_FILE/NL4026902_SI_001.PDF.
82. Quang, N. D.; Kim, D.; Hien, T. T.; Kim, D.; Hong, S. K.; Kim, C. Three-Dimensional Hierarchical Structures of TiO_2/CdS Branched Core-Shell Nanorods as a High-Performance Photoelectrochemical Cell Electrode for Hydrogen Production. *J. Electrochem. Soc.* **2016,** *163*, H434–H439. https://doi.org/10.1149/2.1041606JES/XML.
83. Kargar, A.; Sun, K.; Jing, Y.; Choi, C.; Jeong, H.; Jung, G. Y.; Jin, S.; Wang, D. 3D Branched Nanowire Photoelectrochemical Electrodes for Efficient Solar Water Splitting. *ACS Nano* **2013,** *7*, 9407–9415. https://doi.org/10.1021/NN404170Y/SUPPL_FILE/NN404170Y_SI_001.PDF.
84. Ji, M.; Cai, J.; Ma, Y.; Qi, L. Controlled Growth of Ferrihydrite Branched Nanosheet Arrays and Their Transformation to Hematite Nanosheet Arrays for Photoelectrochemical Water Splitting. *ACS Appl. Mater. Interfaces* **2016,** *8*, 3651–3660. https://doi.org/10.1021/ACSAMI.5B08116/SUPPL_FILE/AM5B08116_SI_001.PDF.
85. Qiu, Y.; Yan, K.; Deng, H.; Yang, S. Secondary Branching and Nitrogen Doping of ZnO Nanotetrapods: Building a Highly Active Network for Photoelectrochemical Water Splitting. *Nano Lett.* **2012,** *12*, 407–413. https://doi.org/10.1021/NL2037326/SUPPL_FILE/NL2037326_SI_001.PDF.
86. Li, Q.; Sun, X.; Lozano, K.; Mao, Y. Facile and Scalable Synthesis of "Caterpillar-Like" ZnO Nanostructures with Enhanced Photoelectrochemical Water-Splitting Effect. *J. Phys. Chem. C* **2014,** *118*, 13467–13475. https://doi.org/10.1021/JP503155C/SUPPL_FILE/JP503155C_SI_001.PDF.
87. Jang, Y. J.; Jang, J. W.; Choi, S. H.; Kim, J. Y.; Kim, J. H.; Youn, D. H.; Kim, W. Y.; Han, S.; Lee, J. S. Tree Branch-Shaped Cupric Oxide for Highly Effective Photoelectrochemical Water Reduction. *Nanoscale* **2015,** *7*, 7624–7631. https://doi.org/10.1039/C5NR00208G.

88. Chen, H.; Wei, Z.; Yan, K.; Yang, B.; Zhu, Z.; Zhang, T.; Yang, S. Epitaxial Growth of ZnO Nanodisks with Large Exposed Polar Facets on Nanowire Arrays for Promoting Photoelectrochemical Water Splitting. *Small* **2014,** *10,* 4760–4769. https://doi.org/10.1002/SMLL.201401298.
89. Zhang, L.; Baumanis, C.; Robben, L.; Kandiel, T.; Bahnemann, D. Bi2WO6 Inverse Opals: Facile Fabrication and Efficient Visible-Light-Driven Photocatalytic and Photoelectrochemical Water-Splitting Activity. *Small* **2011,** *7,* 2714–2720. https://doi.org/10.1002/SMLL.201101152.
90. Gun, Y.; Song, G. Y.; Quy, V. H. V.; Heo, J.; Lee, H.; Ahn, K. S.; Kang, S. H. Joint Effects of Photoactive TiO2 and Fluoride-Doping on SnO2 Inverse Opal Nanoarchitecture for Solar Water Splitting. *ACS Appl. Mater. Interfaces* **2015,** *7,* 20292–20303. https://doi.org/10.1021/ACSAMI.5B05914/ASSET/IMAGES/MEDIUM/AM-2015-05914E_0012.GIF.
91. Lu, Y. R.; Yin, P. F.; Mao, J.; Ning, M. J.; Zhou, Y. Z.; Dong, C. K.; Ling, T.; Du, X. W. A Stable Inverse Opal Structure of Cadmium Chalcogenide for Efficient Water Splitting. *J. Mater. Chem. A Mater.* **2015,** *3,* 18521–18527. https://doi.org/10.1039/C5TA03845F.
92. Quan, L. N.; Jang, Y. H.; Stoerzinger, K. A.; May, K. J.; Jang, Y. J.; Kochuveedu, S. T.; Shao-Horn, Y.; Kim, D. H. Soft-Template-Carbonization Route to Highly Textured Mesoporous Carbon–TiO_2 Inverse Opals for Efficient Photocatalytic and Photoelectrochemical Applications. *Phys. Chem. Chem. Phys.* **2014,** *16,* 9023–9030. https://doi.org/10.1039/C4CP00803K.
93. Moir, J.; Soheilnia, N.; O'Brien, P.; Jelle, A.; Grozea, C. M.; Faulkner, D.; Helander, M. G.; Ozin, G. A. Enhanced Hematite Water Electrolysis Using a 3D Antimony-Doped Tin Oxide Electrode. *ACS Nano* **2013,** *7,* 4261–4274. https://doi.org/10.1021/NN400744D/SUPPL_FILE/NN400744D_SI_001.PDF.
94. Wang, W.; Dong, J.; Ye, X.; Li, Y.; Ma, Y.; Qi, L. Heterostructured TiO2 Nanorod@Nanobowl Arrays for Efficient Photoelectrochemical Water Splitting. *Small* **2016,** *12,* 1469–1478. https://doi.org/10.1002/SMLL.201503553.
95. Yun, G.; Balamurugan, M.; Kim, H. S.; Ahn, K. S.; Kang, S. H. Role of WO_3 Layers Electrodeposited on SnO_2 Inverse Opal Skeletons in Photoelectrochemical Water Splitting. *J. Phys. Chem. C.* **2016,** *120,* 5906–5915. https://doi.org/10.1021/ACS.JPCC.6B00044/ASSET/IMAGES/MEDIUM/JP-2016-00044V_0004.GIF.
96. Lee, J.; Cho, C. Y.; Lee, D. C.; Moon, J. H. Bilayer Quantum Dot-Decorated Mesoscopic Inverse Opals for High Volumetric Photoelectrochemical Water Splitting Efficiency. *RSC Adv.* **2016,** *6,* 8756–8762. https://doi.org/10.1039/C5RA27049A.
97. Das, D.; Veziroğlu, T. N. Hydrogen Production by Biological Processes: A Survey of Literature. *Int. J. Hydrogen Energy* **2001,** *26,* 13–28. https://doi.org/10.1016/S0360-3199 (00)00058-6.
98. Sakurai, H.; Masukawa, H.; Kitashima, M.; Inoue, K. Photobiological Hydrogen Production: Bioenergetics and Challenges for its Practical Application. *J. Photochem. Photobiol. C Photochem. Rev.* **2013,** *17,* 1–25. https://doi.org/10.1016/J.JPHOTOCHEMREV.2013.05.001.
99. Poudyal, R. S.; Tiwari, I.; Koirala, A. R.; Masukawa, H.; Inoue, K.; Tomo, T.; Najafpour, M. M.; Allakhverdiev, S. I.; Veziroğlu, T. N. Hydrogen Production Using Photobiological Methods. In *Compendium of Hydrogen Energy*, 2015; pp 289–317. https://doi.org/10.1016/B978-1-78242-361-4.00010-8.

100. Hydrogen Production: Thermochemical Water Splitting | Department of Energy, (n.d.) [Online]. https://www.energy.gov/eere/fuelcells/hydrogen-production-thermochemical-water-splitting (accessed Sept 2, 2022).
101. Rao, C. N. R.; Dey, S. Solar Thermochemical Splitting of Water to Generate Hydrogen. *Proc. Natl. Acad. Sci. U. S. A.* **2017,** *114,* 13385–13393. https://doi.org/10.1073/PNAS.1700104114/SUPPL_FILE/PNAS.201700104SI.PDF.
102. Wu, X.; Onuki, K. Thermochemical Water Splitting for Hydrogen Production Utilizing Nuclear Heat from an HTGR. *Tsinghua Sci. Technol.* **2005,** *10,* 270–276. https://doi.org/10.1016/S1007-0214(05)70066-3.

CHAPTER 6

Nano Composite PCMs for Thermal Energy Storage Applications

SAURABH PANDEY[1], ABHISHEK ANAND[1], AMRITANSHU SHUKLA[1,2], and ATUL SHARMA[1]

[1]*Non-Conventional Energy Laboratory, Rajiv Gandhi Institute of Petroleum Technology, Jais, Amethi, India*

[2]*Department of Physics, Lucknow University, Lucknow, India*

ABSTRACT

Solar radiation falling on the earth is sufficient to meet the energy needs of mankind whether it is thermal energy or electromagnetic energy. Phase change materials (PCMs) are nicely fitted among other types of base materials to store and then release thermal energy. PCMs have a low thermal conductivity, to improve the thermal properties of PCMs nanoparticles have been affixed. Nanoparticles reduce the phase change temperature range and also modify the phase change process of the PCM nanocomposites. These nanoparticles enhance thermal conductivity along with reducing the supercooling effect. But with it, lots of challenges like instability, cost of production, agglomeration, and abrasion occur. In this chapter, the role of nanocomposite PCMs in thermal energy storage applications would be discussed.

6.1 INTRODUCTION

The vigorous growth of the population rises the demand for energy and which mostly depends on fossil-based fuels. Due to this, limited resource

Technological Advancement in Clean Energy Production: Constraints and Solutions for Energy and Electricity Development. Amritanshu Shukla, Kian Hariri Asli, Neha Kanwar Rawat, Ann Rose Abraham, & A. K. Haghi (Eds.)

is being rapidly eroded and they also increase the level of greenhouse gas which has very bad consequences for both present and future. Rapid climate change is one of the bad consequences related to this. That's why human inclination toward clean and renewable energy on earth is increasing rapidly to avoid these consequences. For which a lot of researchers and scientists are working continuously on renewable energy sources so that the gap between supply and demand of energy can be filled. Direct solar radiation is a prominent source of energy. A study estimates solar energy's global potential of 1600–49,800 EJ (Source: UNDP) is much higher than the energy consumed by the world population (Global primary energy consumption in 2019 is 581.5 EJ).[1] This exciting difference forced us to focus on solar energy which can completely fulfill the world's total energy demand and it is currently estimated that it continuously provides approximately 4 billion years.[2] Solar energy is a promising sustainable renewable energy source whether it is thermal energy or electromagnetic energy need. To utilize solar energy various energy conversion and storage systems are invented in which photovoltaics, DSSCs, rechargeable batteries, thermochemical, sensible heat, and latent heat storage systems are examples of energy conversion and energy storage systems.[3–5] Types of energy storage methods are shown in Figure 6.1, but the problem with solar energy dependency is its intermittency. During night hours and cloudy conditions required energy is not available.[6] These problems are solved by improving energy storage systems. One of the promising techniques which deal with thermal energy storage is the application of PCMs, PCMs are widely scrutinized as latent heat storage systems because of the high heat storage capacity at a constant temperature during phase transition. But lots of diminution is attached with PCMs like low thermal conductivity, supercooling, and stability. These diminutions can be accomplished by nano additives effectively, they also improve the thermal performance of base fluids and molten salts.[7] This chapter focuses on the recent disquisition of thermal energy storage material with different nanomaterials, especially with PCMs.

6.2 NANOMATERIALS

PCMs and liquids are commonly known for thermal energy storage systems, in recent years lots of attempts have been made to enhance heat transfer and many methods are used to enhance the charging-discharging rate, improving the thermal conductivity of TES materials by using bubble agitation, metal matrix insertion, extended surface, metal ring, encapsulation, nanomaterials,

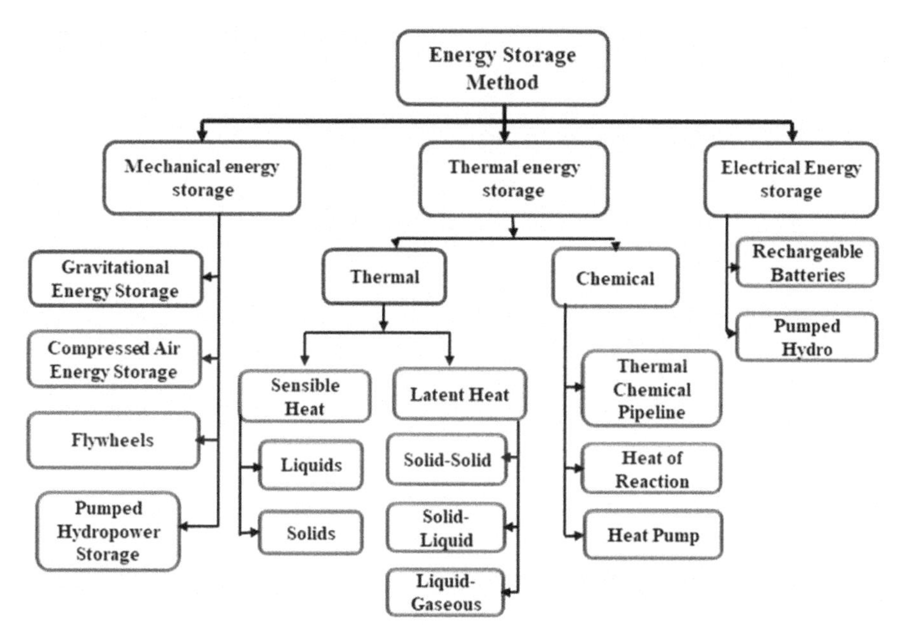

FIGURE 6.1 Types of energy storage methods.

etc. Nanomaterials are widely investigated for energy storage and conversion, whereas in TES it is used to reduce the supercooling effect and increase the thermal conductivity. The precise definition of nanomaterials is not exactly defined, but scientists agree that they are partially characterized by their size, measured in nanometers. Confinement of electrons, photons, and phonons at the nanoscale helm to the feature of new biological, chemical, and physical properties that makes them a new material in itself.[8,9] Nanomaterials have a higher fraction of grain boundary volume which makes them unique from other materials. The small dimensions of NPs affect the melting point, optical properties, magnetic properties, conductivity, and reactivity. For any lattice defects such as vacancies, and dislocations, and for high density of incoherent interfaces, a change in grain size is responsible. Behind structure and morphology, not only their physical properties but also the reactivity as well, whether it is kinetics or thermodynamics of reaction depending on the size of the nanoparticles. Nanomaterials exist naturally as well as humans engineered from materials such as minerals or carbon, etc. Engineered nanomaterials have unique optical, electrical, magnetic, and other properties. Nanomaterials are separated into three categories organics, inorganics, and hybrid as shown in Figure 6.2. Generally, metal, metal oxide, and metalloids-based nanomaterials come under the inorganic category and organics are carbon-based nanomaterials like multi-walled carbon nanotube and single-walled carbon nanotubes, carbon nanotubes, Graphite, fullerenes, etc.,[10] and hybrid are the combinations of organic–organic, inorganic–inorganic, and inorganic–organic nanomaterials.[11]

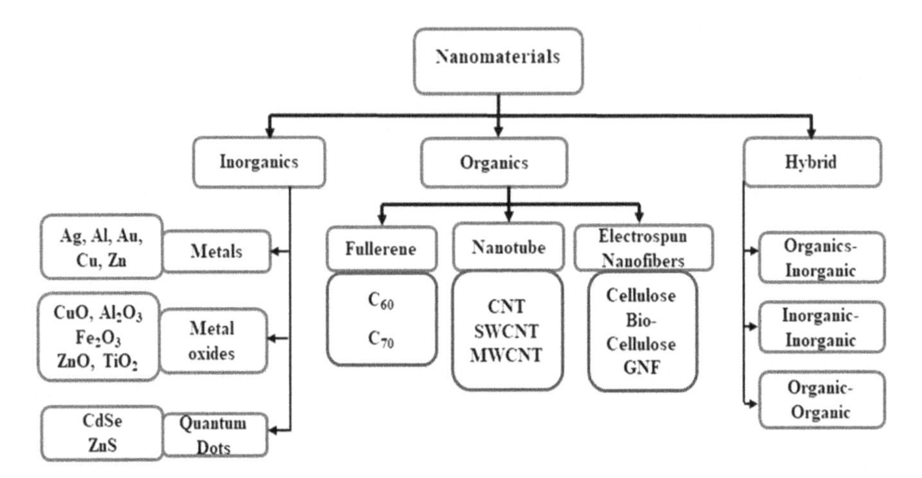

FIGURE 6.2 Types of nanomaterials.

6.2.1 SYNTHESIS OF NANOMATERIALS

There are two synthesis approaches behind these three types of nanomaterials, which we know as top–down and bottom–up as shown in Figure 6.3.

- **Top–down approaches:** In this approach, bulk material is modified to obtain the required nano-size and shape like microcrystals are broken down into nanocrystals. Sputtering and explosion processes are top–down approaches.
- **Bottom–up approaches:** Nanomaterials assemble from atoms and molecules, that is, from base materials comes under the bottom–up approach. Physical–chemical, vapor deposition, sol-gel, and epitaxial growths are some examples.[12,13]

6.3 PHASE CHANGE MATERIALS

Heat or cold in the form of thermal energy is stored as internal energy of the material through thermochemical, sensible, and latent heat storage material.

PCMs, come under latent heat storage material that plays a substantial role in the storage of thermal energy. Energy transfer occurs during the phase transition of material whether it is from liquid to solid or solid to liquid. PCMs, absorb and release heat energy during transition nearly at a constant temperature as compared to other conventional thermal energy storage materials.[14] It absorbs approximately 10–15 times more heat per unit volume than other conventional storage materials such as masonry, water, or rock. As latent heat storage materials, PCMs must possess certain desired thermodynamic, kinetic, and chemical properties and also ease of material availability and economic feasibility kept in mind. Some desirable properties are as follows in Table 6.1.[15]

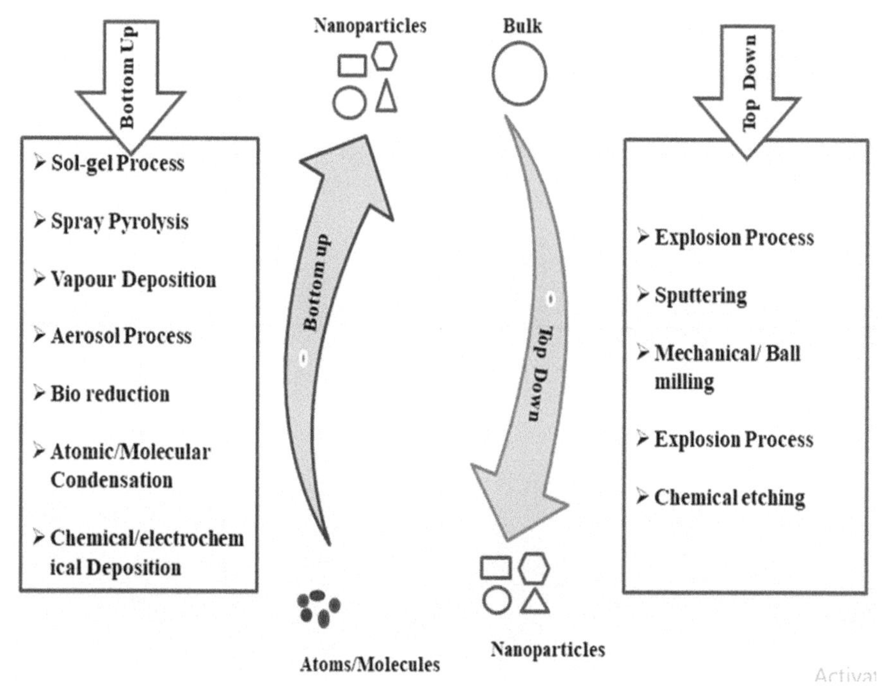

FIGURE 6.3 Different types of synthesis approaches of nanomaterial.

6.3.1 TYPES OF PCMS

Many organic and inorganic chemical materials and eutectics have been identified as PCM from the latent heat of fusion and melting temperature point of view. A classification of PCM is given in Figure 6.4. Majority PCMs

TABLE 6.1 Properties of Phase Change Materials.

Chemical properties	Physical properties	Thermal properties	Kinetic properties	Economics
➢ Chemical stability for a long term ➢ No fire hazards ➢ Compatibility with the formation of materials ➢ No toxicity	➢ Minor volume change ➢ Low vapor Pressure ➢ Compatible phase equilibrium ➢ High density	➢ High latent heat of transition ➢ Convenient phase transition temperature ➢ Good heat transfer	➢ Sufficient crystallization rate ➢ No supercooling	➢ Cost-effective ➢ Large Scale availability

do not have the required yardstick for an appropriate storage media. For an ideal thermal storage media, it is too difficult to find material, so by adequate system design, there is a need for rectification in lousy physical properties of available materials. For example, the supercooling effect is clamped by introducing a cold finger or nucleating agent, and by using suitable thickness incongruent melting can be prevented also many times metallic fins or nanomaterials are used to enhance the thermal conductivity.

6.3.1.1 ORGANIC PHASE CHANGE MATERIAL

Organic PCM is usually non-corrosive and has the property of congruent melting without any phase segregation and latent heat fusion collapse. This is classified as paraffin and non-paraffin as we have seen in Figure 6.4. Kinds of paraffin are a straight chain of n-alkanes, which are used as PCM because of their wide temperature range. Their melting point and heat of fusion increase with increasing the chain of alkanes.[16] Moreover, they are stable below 500°C, chemically inert, and have very little change in their volume after melting with very low vapor pressure, which results in a very long melt-freeze cycle.[15] Paraffin has more desirable properties as it is more reliable, less expensive, non-corrosive, safe, congruent melting, and self-nucleating properties but has some inexpedient properties such as low thermal conductivity and moderately flammable but these inexpedient properties are curtailed by slightly modifying paraffin and storage system.[17] Of all phase transition materials, non-paraffin organics are the most prevalent and have the widest range of features. Unlike paraffin, which has extremely comparable properties, each of these materials will have unique characteristics. The majority of phase change storage candidate materials fall under this category. After conducting a thorough analysis of organic materials, researchers found several esters, fatty acids, alcohols, and glycols that are ideal for storing energy. Fatty acids and other non-paraffin organic compounds are other subgroups of these organic molecules. Due to their flammability, these materials shouldn't be subjected to extreme heat, flames, or oxidizing substances.[18]

The heat of fusion values of fatty acids is high and equivalent to paraffins. Additionally repeatable in their melting and freezing behavior, fatty acids also freeze without supercooling. $CH_3(CH_2)_{2n}$.COOH is the generic formula that describes all fatty acids. However, their main disadvantage is they are 2–2.5 times more expensive than technical grade paraffin. Some characteristics of these organic materials are high heat of fusion, flammability, low thermal conductivity, low flash points, varying levels of toxicity, instability at high temperatures, and slightly low corrosive effect slight corrosive effect.

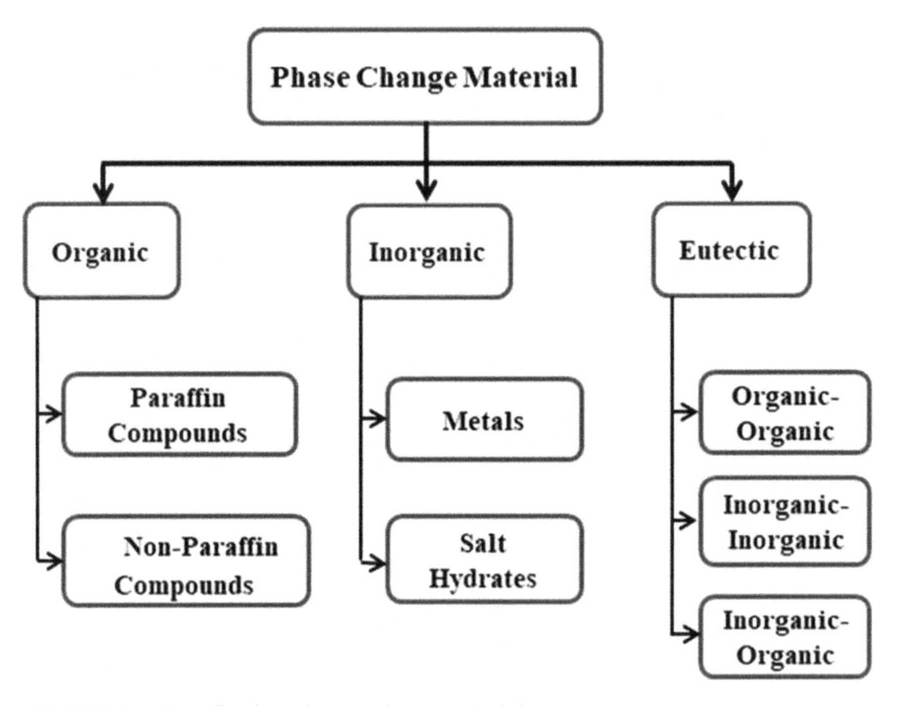

FIGURE 6.4 Classification of phase change materials.

6.3.1.2 INORGANIC PHASE CHANGE MATERIALS

Metallics and salt hydrates are additional categories for inorganic materials. These PCMs do not supercool significantly, and cycling does not affect their temperatures of fusion.

- **Salt Hydrate:** Inorganic salts and water can be combined to form salt hydrates, a characteristic crystalline solid with the usual formula $AB.nH_2O$. Salt hydrates can change from a solid to a liquid state by dehydrating or hydrating the salt, even though this process thermodynamically mimics melting or freezing. Typically, a salt hydrate melts into another salt hydrate with fewer moles of water. The hydrate crystals disintegrate into anhydrous salt and water, or a lower hydrate and water, at the melting point. The fact that the water released during crystallization is insufficient to completely dissolve all of the solid phases present causes incongruent melting, which is one issue with the majority of salt hydrates. The lower hydrate (or anhydrous salt), which has a lower density, sinks to the bottom of the container.[19] The characteristics of salt hydrates that are most appealing are their high latent heat of fusion per unit volume, their relatively high thermal conductivity (nearly twice as high as paraffin's), and minimal volume changes during melting. They are barely harmful, compatible with polymers, and not extremely corrosive. Many salt hydrates are sufficiently affordable for storage applications. The majority of salt hydrates also have weak nucleating capabilities, which causes the liquid to supercool before crystallization starts. The addition of a nucleating agent, which supplies the nuclei on which crystal formation is started, is one approach to solving this issue. Another option is to keep some crystals in a small, cold area so they can act as nuclei. Congruent, incongruent, and semi-congruent melting are three different forms of melting behavior that can be distinguished in molten salts:

i. When the anhydrous salt is fully soluble in its hydrating water at the melting temperature, congruent melting takes place.
ii. Incongruent melting happens when the salt's water of hydration is not completely soluble in it at the melting point.
iii. Due to the hydrate's conversion to a lower-hydrated substance through water loss, the melting composition of the liquid and solid phases that are in equilibrium during a phase transition is different.

The major problem associated with salt hydrates is supercooling, it melts incongruently and the spontaneous formation of salt hydrates with fewer water moles during the discharge process is another issue with salt hydrates. Chemicals can stop lower salt hydrates from nucleation, which enhances their solubility more than the original salt hydrates with greater water molecular weight.[20]

- **Metallics:** The metal eutectics and low melting metals fall within this category. Due to weight penalties, these metallics have not yet been given substantial consideration for PCM technology. They are plausible candidates, nevertheless, when the volume is taken into account due to the high fusion heat output per unit volume. Due to their high thermal conductivities, fillers that increase weight are not necessary. The employment of metallics brings forth a variety of peculiar technical issues. The strong heat conductivity of the metallics distinguishes them significantly from other PCMs.

6.3.1.3 EUTECTICS

A eutectic is a minimum-melting mixture of two or more components, each of which melts and freezes concurrently during crystallization to produce a mixture of the component crystals.[21] Because they freeze to a close-knit combination of crystals, eutectic materials rarely melt or freeze without the components segregating. Both components simultaneously liquefy when heated, making separation unlikely. Since they are minimum melting, some segregated PCM compositions have been wrongly referred to as eutectics. But it would be more accurate to refer to them as peritectic as they go through a peritectic reaction during phase change.[22]

6.4 CLASSIFICATION OF NANOCOMPOSITES

The researchers are emphasizing making the use of different nanomaterials in PCMs-based TES as effective as possible, to bring about a paradigm shift in various parameters. These include thermal conductivity, thermal diffusivity/reliability/stability/durability, efficiency, melting/solidification, and degree of supercooling time as shown in Tables 6.2–6.4 where augmentation/mitigation of thermal conductivity, melting and solidification, respectively, are listed with different composition of nanomaterials. Nanocomposites are a kind of mixture obtained as the result of the incorporation of TES base materials and nanomaterials. It can be organic or inorganic both. Organics nanocomposites have non-corrosive in nature, chemically and thermally stable whereas inorganics nanocomposites have greater phase change enthalpy. Paraffin blended, HDPE blended, and fatty acid blended nanocomposites are some examples of organic nanocomposites and some salt hydrates blended nanocomposites are an example of inorganic nanocomposites.[23,24]

6.4.1 SYNTHESIS TECHNIQUE

Finding the desired size and dimension of nanocomposite is one of the major difficulties in the field of nanotechnology. Innumerable efforts were made by researchers in the early stages, as a result of which many synthesis methods are developed and considered as standard synthesis processes of nanocomposites. In this sequence, microencapsulation, blending, and impregnation are developed nanocomposite synthesis techniques.

TABLE 6.2 Summary of Different Nanomaterials in Improving PCMs' Thermal Conductivity.

PCM	Nanomaterials	Nanomaterial concentration	Augmentation in thermal conductivity	References
Cyclohexane (C_6H_{12})	CuO	1, 2, and 4 wt%	Around 5%	[25]
Paraffin wax	Al_2O_3	5 and 10 vol%	Around 30%	[26]
Oleic acid	CuO	0.5, 1.0, 1.5, and 2.0 mass%	Up to 98.66%	[27]
Paraffin wax	Hexagonal Boron nitrode sheets	1–10 mass%	Up to 60%	[28]
Saturated $BaCl_2$ aqueous solution	TiO_2	0.167, 0.283, 0.565, and 1.13 vol%	Up to 16.67%	[29]
Palmitic acid	Functionalized MWCNT	1, 2, 3, 5, and 7 wt%	About 60%	[30]
Paraffin wax	CuO	2, 5, and 10 wt%	3.77%, 6.92%, and 13.21%, respectively	[31]
Paraffin RT-20	MWCNT	0.15, 0.3, 0.45, and 0.6 wt%	40–50% for 0.6 mass %	[28]
Water	Cu	0–0.2 vol%	74.5%	[32]
Paraffin wax	Cu	1 and 2 wt%	12.24% and 31.29%, respectively	[33]
n-Octadecane	Al_2O_3	5 and 10 wt%	6% and 17% at 30°C and 60°C, respectively	[34]
Paraffin wax	MWCNTs (short and long)	1, 3, and 5 wt%	For short MWCNTs 8.4%, and for long MWCNTs 33.7%	[35]
Tetradecanol (TD)	Ag	1:64–16:1	Increase in thermal conductivity	[7]
Paraffin wax	xGnP	1, 2, 3, 5, and 7 wt%	2 times	[36]
Paraffin wax	TiO_2	1, 3, and 5 wt%	About 10%	[37]
Tetradecanol (TD)	Cu nanowires	0–11.9 vol%	The thermal conductivity increases as particle loading increases and fluctuates significantly at 1.5 vol% and reaches nine times greater thermal conductivity at 11.9 vol%	[38]
Paraffin wax	CNF	4 wt%	Around 40%	[39]

TABLE 6.2 *(Continued)*

PCM	Nanomaterials	Nanomaterial concentration	Augmentation in thermal conductivity	References
Palmitic acid	MWCNT	0.2, 0.5, and 1 wt%	In solid state around 45% and in liquid state around 35%	[24]
Paraffin wax	Al_2O_3, TiO_2, SiO_2, and ZnO	1, 2, and 3 wt%	TiO_2 is superior to others for the enhancement of thermal conductivity	[40]
n-Octadecane	Mesoporous silica ($MPSiO_2$)	1, 3, and 5 wt%	Around 6%	[41]
Paraffin wax	Al_2O_3 and CuO	0–10 vol%	At 27°C About 30% rise and with the rise in temperature thermal conductivity rises	[26]
Paraffin wax	Cu	0.025 and 0.05 vol%	7.64% and 1.75%, respectively	[33]
Organic ester	Ag	0.1–5 wt%	Approximately 10–67%	[42]
Paraffin wax	xGnP and Graphene	1, 2, 5, and 10 wt%	The thermal conductivity of xGnP is superior to that of graphene, and it is increased tenfold by adding 10 mass%.	[43]
Paraffin wax	Nano-graphite	1, 4, 7, and 10 wt%	7.41 times	[44]

TABLE 6.3 Role of Nanomaterial in the Enhancement of Melting Rate..

PCM	Nanomaterial	Nanomaterial concentration	Augmentations	Phase change process	References
Paraffin wax	Al_2O_3 and TiO_2	0.075 wt%	5.92% for 0.05 mass% TiO_2	Melting	[35]
Paraffin wax	Al_2O_3 and CuO	1 wt%	20 min decrease in melting time for Al_2O_3 and by 15.6 min for CuO	Melting	[45]
Palmitic acid–	TiO2	0.3 wt%	21.05% decrease in melting time	Melting	[27]
Paraffin wax	Al_2O_3	5 wt%	29% decrease in melting time	Melting	[46]
Paraffin wax	Cu	5 vol%	15% decrease in melting time	Melting	[47]
Paraffin wax	Cu	1 wt%	33% decrease in melting time	Melting	[48]
Paraffin wax	Cu	5 vol%	Melting time reduced by 14.6%	Melting	[49]
Paraffin wax	Cu	5 vol%	27.7% decrease in melting time	Melting	[50]
Paraffin wax	Al_2O_3	5 vol%	Time for 50% melting is reduced by 9%	Melting	[51]
Paraffin wax	Al_2O_3, TiO_2, and CuO	3 vol%	The melting rate is slower for 3 vol% in comparison with pure PCM	Melting	[52]
Paraffin wax	MWCNT	0.9 wt%	Melting time reduced by 29%	Melting	[53]
Paraffin wax	MWCNT	0.9 wt%	30% decrease in Melting time	Melting	[54]

TABLE 6.4 Role of Nanomaterial in Changing Solidification Rate.

PCM	Nanomaterial	Nanomaterial concentration	Change in solidification	Phase change process	Reference
Water (base fluid)	Cu	0.05 vol%	16% decrease in solidification time	Solidification	[55]
Water (base fluid)	Cu	10 vol%	Solidification time reduced by 16%	Solidification	[56]
Water (base fluid)	CuO Al_2O_3, TiO_2	20 vol%	16% decrease in solidification time	Solidification	[57]
Water (base fluid)	Cu	20 vol%	57% decrease in solidification time	Solidification	[58]
Water	CuO	10 wt%	Solidification rate reduced with particle size	Solidification	[59]
Cyclohexane	CuO	4 wt%	5.2% decrease in solidification time for 2 mass% of CuO experimentally and by 8.09% in numerical prediction	Solidification	[32]
Eicosane	xGNPs	4.5 wt%	51% decrease in solidification time for small spheres and 53%for large sphere	Solidification	[35]
Deionized (DI) water	GNP	0.6 wt%	Solidification time decreased by 24%	Solidification	[60]
Paraffin wax	Cu	5 vol%	Solidification time decreased by18%	Solidification	[61]
Deionized (DI) water	CuO	0.1 wt%	Solidification time decreased by 35%	Solidification	[62]
Oleic acid	CuO	0.6 wt%	27.6% decrease in solidification time	Solidification	[63]

6.4.1.1 MICROENCAPSULATION

PCM and nanoparticles can be combined to create the core of microcapsules. PCMs based on long-chain fatty acids have supercooling concerns. The thermal conductivity of these organic PCMs is poor. To lower the super-cooling degree, a research group created a eutectic mixture of these PCMs. Ag-doped ZnO nanoparticles were added to PCMs to improve their poor heat conductivity.[64] To create P-type conductive ZnO and produce a synergistic surface plasmon effect, Ag substitution was chosen. The eutectic nanocomposite that had undergone ultrasonication had been further microencapsulated with a melamine-formaldehyde (MF) shell using in situ polymerization. The microcapsules with a size of 3–5 μm could store 75–80 J/g of heat. This technique increased the eutectic's thermal conductivity from 0.2513 W/mK to 0.3735 W/mK. The eutectic PCM's thermal stability and thermal cycling capabilities were improved by the shell and nanoparticles. Erythritol and sugar alcohol have a higher heat storage enthalpy than commonly used PCMs composed of paraffin wax and fatty acids. But a high supercooling temperature of roughly 100°C makes it unusable for practical uses. Erythritol's supercooling properties were increased by 83.6% through the creation of erythritol nanocomposites. There were two primary steps in the preparation procedure. The first step was creating an emulsion with erythritol as the main ingredient and alumina nanoparticles as the nucleating agent. With the help of a polysiloxane precursor created by hydration, the emulsion's interface was adsorbed. The second stage involved using ultraviolet (UV) light to polymerize the polysiloxane precursor to construct a shell around the PCM core. The thermal conductivity of the produced microencapsulated PCM (MPCM) was 0.84 W/mK, which was 29.2% higher than erythritol. In MPCM, PCM's crystal structure stayed the same and provided a respectable 203 J/g of latent heat storage. But enhanced crystallization kinetics resulted from greater heat conductivity. As a result, MPCM's heat release ratio was 52.2% higher than erythritol. The organic shell materials that are employed in commerce have high-pressure sensitivity and low thermal conductivity. They can be used to encase PCM in a shell of material and are suited for use in potential implementation because of their great strength. The suitability of the $CaCO_3$ shell for encasing PCM via the self-assembly method was examined by Zhang et al.[65] 800 rpm, 45°C, and 15 mmol/L were found to be the ideal stirring speed, temperature, and emulsifier concentration, respectively. The ratio of the core to the shell ranged from half, one, two, three to four, correspondingly. The optimal core-to-shell ratio was 3:1. Beyond this number, core content increases resulted in agglomerated particles. Rough shell surfaces were formed by lowering the ratio below three. Large $CaCO_3$ crystals that are unable to enclose the core were formed under the influence of extremely high shell concentration. This might cause PCM to leak. The presence of a $CaCO_3$ shell for an optimal formulation was confirmed by FTIR, SEM, and TEM investigation. MPCMs with the ability to store thermal energy were revealed to have crystalline structures using XRD and DSC examination. The MPCMs have a 134.83 J/g latent heat storage capacity. A leakage test revealed that PCM was stabilized by $CaCO_3$ shell formation. The inorganic shell enhanced PCM's thermal stability. Thermal conductivity rose from 0.152 W/mK to 0.542 W/mK as a result. MPCM dispersion was stirred for 2 h at 2000 rpm to assess the mechanical durability of the MPCMs. MPCMs that had been agitated had the same particle size. The SEM investigation found the shell structure to be rough. Only a 4% reduction in the heat storage enthalpy occurred. These findings suggest that the $CaCO_3$ shell is resilient. MPCMs made of paraffin and graphene were assembled electrostatically.[66] By using a high-speed homogenizer, the paraffin PCM was emulsified with stearic acid and disseminated in water. There was a negative charge in this dispersion. Through the use of ultrasonication, cetyltrimethylammonium bromide disperses with graphene nanoparticles. This positively charged dispersion was dropped into the negatively charged paraffin dispersion while being stirred. The graphene shell that surrounds the paraffin core is created by the electrostatic rearrangement of the mixture. In SEM pictures, the smooth surface of paraffin and the wrinkled surface of MPCMs indicate the creation of a core/shell structure. The characteristic paraffin peaks visible in the MPCM structure confirms microencapsulation.

For microencapsulating oleic acid PCM, Hussain and Kalaiselvam created a hybrid Ag_2O-urea formaldehyde shell.[67] Cationic, anionic, and nonionic surfactants were used to carry out the in-situ polymerization. Over other surfactants, the cationic surfactant had a higher PCM encapsulation ratio. The creation of the Ag_2O-urea formaldehyde shell over the surfactant stabilized core PCM particles further improves the mechanical, thermal, and surface morphological properties. FTIR, XRD, and UV-visible absorption measurements can all be used to confirm the presence of Ag_2O particles.

6.4.1.2 BLENDING

Aluminum, silver, gold, copper, CuO, AlN, SiO_2, Al_2O_3, SiC, TiO_2, Fe_2O_3, Fe_3O_4, ZnO, BN, carbon nanofiber (CNF), carbon nanotubes (CNT), and graphene nanoplatelets, graphene oxide (GO), MWCNT, mesoporous silica

($MPSiO_2$), Cu nanowires, expanded graphite (EG) nanoparticles, etc.,[35] are some nanomaterials used in most of the blending decipher. The purest metal nanoparticles are crystalline and have good thermal conductivity. They are less expensive than other manufactured nanomaterials like CNTs and CNFs. By contemplating this, a group of researchers synthesize nanocomposite in PCM matrix by dispersing 0.1–5% Al and Cu nanoparticles. This study showed that after a certain level, excessive nanomaterial loading decreases PCM's ability to store latent heat and increases its dynamic viscosity. By taking into account all of these parameters, the optimal concentration of Al and Cu nanoparticles was determined to be 2% and 1%, respectively. The advantages and disadvantages of nanomaterial dispersion are balanced by nanocomposites made at an optimal concentration level.

The economic and environmental effects of incorporating Fe_3O_4, CNT, and Al_2O_3 on PCMs were examined. Fe_3O_4, CNT, and Al_2O_3 nanoparticles have thermal conductivities of 6 W/mK, 3400 W/mK, and 35 W/mK, respectively.[68] Due to the better thermal conductivity of CNT-based nanocomposite, the heat transmission is increased by a greater amount. The PCM–CNT nanocomposite provided the latent heat storage tank (LHST) with the lowest operating costs and energy consumption. The rationale for lower energy use with nanocomposite is the increase in heat flow.

Older bending research solely used commercially available nanoparticles to disperse them. Recent studies have used chemical reactions to change the structural skeleton of nanomaterials. The composite's thermal properties are enhanced by the nanomaterial's changed shape. Titania and copper acetate were combined to create a copper–titania hybrid nanomaterial with Fin-like geometry. The phase transition process of the produced nanomaterial improved when it was redispersed in methyl cinnamate PCM. Composites with hybrid nanomaterial incorporation lowered the amount of supercooling and the time it took for heat to transfer. The thermal conductivity of the nanocomposite increased to 0.347 W/mK.

TiO_2 nanoparticle dispersion was used to create silver nanoparticles from the $AgNO_3$ solution.[69] The adsorption of silver nanoparticles on the titanium layer is confirmed by the XRD and SEM analysis. Sodium dodecyl sulfate (SDS) surfactant was used to disperse the produced hybrid nanoparticles in paraffin wax. Surfactant enhances PCM matrix and hybrid nanoparticle's physical interaction. Physical contact between components is confirmed by FTIR analysis and TGA revealed that the presence of SDS increased the thermal stability of the nanocomposite.[68]

Leakage is reduced in blended nanocomposite PCMs, complex structural pathways included in the blending of nanomaterials tether PCM leakage.

The features of a composite made up of polyethylene glycol (PEG), epoxy, ethylene glycol (EG), and gopher oil (GO) were studied.[70] EG and GO were immersed in molten PCM for the development of the composite by stirring in the solvent. The solvent was thereafter heated and evaporated. The curing agent and epoxy resin were poured at 70°C. To prepare the composite, the prepared mixture was molded and successively dried for 24 h at 30°C and 4 h at 120°C. PCM leakage over its melting point was prevented by the capillary force of cured epoxy, the hydrogen bonding between PEG and epoxy, and the high surface area to volume ratio of EG. Due to the polymer chain trap between GO platelets, capillary force, and hydrogen bonding contact, GO inhibits PEG leakage. The amount of epoxy in the sample decreases leakage by 43%. Furthermore, 15% lowers the normalized enthalpy that was calculated using the percentage of PCM in the composite. As epoxy concentration rises, there is less nucleation and more encapsulation. On the other hand, samples with less epoxy showed a 2% decrease in normalized enthalpy. When compared to pure PEG, this sample's leakage test showed only 8% leakage.

6.4.1.3 IMPREGNATION

By using the impregnation approach, PCM can accumulate in the nanopores created in the nanomaterial network. The physical connections that are created between PCM and nanomaterials prevent PCM leakage. To prevent leakage, additional adhesive coating of the impregnated PCM nanocomposites is an option. The maximal paraffin content for impregnating into porous EG was 94%, according to the leakage-proof test of vacuum-impregnated paraffin wax into dried EG-produced nanocomposite.[71] In addition to structural support, this nanocomposite needs defense against paraffin volatilization. The improved nanocomposite tablet composition was coated in the epoxy adhesive to accomplish both goals. The optimal concentration assured that epoxy adhesive did not enter the EG pores. The nanocomposite's thermal resistance can be increased by such epoxy passing. The effectiveness of heat storage would suffer as a result. The adhesive was applied to a PCM nanocomposite at an optimal concentration. The synthesized nanocomposite has a latent heat capacity of 143.8 J/g. It now has a thermal conductivity of 2.141 W/mK.

Paraffin wax was infused into the porous structure of developed porous oil palm kernel shell-activated carbon from waste bioproduct. By applying two coats of epoxy-nano alumina coat, the structural stability and thermal conductivity of the impregnated composite were enhanced. 60% alumina nanoparticles added to epoxy resin increased thermal conductivity by 271%.

With the use of alumina nanoparticles, the thermal resistance related to heat transmission through epoxy resin was reduced.

Synthetic microporous foams have good structural stability and can be made in any dimension. Other nanoporous materials, which demand extra coating for structural stability, make it impossible. However, the microporous foam can hold less PCM without leaking than nanoporous materials. Nanoparticles can be added to microporous foams to improve their leakage performance. Leakage of molten PCM is prevented by the high surface area and reactivity of nanoparticles. Shape stability is improved by the synergistic interaction of nanoparticles with microporous materials. Additionally, it can give composites additional material-specific features. With the use of polydopamine, Wang and his company created flexible carbon foam (CF) percolated with Ti_2O_3 nanoparticles (PDA).[72] The conglutination of the Ti_2O_3/PDA network in CF is made possible by PDA's strong adhesive properties. CF foam has been precoated on the Ti_2O_3/PDA network, creating a microporous substrate ideal for PCM impregnation. The capillary forces connected to Ti_2O_3 nanoparticles and the lipophilic nature of PDA increased CF foam's leaking properties. The highest amount of PCM that can be included in substrates that have been produced is 84%. The prepared composite was capable of photothermal conversion and was bending and compression-resistant. Because of the synergistic influence of composite components, certain characteristics are produced. In the hybrid structure of impregnating PCM prepared by Yang,[73] less amount of PCM may be included in GF without leaking due to its high pore size. The GF's covalent bonds enhance phonon transport. However, the pore size of graphene aerogel (GA) is tiny. Without leakage, it can contain a lot of PCM. However, the physical crosslinks in GA increase heat resistance and lower phonon transit rate. By combining GF and GA structures, it is possible to improve the thermal characteristics and decrease PCM leakage. Self-assembly and chemical vapor deposition techniques were used to create this structure. When paraffin wax was infused into the created hybrid microstructure, heat conductivity was increased by 574%.

6.5 APPLICATION OF NANOCOMPOSITE PCM

6.5.1 SOLAR WATER HEATING SYSTEM

In homes and workplaces where low temperatures are frequently needed, water heating is a significant energy consumer. The main non-renewable energy sources utilized to produce hot water are fossil fuels. Unfortunately, despite attempts to develop and promote the use of renewable, solar thermal water heaters for more than a century, this issue persists. By far, solar thermal collector systems are the most inefficient way to capture solar radiation for heating water. There are many different kinds of solar collectors, including integrated collectors, focused collectors, evacuated tube collectors, and flat plate collectors.[74] The optimal thermal collector should effectively capture solar energy, transform it into thermal energy, and reduce system heat losses. The main drawback of thermal collectors is that they need a storage medium because solar energy is sporadic. While some integrated collectors employ water as the storage medium, PCMs have been used to store and release latent heat for integrated solar collectors. To improve the heat transfer rate and thermal conductivity of PCMs, nanomaterials have been added. Advanced heat transfer fluid and nanocomposites cascaded with well-transitioning temperatures in diverse designs have increased the thermal storage capacity of solar thermal systems. An experiment shows that dispersing 1% of the micro-Cu powder in paraffin wax as a foundation material increases thermal conductivity by 12.2%. The dispersed nano Cu concentration is increased to 2% by weight, which results in a 24% increase in thermal conductivity for the nanocomposite. Examples of nanocomposites for the enhancement of solar collectors are shown in Table 6.5.

6.5.2 FOOD INDUSTRY

PCMs have undergone testing as heat storage and transit systems for heat processing sections, chilling storage units, as well as packaging applications in the food industry.[79] It has been discovered that using paraffin wax to dry sweet potatoes could result in energy savings.[80] Micro/nanostructures containing dodecane and zein may be employed in smart packaging systems for food applications, according to Pérez-Masiá and López-Rubio.[81] In a different study, they used electrospinning technology to increase the heat storage capacity using polylactide (PLA),[82] polycaprolactone (PCL), and dodecane. The distribution of food temperature can be significantly[80] enhanced by employing heat pipe prototypes containing PCMs, according to Lu and Zhang.[83] Additionally, it has been demonstrated that Rubitherm RT6-loaded nanoscale calcium silicate (NCS) maintained a container's temperature of 10°C for 5 h while the temperature outside increased. Additionally Ref. [84], Chalco-Sandoval and Fabra found that encapsulated RT5 could change the quantity of heat retention to about 34 weight percent of

TABLE 6.5 Nanocomposites for the Enhancement of Solar Collectors.

Nanomaterial composites	Applications	Remarks	References
Single-wall carbon nanohorns (SWCNHs) in an aqueous suspension	Solar collector heat exchanger	The investigated concentration resulted in a 10% increase in thermal conductivity. The spectral transmission demonstrated that single-wall carbon nanohorns (SWCNHs) enhanced the fluid's platonic characteristics, resulting in a notable rise in the amount of light extinction even at relatively low concentrations.	[75]
Carbon nanowalls (CNWs) in an aqueous solution	Solar thermal collector	The largest CNS mass content (0.04 g) nanofluid had the largest temperature enhancement of 8.1°C, which demonstrates the efficient absorption capabilities of the CNS nanofluids toward solar irradiation The results indicate that functionalized CNS nanofluids have the potential to effectively improve the solar absorption capabilities of direct-absorption solar collectors.	[76]
Multi-walled carbon nanotube (MWCNT) nanofluid	Flat plate solar collector	0.2% of the MWCNTs were investigated with water as the base fluid. Different pH ranges were selected. The nanofluid with a pH of 6.5 achieved the highest heat absorption coefficient, $F_R U_L$ (38.84), while the nanofluid with a pH of 9.5 (FRUL = 30.2). The enhancement increased the collector efficiency by 25%.	[77]
CuO nanofluid	Evacuated solar collector	The CuO nanofluid concentration range of 0.8–1.5% was tested. 1.2% was found to provide the highest thermal conductivity. The performance of the evaporator improved by 30% as compared to ionized water.	[78]

all PCMs.[85] Additionally, PCMs have been employed to optimize thermal processes (such as chilling bins and applications for super-cooling food) and reduce thermal energy.[86]

6.5.3 TEXTILES

For heat management, PCMs are also used in the design of various textiles, apparel, and footwear. For instance, the PCMs can absorb heat, which increases the operator's comfort level in warm situations. PCMs can be integrated directly into fabric fibers, allaying concerns regarding coating durability.[87] Through the use of single and coaxial electrospinning techniques, Dang and Nguyen have also created nanofibers based on PEG (PEG600 and PEG1000) as PCMs and PVDF as supporting polymers.[88] They employed SiO_2 throughout a single electrospinning procedure to prevent altering the nanofiber shape and PEG leakage. According to their findings, the non-woven mats could be used in the production of smart fabrics and TES. Through the use of the in situ polymerization process, PEG nanocapsules have been created. Following the pad-dry-cure technique, the 141 nm-sized nanocapsules that were produced were subsequently coated on the fabric, demonstrating the treated fabric's high thermal stability.[89] To enhance the thermoregulating properties, Yoo and Lim coated cotton materials with nanosilver-coated nonadecane.[90] According to their findings, the PCMs might be applied to the outside layer of clothes to enhance their capacity to regulate body temperature. Coaxial electrospinning has also created ultrafine PCFs based on PEG and CA. The findings showed that the compound fiber's ultimate strain and strength were inferior to those of CA fibers.[91]

6.5.4 BUILDINGS

Space cooling is a major scientific difficulty that affects many different production sectors, such as transportation, manufacturing, microelectronics, and sports arenas. Heating and cooling a building or area has been a common activity for many years. Buildings can be built among trees to help keep cool in hot weather, or walls can be constructed out of bricks to do the same. To effectively utilize TES, PCM is now used in a variety of building technologies to assist with cooling and heating according to the desired level of comfort. Because PCMs have a great ability to store latent heat, they have been taken into consideration for thermal storage in building applications.

Now that PCM is impregnated in trombe, the building's walls, wallboards, shutters, under-floor heating systems, and ceiling boards can all be used for heating and cooling purposes. Since most PCMs have relatively low thermal conductivities, integrating nanomaterials into PCMs to improve their thermophysical properties has only recently been necessary to maximize utilization. Numerous academic studies have demonstrated the potential for the creation and implementation of PCM nanocomposites to improve TES in construction applications. The usage of PCM-augmented nanocomposites in buildings or spaces mostly focuses on exploiting solar radiation, which is a natural source of heat, for heating in cold weather and for cooling in hot weather using artificial heat or cold sources. In any case, thermal energy storage is crucial to balancing supply and demand in terms of time as well as power. The reported studies on the use of nanoparticles in TES for building heating and cooling are summarized in Table 6.6.

TABLE 6.6 Nanomaterial Composite Application in Building Heating and Cooling.

Nanomaterial composites	Applications	Remarks	References
Electro-chromic material with nanostructure photocatalyst	Smart material for a benign indoor environment	Solar photocatalysts are used in air purification systems, and the nanostructured control characteristics of smart windows reduce heat absorption, which reduces the demand for ventilation	[92]
Graphite matrix composite	Cylindrical shell thermal storage unit for space heating	A 15% graphite mixing fraction improved the paraffin-based (HTF thermal)'s conductivity. Without causing any physical harm to the (LHTES) units' medium, the charging behavior was enhanced and the extendibility of the discharge was seen	[47]
Paraffin graphite composite	Piping network for district heating	To enhance the PCM, thermal conductance was raised by 10%; system efficiency rose by 28% as a result.	[93]

6.6 ELECTRONICS

The increase in computing power presents a difficulty for electronic device temperature management. For desktops and laptops, a heat sink with a fan can be used to dissipate the heat loads. However, due to the form factors of portable electronic devices (such as smartphones), PCMs can be employed for temperature management.[94] In this instance, the heat that is emitted melts the PCMs, which subsequently release heat into the environment. If the electronic gadgets are left on heat after the PCMs have completely melted, the temperature will rise due to sensible heating. The use of paraffin is common in systems where the temperature must be maintained between 40 and 45°C.[94] Combining a volume proportion of Al_2O_3 nanoparticles with tricosane, Krishna, and Kishore created nano-enhanced PCMs as an energy storage medium for electronic cooling applications.[95] Their findings demonstrated that the heat pipes with nano-enhanced PCMs had a considerable decrease in evaporator temperature of about 25.75%, which may save about 53% of the fan power. The paraffin and expanded graphite composites were used by Ling and Chen for a battery heat management system.[96] The outcomes showed that paraffin-filled heat pipe modules could reduce fan power consumption. Thermal conductivity of silver-coated PCM nanocapsules with n-octacosane as the core and silica as the shell rose from 0.246 W/m K to 1.346 W/m K, according to Zhu and Chi.[97] They suggested using the PCM nanocapsules they were cool the electronic equipment.

6.7 CONCLUSION

The storage of thermal energy is developing quickly. By altering the thermal and optical properties of heat transfer fluids, nanomaterials in the form of nanofluids, nanocomposites, and nanofluid PCMs can enhance the heat transfer phenomena. The combination of nanoparticles with base fluids can significantly increase the performance and efficiency of solar collectors, as shown by all theoretical and practical experiments. The parameters for optimal output depend on the characteristics of the nanofluid, which in turn depend on the operating temperature. A few scattered investigations present conflicting findings. It is still unknown how to prepare nanofluids according to standards. Numerous sources have reported that particle size affects heat conductivity. However, it is not clear how it affects the heat transfer coefficient in laminar and turbulent zones. Using nanofluids for solar thermal energy storage comes with several difficulties, including high production costs, instability, agglomeration, and erosion. As a result, techno-economic optimization is required before moving further with new solar thermal energy storage projects. Key factors such as nanoparticle size, concentration, shape, and dispersion technology should be considered during optimization. It's crucial to identify the variables that will benefit solar energy storage applications the most.

KEYWORDS

- **solar radiation**
- **thermal energy storage**
- **phase change materials**
- **nanoparticles**
- **nanocomposite**

REFERENCES

1. Moriarty, P.; Honnery, D. What is the Global Potential for Renewable Energy? *Renew. Sustain. Energy Rev.* **2012,** *16*, 244–252. https://doi.org/10.1016/j.rser.2011.07.151.
2. Mohana, Y.; Mohanapriya, R.; Anukiruthika, T.; Yoha, K. S.; Moses, J. A.; Anandharamakrishnan, C. Solar Dryers for Food Applications: Concepts, Designs, and Recent Advances. *Sol. Energy* **2020,** *208*, 321–344. https://doi.org/10.1016/j.solener. 2020.07.098.
3. Earis, P.; Nugent, N.; Martinez-fresno, M.; Wilkes, S.; Spring, S.; Dean, J.; Lewis, N.; Berkeley, U. C.; Bolt, H. Energy & Environmental Science Energy & Environmental Science. *Energy* **2009,** 3–10.
4. Li, M.; Lu, J.; Chen, Z.; Amine, K. 30 Years of Lithium-Ion Batteries. *Adv. Mater.* **2018,** *30*, 1–24. https://doi.org/10.1002/adma.201800561.
5. Douvi, E.; Pagkalos, C.; Dogkas, G.; Koukou, M. K.; Stathopoulos, V. N.; Caouris, Y.; Vrachopoulos, M. G. Phase Change Materials in Solar Domestic Hot Water Systems: A Review. *Int. J. Thermofluids* **2021,** *10*, 100075. https://doi.org/10.1016/j.ijft.2021.100075.
6. Sudhakar, P. A Review on Performance Enhancement of Solar Drying Systems. *IOP Conf. Ser. Mater. Sci. Eng.* **2021,** *1130*, 012042. https://doi.org/10.1088/1757-899x/1130/1/012042.
7. Zeng, J. L.; Sun, L. X.; Xu, F.; Tan, Z. C.; Zhang, Z. H.; Zhang, J.; Zhang, T. Study of a PCM Based Energy Storage System Containing Ag Nanoparticles. *J. Therm. Anal. Calorim.* **2007,** *87*, 371–375. https://doi.org/10.1007/s10973-006-7783-z.
8. Kumar, N.; Kumbhat, S. Unique Properties, Essentials Nanosci. *Nanotechnology* **2016,** 326–360. https://doi.org/10.1002/9781119096122.ch8.
9. Tiwari, J. N.; Tiwari, R. N.; Kim, K. S. Zero-Dimensional, One-dimensional, Two-dimensional and Three-dimensional Nanostructured Materials for Advanced Electrochemical Energy Devices. *Prog. Mater. Sci.* **2012,** *57*, 724–803. https://doi.org/10.1016/j.pmatsci.2011.08.003.
10. Lan, Y.; Wang, Y.; Ren, Z. F. Physics and Applications of Aligned Carbon Nanotubes. *Adv. Phys.* **2011,** *60*, 553–678. https://doi.org/10.1080/00018732.2011.599963.
11. Al-Kayiem, H.; Lin, S.; Lukmon, A. Review on Nanomaterials for Thermal Energy Storage Technologies. *Nanosci. Nanotechnol. Asia* **2013,** *3*, 60–71. https://doi.org/10.2174/22113525113119990011.

12. Kolahalam, L. A.; Kasi Viswanath, I. V.; Diwakar, B. S.; Govindh, B.; Reddy, V.; Murthy, Y. L. N. Review on Nanomaterials: Synthesis and Applications. *Mater. Today Proc.* **2019,** *18*, 2182–2190. https://doi.org/10.1016/j.matpr.2019.07.371.
13. Baig, N.; Kammakakam, I.; Falath, W.; Kammakakam, I. Nanomaterials: A Review of Synthesis Methods, Properties, Recent Progress, and Challenges. *Mater. Adv.* **2021,** *2*, 1821–1871. https://doi.org/10.1039/d0ma00807a.
14. Nazir, H.; Batool, M.; Bolivar Osorio, F. J.; Isaza-Ruiz, M.; Xu, X.; Vignarooban, K.; Phelan, P.; Inamuddin; Kannan, A. M. Recent Developments in Phase Change Materials for Energy Storage Applications: A Review. *Int. J. Heat Mass Transf.* **2019,** *129*, 491–523. https://doi.org/10.1016/j.ijheatmasstransfer.2018.09.126.
15. Sharma, A.; Tyagi, V. V.; Chen, C. R.; Buddhi, D. Review on Thermal Energy Storage with Phase Change Materials and Applications. *Renew. Sustain. Energy Rev.* **2009,** *13*, 318–345. https://doi.org/10.1016/j.rser.2007.10.005.
16. Kenisarin, M. M. Thermophysical Properties of Some Organic Phase Change Materials for Latent Heat Storage: A Review. *Sol. Energy* **2014,** *107*, 553–575. https://doi.org/10.1016/j.solener.2014.05.001.
17. Gulfam, R.; Zhang, P.; Meng, Z. Advanced Thermal Systems Driven by Paraffin-Based Phase Change Materials – A Review. *Appl. Energy* **2019,** *238*, 582–611. https://doi.org/10.1016/j.apenergy.2019.01.114.
18. Yuan, Y.; Zhang, N.; Tao, W.; Cao, X.; He, Y. Fatty Acids as Phase Change Materials: A Review. *Renew. Sustain. Energy Rev.* **2014,** *29*, 482–498. https://doi.org/10.1016/j.rser.2013.08.107.
19. Xie, N.; Huang, Z.; Luo, Z.; Gao, X.; Fang, Y.; Zhang, Z. Inorganic Salt Hydrate for Thermal Energy Storage. *Appl. Sci.* **2017,** *7*. https://doi.org/10.3390/app7121317.
20. Readers, D.; Researches, R.; Studies, E. e c e n t s, 2021.
21. Tyagi, V. V.; Buddhi, D. PCM Thermal Storage in Buildings: A State of Art. *Renew. Sustain. Energy Rev.* **2007,** *11*, 1146–1166. https://doi.org/10.1016/j.rser.2005.10.002.
22. Zhou, D.; Zhou, Y.; Liu, Y.; Luo, X.; Yuan, J. Preparation and Performance of Capric-Myristic Acid Binary Eutectic Mixtures for Latent Heat Thermal Energy Storages. *J. Nanomater.* **2019,** *2019*. https://doi.org/10.1155/2019/2094767.
23. Socher, R.; Krause, B.; Hermasch, S.; Wursche, R.; Pötschke, P. Electrical and Thermal Properties of Polyamide 12 Composites with Hybrid Fillers Systems of Multiwalled Carbon Nanotubes and Carbon Black. *Compos. Sci. Technol.* **2011,** *71*, 1053–1059. https://doi.org/10.1016/j.compscitech.2011.03.004.
24. Wang, J.; Xie, H.; Xin, Z.; Li, Y.; Chen, L. Enhancing Thermal Conductivity of Palmitic Acid Based Phase Change Materials with Carbon Nanotubes as Fillers. *Sol. Energy* **2010,** *84*, 339–344. https://doi.org/10.1016/j.solener.2009.12.004.
25. Fan, L.; Khodadadi, J. M. An Experimental Investigation of Enhanced Thermal Conductivity and Expedited Unidirectional Freezing of Cyclohexane-Based Nanoparticle Suspensions Utilized as Nano-Enhanced Phase Change Materials (NePCM). *Int. J. Therm. Sci.* **2012,** *62*, 120–126. https://doi.org/10.1016/j.ijthermalsci.2011.11.005.
26. valan Arasu, A.; Sasmito, A. P.; Mujumdar, A. S. Numerical Performance Study of Paraffin Wax Dispersed with Alumina in a Concentric Pipe Latent Heat Storage System. *Therm. Sci.* **2013,** *17*, 419–430. https://doi.org/10.2298/TSCI110417004A.
27. Harikrishnan, S.; Kalaiselvam, S. Experimental Investigation of Solidification and Melting Characteristics of Nanofluid as PCM for Solar Water Heating Systems. *Int. J. Emerg. Technol. Adv. Eng.* **2013,** *3*, 628–635. http://www.ijetae.com/files/Conference ICERTSD-2013/IJETAE_ICERTSD_0213_95.pdf.

28. Kumaresan, V.; Velraj, R.; Das, S. K. The Effect of Carbon Nanotubes in Enhancing the Thermal Transport Properties of PCM During Solidification. *Heat Mass Transf. Und Stoffuebertragung* **2012,** *48*, 1345–1355. https://doi.org/10.1007/s00231-012-0980-3.
29. He, Q.; Wang, S.; Tong, M.; Liu, Y. Experimental Study on Thermophysical Properties of Nanofluids as Phase-Change Material (PCM) in Low Temperature Cool Storage. *Energy Convers. Manag.* **2012,** *64*, 199–205. https://doi.org/10.1016/j.enconman.2012.04.010.
30. Ji, P.; Sun, H.; Zhong, Y.; Feng, W. Improvement of the Thermal Conductivity of a Phase Change Material by the Functionalized Carbon Nanotubes. *Chem. Eng. Sci.* **2012,** *81*, 140–145. https://doi.org/10.1016/j.ces.2012.07.002.
31. Jesumathy, S.; Udayakumar, M.; Suresh, S. Experimental Study of Enhanced Heat Transfer by Addition of CuO Nanoparticle. *Heat Mass Transf. Und Stoffuebertragung* **2012,** *48*, 965–978. https://doi.org/10.1007/s00231-011-0945-y.
32. Khodadadi, J. M.; Hosseinizadeh, S. F. Nanoparticle-Enhanced Phase Change Materials (NEPCM) with Great Potential for Improved Thermal Energy Storage. *Int. Commun. Heat Mass Transf.* **2007,** *34*, 534–543. https://doi.org/10.1016/j.icheatmasstransfer.2007.02.005.
33. Lin, S. C.; Al-Kayiem, H. H. Thermophysical Properties of Nanoparticles-Phase Change Material Compositions for Thermal Energy Storage. *Appl. Mech. Mater.* **2012,** *232*, 127–131. https://doi.org/10.4028/www.scientific.net/AMM.232.127.
34. Ho, C. J.; Gao, J. Y. Preparation and Thermophysical Properties of Nanoparticle-in-Paraffin Emulsion as Phase Change Material. *Int. Commun. Heat Mass Transf.* **2009,** *36*, 467–470. https://doi.org/10.1016/j.icheatmasstransfer.2009.01.015.
35. Jegadheeswaran, S.; Sundaramahalingam, A.; Pohekar, S. D. High-Conductivity Nanomaterials for Enhancing Thermal Performance of Latent Heat Thermal Energy Storage Systems. *J. Therm. Anal. Calorim.* **2019,** *138*, 1137–1166. https://doi.org/10.1007/s10973-019-08297-3.
36. Kim, S.; Drzal, L. T. High Latent Heat Storage and High Thermal Conductive Phase Change Materials Using Exfoliated Graphite Nanoplatelets. *Sol. Energy Mater. Sol. Cells* **2009,** *93*, 136–142. https://doi.org/10.1016/j.solmat.2008.09.010.
37. Nabhan, B. J. Using Nanoparticles for Enhance Thermal Conductivity of Latent Heat Thermal Energy Storage. *J. Eng.* **2015,** *21*, 37–51.
38. Zeng, J. L.; Zhu, F. R.; Yu, S. B.; Zhu, L.; Cao, Z.; Sun, L. X.; Deng, G. R.; Yan, W. P.; Zhang, L. Effects of Copper Nanowires on the Properties of an Organic Phase Change Material. *Sol. Energy Mater. Sol. Cells* **2012,** *105*, 174–178. https://doi.org/10.1016/j.solmat.2012.06.013.
39. P.J.S. Jr, B.I. Farah, A. Mosley, IMECE2011-65070, (2014) 4–6.
40. Teng, T. P.; Yu, C. C. The Effect on Heating Rate for Phase Change Materials Containing MWCNTs. *Int. J. Chem. Eng. Appl.* **2012,** *3*, 340–342. https://doi.org/10.7763/ijcea.2012.v3.214.
41. Motahar, S.; Nikkam, N.; Alemrajabi, A. A.; Khodabandeh, R.; Toprak, M. S.; Muhammed, M. A Novel Phase Change Material Containing Mesoporous Silica Nanoparticles for Thermal Storage: A Study on Thermal Conductivity and Viscosity. *Int. Commun. Heat Mass Transf.* **2014,** *56*, 114–120. https://doi.org/10.1016/j.icheatmasstransfer.2014.06.005.
42. Parameshwaran, R.; Jayavel, R.; Kalaiselvam, S. Study on Thermal Properties of Organic Ester Phase-Change Material Embedded with Silver Nanoparticles. *J. Therm. Anal. Calorim.* **2013,** *114*, 845–858. https://doi.org/10.1007/s10973-013-3064-9.
43. Shi, J. N.; Der Ger, M.; Liu, Y. M.; Fan, Y. C.; Wen, N. T.; Lin, C. K.; Pu, N. W. Improving the Thermal Conductivity and Shape-Stabilization of Phase Change Materials Using

Nanographite Additives. *Carbon N. Y.* **2013,** *51*, 365–372. https://doi.org/10.1016/j.carbon.2012.08.068.

44. Li, M. A Nano-Graphite/Paraffin Phase Change Material with High Thermal Conductivity. *Appl. Energy* **2013,** *106*, 25–30. https://doi.org/10.1016/j.apenergy.2013.01.031.
45. Jameel Zaidan, M.; Hamed Alhamdo, M. Effect of Perforated and Smooth Fins on Thermal Performance of a Latent Heat Energy System. *J. Eng. Sustain. Dev.* **2019,** *23*, 109–127. https://doi.org/10.31272/jeasd.23.6.9.
46. Hayder, A. M.; Bin Sapit, A.; Abed, Q. A.; Abbas, M. S.; Saheb, B. A.; Mohsin, N. M. B. Thermal Performance of Corrugated Solar Air Heater Integrated with Nanoparticles to Enhanced the Phase Change Material (PCM). *Int. J. Mech. Mechatron. Eng.* **2019,** *19*.
47. Sarkar, J.; Bhattacharyya, S. Application of Graphene and Graphene-Based Materials in Clean Energy-Related Devices Minghui. *Arch. Thermodyn.* **2012,** *33*, 23–40. https://doi.org/10.1002/er.
48. Wu, S. Y.; Wang, H.; Xiao, S.; Zhu, D. S. An Investigation of Melting/Freezing Characteristics of Nanoparticle-Enhanced Phase Change Materials. *J. Therm. Anal. Calorim.* **2012,** *110*, 1127–1131. https://doi.org/10.1007/s10973-011-2080-x.
49. Hosseini, S. M. J.; Ranjbar, A. A.; Sedighi, K.; Rahimi, M. Melting of Nanoprticle-Enhanced Phase Change Material Inside Shell and Tube Heat Exchanger. *J. Eng.* **2013,** *2013*. https://doi.org/10.1155/2013/784681.
50. Hosseinizadeh, S. F.; Darzi, A. A. R.; Tan, F. L. Numerical Investigations of Unconstrained Melting of Nano-Enhanced Phase Change Material (NEPCM) Inside a Spherical Container. *Int. J. Therm. Sci.* **2012,** *51*, 77–83. https://doi.org/10.1016/j.ijthermalsci.2011.08.006.
51. Yanuar, F. The Health Status Model in Urban and Rural Society in West Sumatera, Indonesia: An Approach of Structural Equation Modeling. *Indian J. Sci. Technol.* **2016,** *9*, 1–8. https://doi.org/10.17485/ijst/2016/v9i4/72601.
52. Yanuar, F. The Health Status Model in Urban and Rural Society in West Sumatera, Indonesia: An Approach of Structural Equation Modeling. *Indian J. Sci. Technol.* **2016,** *9*. https://doi.org/10.17485/ijst/2016/v9i4/72601.
53. Lokesh, S.; Murugan, P.; Sathishkumar, A.; Kumaresan, V.; Velraj, R. Melting/Solidification Characteristics of Paraffin Based Nanocomposite for thermal energy storage applications, Therm. Sci. 21 (2017) 2517–2524. https://doi.org/10.2298/TSCI150612170L.
54. Murugan, P.; Ganesh Kumar, P.; Kumaresan, V.; Meikandan, M. K. Malar Mohan, R. Velraj, Thermal energy storage behaviour of nanoparticle enhanced PCM during freezing and melting, Phase Transitions. 91 (2018) 254–270. https://doi.org/10.1080/01411594.2017.1372760.
55. Irwan, M. A. M.; Nor Azwadi, C. S.; Asako, Y. Review on Numerical Simulations for Solidification & Melting ofNano-Enhanced Phase Change Materials (NEPCM). *IOP Conf. Ser. Earth Environ. Sci.* **2019,** *268*. https://doi.org/10.1088/1755-1315/268/1/012114.
56. Kashani, S.; Ranjbar, A. A.; Abdollahzadeh, M.; Sebti, S. Solidification of Nano-Enhanced Phase Change Material (NEPCM) in a Wavy Cavity. *Heat Mass Transf. Und Stoffuebertragung* **2012,** *48*, 1155–1166. https://doi.org/10.1007/s00231-012-0965-2.
57. Hosseini, M.; Shirvani, M.; Azarmanesh, A. Solidification of Nano-Enhanced Phase Change Material (nepcm) In An Enclosure. *J. Math. Comput. Sci.* **2014,** *08*, 21–27. https://doi.org/10.22436/jmcs.08.01.02.
58. Abdelrazik, A. S.; Al-Sulaiman, F. A.; Saidur, R.; Ben-Mansour, R. A Review on Recent Development for the Design and Packaging of Hybrid Photovoltaic/Thermal

(PV/T) Solar Systems. *Renew. Sustain. Energy Rev.* **2018,** *95*, 110–129. https://doi.org/10.1016/j.rser.2018.07.013.
59. El Hasadi, Y. M. F.; Khodadadi, J. M. Numerical Simulation of the Effect of the Size of Suspensions on the Solidification Process of Nanoparticle-Enhanced Phase Change Materials. *J. Heat Transfer.* **2013,** *135*, 1–11. https://doi.org/10.1115/1.4023542.
60. Sathishkumar, A.; Kathirkaman, M. D.; Ponsankar, S.; Balasuthagar, C. Experimental Investigation on Solidification Behaviour of Water Base Nanofluid PCM for Building Cooling Applications. *Indian J. Sci. Technol.* **2016,** *9*. https://doi.org/10.17485/ijst/2016/v9i39/94966.
61. Mahato, A.; Kumar, A. Modelling of Melting/Solidification Behaviour of Nanoparticle-enhanced Phase, 2013.
62. Chandrasekaran, P.; Cheralathan, M.; Kumaresan, V.; Velraj, R. Enhanced Heat Transfer Characteristics of Water Based Copper Oxide Nanofluid PCM (Phase Change Material) in a Spherical Capsule During Solidification for Energy Efficient Cool Thermal Storage System. Energy **2014,** *72*, 636–642. https://doi.org/10.1016/j.energy.2014.05.089.
63. Harikrishnan, S.; Kalaiselvam, S. Preparation and Thermal Characteristics of CuO-oleic Acid Nanofluids as a Phase Change Material. *Thermochim. Acta.* **2012,** *533*, 46–55. https://doi.org/10.1016/j.tca.2012.01.018.
64. Dhivya, S.; Hussain, S. I.; Jeya Sheela, S.; Kalaiselvam, S. Experimental Study on Microcapsules of Ag Doped ZnO Nanomaterials Enhanced Oleic-Myristic Acid Eutectic PCM for Thermal Energy Storage. *Thermochim. Acta.* **2019,** *671*, 70–82. https://doi.org/10.1016/j.tca.2018.11.010.
65. Zhang, Q.; Liu, C.; Rao, Z. Preparation and Characterization of n-Nonadecane/CaCO3 Microencapsulated Phase Change Material for Thermal Energy Storage. *ChemistrySelect.* **2019,** *4*, 8482–8492. https://doi.org/10.1002/slct.201901436.
66. Deng, H.; Guo, Y.; He, F.; Yang, Z.; Fan, J.; He, R.; Zhang, K.; Yang, W. Paraffin@graphene/Silicon Rubber Form-Stable Phase Change Materials for Thermal Energy Storage. *Fullerenes Nanotub. Carbon Nanostruct.* **2019,** *27*, 626–631. https://doi.org/10.1080/1536383X.2019.1624539.
67. Hussain, S. I.; Kalaiselvam, S. Nanoencapsulation of Oleic Acid Phase Change Material with Ag2O Nanoparticles-Based Urea Formaldehyde Shell for Building Thermal Energy Storage. *J. Therm. Anal. Calorim.* **2020,** *140*, 133–147. https://doi.org/10.1007/s10973-019-08732-5.
68. Granqvist, C. G.; Azens, A.; Heszler, P.; Kish, L. B.; Österlund, L. Nanomaterials for Benign Indoor Environments: Electrochromics for "Smart Windows", Sensors for Air Quality, and Photo-Catalysts for Air Cleaning. *Sol. Energy Mater. Sol. Cells* **2007,** *91*, 355–365. https://doi.org/10.1016/j.solmat.2006.10.011.
69. Bose, P.; Amirtham, V. A. Effect of Titania-Silver Nanocomposite Particle Concentration and Thermal Cycling on Characteristics of Sodium Dodecyl Sulfate Added Paraffin Wax Thermal Energy Storage Material. *Energy Storage* **2021,** *3*. https://doi.org/10.1002/est2.192.
70. Sabagh, S.; Bahramian, A. R.; Madadi, M. H. Improvement in Phase-Change Hybrid Nanocomposites Material Based on Polyethylene Glycol/Epoxy/Graphene for Thermal Protection Systems. *Iran. Polym. J.* **2020,** *29*, 161–169. https://doi.org/10.1007/s13726-020-00783-y.
71. Ren, X.; Shen, H.; Yang, Y.; Yang, J. Study on the Properties of a Novel Shape-Stable Epoxy Resin Sealed Expanded Graphite/Paraffin Composite PCM and its Application

in Buildings. *Ph. Transit.* **2019,** *92*, 581–594. https://doi.org/10.1080/01411594.2019.1610174.

72. Wang, W.; Cai, Y.; Du, M.; Hou, X.; Liu, J.; Ke, H.; Wei, Q. Ultralight and Flexible Carbon Foam-Based Phase Change Composites with High Latent-Heat Capacity and Photothermal Conversion Capability. *ACS Appl. Mater. Interfaces* **2019,** *11*, 31997–32007. https://doi.org/10.1021/acsami.9b10330.
73. Yang, J.; Qi, G. Q.; Bao, R. Y.; Yi, K.; Li, M.; Peng, L.; Cai, Z.; Yang, M. B.; Wei, D.; Yang, W. Hybridizing Graphene Aerogel into Three-Dimensional Graphene Foam for High-Performance Composite Phase Change Materials. *Energy Storage Mater.* **2018,** *13*, 88–95. https://doi.org/10.1016/j.ensm.2017.12.028.
74. Tian, Y.; Zhao, C. Y. A Review of Solar Collectors and Thermal Energy Storage in Solar Thermal Applications. *Appl. Energy* **2013,** *104*, 538–553. https://doi.org/10.1016/j.apenergy.2012.11.051.
75. Sani, E.; Barison, S.; Pagura, C.; Mercatelli, L.; Sansoni, P.; Fontani, D.; Jafrancesco, D.; Francini, F. Carbon Nanohorns-Based Nanofluids as Direct Sunlight Absorbers. *Opt. Express.* **2010,** *18*, 5179. https://doi.org/10.1364/oe.18.005179.
76. Poinern, G. E. J.; Brundavanam, S.; Shah, M.; Laava, I.; Fawcett, D. Photothermal Response of CVD Synthesized Carbon (nano)Spheres/Aqueous Nanofluids for Potential Application in Direct Solar Absorption Collectors: A Preliminary Investigation. *Nanotechnol. Sci. Appl.* **2012,** *5*, 49–59. https://doi.org/10.2147/nsa.s34166.
77. Yousefi, T.; Shojaeizadeh, E.; Veysi, F.; Zinadini, S. An Experimental Investigation on the Effect of pH Variation of MWCNT-H 2O Nanofluid on the Efficiency of a Flat-Plate Solar Collector. *Sol. Energy* **2012,** *86*, 771–779. https://doi.org/10.1016/j.solener.2011.12.003.
78. Lu, L.; Liu, Z. H.; Xiao, H. S. Thermal Performance of an Open Thermosyphon Using Nanofluids for High-Temperature Evacuated Tubular Solar Collectors. Part 1: Indoor Experiment. *Sol. Energy* **2011,** *85*, 379–387. https://doi.org/10.1016/j.solener.2010.11.008.
79. Lu, W.; Tassou, S. A. Characterization and Experimental Investigation of Phase Change Materials for Chilled Food Refrigerated Cabinet Applications. *Appl. Energy* **2013,** *112*, 1376–1382. https://doi.org/10.1016/j.apenergy.2013.01.071.
80. Alehosseini, E.; Jafari, S. M. Micro/Nano-Encapsulated Phase Change Materials (PCMs) as Emerging Materials for the Food Industry. *Trends Food Sci. Technol.* **2019,** *91*, 116–128. https://doi.org/10.1016/j.tifs.2019.07.003.
81. Pérez-Masiá, R.; López-Rubio, A.; Lagarón, J. M. Development of Zein-Based Heat-Management Structures for Smart Food Packaging. *Food Hydrocoll.* **2013,** *30*, 182–191. https://doi.org/10.1016/j.foodhyd.2012.05.010.
82. Chalco-Sandoval, W.; Fabra, M. J.; López-Rubio, A.; Lagaron, J. M. Use of Phase Change Materials to Develop Electrospun Coatings of Interest in Food Packaging Applications. *J. Food Eng.* **2017,** *192*, 122–128. https://doi.org/10.1016/j.jfoodeng.2015.01.019.
83. Lu, Y. L.; Zhang, W. H.; Yuan, P.; Xue, M. D.; Qu, Z. G.; Tao, W. Q. Experimental Study of Heat Transfer Intensification by Using a Novel Combined Shelf in Food Refrigerated Display Cabinets (Experimental Study of a Novel Cabinets). *Appl. Therm. Eng.* **2010,** *30*, 85–91. https://doi.org/10.1016/j.applthermaleng.2008.10.003.
84. Johnston, J. H.; Grindrod, J. E.; Dodds, M.; Schimitschek, K. Composite Nano-Structured Calcium Silicate Phase Change Materials for Thermal Buffering in Food Packaging. *Curr. Appl. Phys.* **2008,** *8*, 508–511. https://doi.org/10.1016/j.cap.2007.10.059.

85. Chalco-Sandoval, W.; Fabra, M. J.; López-Rubio, A.; Lagaron, J. M. Optimization of Solvents for the Encapsulation of a Phase Change Material in Polymeric Matrices by Electro-Hydrodynamic Processing of Interest in Temperature Buffering Food Applications. *Eur. Polym. J.* **2015,** *72*, 23–33. https://doi.org/10.1016/j.eurpolymj.2015.08.033.
86. Oró, E.; De Gracia, A.; Cabeza, L. F. Active Phase Change Material Package for Thermal Protection of Ice Cream Containers. *Int. J. Refrig.* **2013,** *36*, 102–109. https://doi.org/10.1016/j.ijrefrig.2012.09.011.
87. Sarier, N.; Onder, E. Organic Phase Change Materials and Their Textile Applications: An Overview. *Thermochim. Acta.* **2012,** *540*, 7–60. https://doi.org/10.1016/j.tca.2012.04.013.
88. Dang, T. T.; Nguyen, T. T. T.; Chung, O. H.; Park, J. S. Fabrication of Form-Stable Poly(ethylene glycol)-Loaded Poly(vinylidene fluoride) Nanofibers via Single and Coaxial Electrospinning. *Macromol. Res.* **2015,** *23*, 819–829. https://doi.org/10.1007/s13233-015-3109-y.
89. Karthikeyan, M.; Ramachandran, T.; Sundaram, O. L. S. Nanoencapsulated Phase Change Materials Based on Polyethylene Glycol for Creating Thermoregulating Cotton. *J. Ind. Text.* **2014,** *44*, 130–146. https://doi.org/10.1177/1528083713480378.
90. Yoo, H.; Lim, J.; Kim, E. Effects of the Number and Position of Phase-Change Material-Treated Fabrics on the Thermo-Regulating Properties of Phase-Change Material Garments. *Text. Res. J.* **2013,** *83*, 671–682. https://doi.org/10.1177/0040517512461700.
91. Chen, C.; Zhao, Y.; Liu, W. Electrospun Polyethylene Glycol/Cellulose Acetate Phase Change Fibers with Core-Sheath Structure for Thermal Energy Storage. *Renew. Energy* **2013,** *60*, 222–225. https://doi.org/10.1016/j.renene.2013.05.020.
92. Granqvist, C. G.; Azens, A.; Heszler, P.; Kish, L. B.; Österlund, L. Nanomaterials for Benign Indoor Environments: Electrochromics for "Smart Windows", Sensors for Air Quality, and Photo-Catalysts for Air Cleaning. *Sol. Energy Mater. Sol. Cells* **2007,** *91*, 355–365. https://doi.org/10.1016/j.solmat.2006.10.011.
93. Colella, F.; Sciacovelli, A.; Verda, V. Numerical Analysis of a Medium Scale Latent Energy Storage Unit for District Heating Systems. *Energy* **2012,** *45*, 397–406. https://doi.org/10.1016/j.energy.2012.03.043.
94. Chougule, S. S.; Sahu, S. K. Thermal Performance of Nanofluid Charged Heat Pipe With Phase Change Material for Electronics Cooling. *J. Electron. Packag. Trans. ASME.* **2015,** *137*, 1–7. https://doi.org/10.1115/1.4028994.
95. Krishna, J.; Kishore, P. S.; Solomon, A. B. Heat Pipe with Nano Enhanced-PCM for Electronic Cooling Application. *Exp. Therm. Fluid Sci.* **2017,** *81*, 84–92. https://doi.org/10.1016/j.expthermflusci.2016.10.014.
96. Ling, Z.; Chen, J.; Fang, X.; Zhang, Z.; Xu, T.; Gao, X.; Wang, S. Experimental and Numerical Investigation of the Application of Phase Change Materials in a Simulative Power Batteries Thermal Management System. *Appl. Energy* **2014,** *121*, 104–113. https://doi.org/10.1016/j.apenergy.2014.01.075.
97. Zhu, Y.; Chi, Y.; Liang, S.; Luo, X.; Chen, K.; Tian, C.; Wang, J.; Zhang, L. Novel Metal Coated Nanoencapsulated Phase Change Materials with High Thermal Conductivity for Thermal Energy Storage. *Sol. Energy Mater. Sol. Cells* **2018,** *176*, 212–221. https://doi.org/10.1016/j.solmat.2017.12.006.

CHAPTER 7

Plasmonics Method of Improving Solar Cell Efficiency

CHANCHAL LIZ GEORGE[1,2], ROHINI MANOJ[1,2], RINSY THOMAS[1,2], and K. V. ARUN KUMAR[1,2]

[1]*Department of Physics, CMS College (Autonomous), Kottayam, Kerala, India*

[2]*Nanotechnology and Advanced Materials Research Centre, CMS College (Autonomous), Kottayam, Kerala, India*

ABSTRACT

Photovoltaic devices play a crucial role in converting renewable solar energy into electricity. Efficiency improvement of the photovoltaic (PV) cell is one of the interesting research fields in the future for solving the problem of energy crisis and environmental concerns.

Potential applications of plasmonic nanoparticles in solar cells have tremendous applications in future, so researchers are impressed in field. In plasmonic solar cells, we make use of the nanoscale properties of noble metal particles, so it utilizes the surface plasmon resonance properties of nanoparticles. Plasmonic structure increases light absorption and scattering, thereby excites more electrons, followed by trapping of more plasmon-polaritons within the structure. Comparing solar cells with plasmonic solar cells, these factors favor the efficiency enhancement. We discussed here the applications of plasmonic with different types of solar cells like silicon solar cell, dye- sensitized solar cell, and organic solar cell.

Technological Advancement in Clean Energy Production: Constraints and Solutions for Energy and Electricity Development. Amritanshu Shukla, Kian Hariri Asli, Neha Kanwar Rawat, Ann Rose Abraham, & A. K. Haghi (Eds.)

7.1 INTRODUCTION

Several conventional sources of energy including coal, fossil fuels, geothermal, natural gas, nuclear, stalk of agriculture and considerably more have been used by mankind for a long time. Repositories of these conventional sources of energy are insufficient and due to various environmental threats caused by them, we have chosen nonconventional energy resources which are sustainable and clean energy resources. Sustainable energy sources include solar, wind, tidal, biomass as well as other sources, which can be used to generate energy repeatedly. Due to the global availability of solar energy, it is highly desirable and viable alternative to fossil fuels. Our constantly growing energy needs could be satisfied by the sunlight that reaches the surface of the earth. Incident solar energy gets converted into electrical energy by solar photovoltaic technology. The capability of solar cell is determined by the effectiveness of the process of absorbing light that produces electron-hole pairs and the following extraction of charge carriers.[1] The majority of commercial photovoltaic (PV) cells are built on Si wafers which are becoming thinner every year and reduce the consumable cost. Relatively poor solar absorption is the main disadvantage of lowering wafer thickness. This drawback can be overcome by a method known as light trapping. New technologies like organic photovoltaic cells (OPVs), dye-sensitized solar cells (DSSCs), quantum dot solar cells (QDSCs), and perovskite solar cells (PSCs) have emerged to improve or even replace conventional silicon solar cells. However, problems with photostability afflict both organic dyes and quantum dots solar cells, particularly in applications involving photocatalysis or photoelectrochemical cells (PECs).[2] It has been suggested that using plasmonic effects could enhance light absorption in solar cells active layers[3] by means of integrating metallic nanostructures in solar cells. Localized surface plasmon resonance (LSPR) is a significant feature of metal nanoparticles. In metal nanoparticles, the collective oscillation of charge induced by an applied field is in resonance with the periodic electron displacement versus positive nuclei which are known as LSPR. Plasmonic structures are an efficient approach for improving energy conversion efficiency and reducing the ultimate solar thickness. The effectiveness of solar energy conversion can be increased by plasmonic nanostructures by means of the following methods: (i) lengthening the optical path and increasing the light absorption in semiconductor by focusing the incident field through photonic enhancement. (ii) direct electron transfer (DET) which results in separation of charges in the semiconductor is a method of transfer of plasmonic energy from the metal to the semiconductor and another method is resonant energy transfer induced by plasmon.[4] Compared to dyes and quantum dots, LSPR has a wider resonance (typically 100 nm or larger). By varying the particle's shape, composition, and surroundings, the range of LSPR is fine tunable across the full range of ultraviolet/visible/near infrared (UV-NIR). The unexplored optical world at the subwavelength scale is being opened up by plasmonics. Plasmonics' ability to create thin cells results in predicted material savings. New photovoltaic opportunities may be provided by high-index, non-absorbing dielectrics, and the developing field of metamaterial plasmonics. It also enables us the effective use of advanced ideas with great performance, like hot carrier cells.[5]

7.2 SOLARCELL/PLASMONIC SOLAR CELLS

7.2.1 SOLAR CELL

Solar energy can make a significant contribution to combating climate change. For photovoltaic electricity to compete with fossil fuel-based technology, the cost must be decreased by 2–5 times (depending on local prices for fossil fuel electricity). Photovoltaic has the potential to make a great contribution to addressing climate change. Currently, 90% of the global market for solar cells is centered on crystalline silicon wafers with thickness between 200 and 300 μm. About 40% of the price of a solar module based on crystalline silicon is the price of the silicon wafers. Thin film solar cells are usually about 1–2 μm thickness and are deposited on substrates which are low-cost materials such as plastic, glass, or stainless steel. A main drawback of all thin-film solar cell-based technology is the inefficient band-pass light absorption, especially for indirect-band semiconductor silicon. So, it is crucial to design the solar cell in such a way that light is trapped inside in order to increase its absorption capacity. Because silicon is a feeble absorber, light traps are also applied in wafer-based cells. In the case of wafer-based cells, a pyramid sized from 2 to 10 μm is engraved on the surface. Surface arrays of this size are not fit for thin film cells of micrometer range thickness, requiring the development of a new process. A large increase in photocurrent was achieved due to the fact that light could be captured by depositing a thin film solar cell on the substrate after texture formation in the wavelength range. However, rough semiconductor surfaces increase recombination of surface, and generally the material quality of semiconductors deposited on rough surfaces are generally poor. Recent approach has appeared to increase absorption of light by utilization of noble metal nanoparticles scattering caused by surface plasmon resonance.[6]

7.2.2 PLASMONIC SOLAR CELLS

One of the methods of improving the efficiency of thin-film solar cells is trapping powder with a metallic substance. Metal and the insulating semi-conductor interface help to support the surface plasmon. Surface plasmons can be generated in metal nanoparticles (locally and in a constrained spatial elongation) or along dielectric metal surfaces known as surface plasmonic polaritons (SPPs), where they transmit and limit light to the interface. The use of plasmonic elements improves the efficiency of solar cells by focusing or trapping light in the absorbing layer and also acts as a low-cost anti-reverse or anti-glare contact. Sudden light at the resonant frequency of the plasmon results in electronic oscillations on the nanoparticles surface. The oscillating electrons are captured by the conductive layer, thus creating an electric current. The voltage developed depends on the potential of the electrolyte in contact with the nanoparticles and on the band gap of the conductive layer.[7]

Direct plasmonic solar cells are solar cells in which light is converted into electrical energy by plasmons as the active photovoltaic material. The plasmonic solar cells thickness can differ from typical silicon cell thickness to less than 2 μm. Substrates that are more inexpensive than silicon, such as plastic, glass, or steel, can be used. Thin film solar cells cannot absorb light as much as thick solar cells manufactured from materials with the same value of absorption coefficient, is one of its major drawback. This is why all light harvesting techniques are so important in thin film solar cell technology and hence plasmonic solar cells are so significant.[7]

7.3 PLASMONIC STRUCTURES IN DIFFERENT TYPES OF SOLAR CELLS

7.3.1 PLASMONICS IN SILICON SOLAR CELLS

The primary element of solar cells is pure silicon, and it can be used as a component of an electrical system for hundreds of years. Silicon solar panels are frequently called as "first generation" panels, because silicon solar cell-based technology was developed in 1950s. Now, greater than 90% of the market of solar cells today depends on silicon. Pure crystalline silicon is a poor conductor of electricity since it has a semiconductor material within it. To solve this issue, the silicon in a solar cell has impurities, which means that distinct atoms are purposefully mixed with the atoms of silicon, thereby enhancing the efficiency of silicon to absorb solar energy and transform light energy into electricity.[8]

Plasmonic nanostructures have recently been explored as an efficient way to improve light absorption by solar cells. Small metallic nanostructures and the strong interaction of light allow control of light propagation at the nanoscale, allowing the production of ultra-thin solar cells in which light is efficiently absorbed by trapping in the active layer. Usually, two methods are used for incorporating metal nanoparticles into solar cells. In the first method, the crystalline and amorphous silicon solar cells are used designed and manufactured silver nanoparticle coatings that improve the anti-reflective coatings. In the second method, the ordered and disordered arrangement of metal nanostructures is designed to convert light into waveguide modes in thin layers of semiconductors. Employing a relatively inexpensive wide-ranging nano imprinting technique, a method is developed a – Si:H solar cell back-contact light harvesting surface that exhibits improved efficiency over standard random texture cells.[9]

Currently, a large number of solar cells are developed using silicon with a wafer of 150 to 250 microns thickness. The price of solar cells must be decreased if this technology is to be competitive with other forms of generating electricity. For the cost reduction of solar panels, it is essential to reduce the production cost. This can be attained by decreasing the thickness of the silicon in solar cell which reduces the price of the material. But this enhances the penetration of sunlight across the cells. This enhanced transmission can be reduced by using a light harvesting system such as silver nanocrystals embedded in the solar cell. In the case of traditional thick silicon solar cells, the light path length is increased by trapping the light by the pyramidal structure on the surface and scattering the light inside the solar cell. This is especially true for crystalline silicon, which has an indirect band gap and low value of absorption coefficient. For thin silicon wafers, geometric texture is a big problem. One way to solve the issue of light capture is to employ the effect of surface plasmon resonance of metal nanoparticles.[10]

Solar cells made of crystalline silicon may utilize photons with energy as high as 1100 nm. However, less long wavelength light is absorbed and a large amount of light passes through the solar cell as the silicon wafer gets thinner. To scatter a proportion of light that is not ordinarily absorbed when particles are positioned in front of a solar cell, it is preferable to have a wide scattering peak at a wavelength between 700 and 1150 nm. Surface plasmon resonance depends on surface features such as size, shape, and surrounding substance. Scattering properties can be dramatically changed by incorporating silver nanoparticles into other materials. It is found that silver nanoparticles with a various diameter range shift scattering peak to 700 nm in the matrices like SiO_2, Al_2O_3, or SiN. This cannot be achieved by chemical synthesis. Therefore, silver sputtering was proposed. Metal nanoparticles can be obtained in various shapes and sizes using different processes. It is observed that silver nanoparticles are effectively produced by thermal evaporation on a thin wafer of silicon. This is a popular method for obtaining silver nanoparticles employed for the photovoltaic applications. To completely make out the process of deposition, elaborate studies have been carried out on the parameters of evaporation and annealing, as well as the influence of various substrates. The reflectance of the plates is also measured with and without particles.[10]

Incorporating plasmon-like metal nanoparticles into device structures can improve device performance through placement-dependent mechanisms. There are several ways to distribute metal nanoparticles, for example, in an absorbing layer, a charge transfer layer, applying them to an electrode, or embedding them between layers in a sandwich structure. Basically, Ag, Au, and Ag/Au (bimetal) nanoparticles are incorporated into organic PVs to obtain light energy additionally. It has been proven that Gold is superior than silver in the active layer as plasmonic nanoparticles. Bimetallic Au/Ag HTL nanoparticles too increase PCE. Due to the different composition of organic solar cells, there is no clear mechanism for specific nanomaterials such as plasmonic nanoparticles.[11]

The plasmonic effect in thin-film silicon solar cells is a new technology with encouraging applications in solar cell manufacturing. Take advantage of the nanoscale properties of Al, Au, Ag, Ti, SiO_2, Cu, etc., nanoparticles embedded at the interface between dielectric and metal contacts improve the light harvesting features of thin-film silicon solar cells by enhancing absorptivity and generating hot electrons and thereby increases the photocurrent of the solar cell. Plasmonic nanoparticles deposited on the front surface of a silicon solar cell are the most efficient and easy to apply.[11]

7.3.2 PLASMONIC DYE-SENSITIZED SOLAR CELL

Dye-sensitized solar cells (DSC, DSSC, DYSC, or Grätzel cells) are the lowest cost solar cells of the various thin film solar cells. Photoelectrochemical system centered on a semiconductor device between an electrolyte and a photosensitive anode. The current generation of the dye solar cell, further known as the Grätzel element, was initially discovered by Brian O'Regan and Michael Grätzel of the University of California at Berkeley in 1988.[12]

In the late 1960s, it was invented that an organic dye ignited at the oxide electrode of an electrochemical cell could generate electricity. To understand and model the key process of photosynthesis, this phenomenon was studied at UC Berkeley using chlorophyll extracted from spinach (a biomimetic or bioengineering approach). In 1972, based on these experiments, electricity generation using the DSSC principle was showed and discussed. The dye solar cells are unstable and have been identified as a serious problem. Efficiency can be improved over the next 20 years by optimizing the porosity of fine oxide powder electrodes, but instability is still a problem.[12]

Every year, the Earth collects a significant amount of solar light energy, and using this energy to meet our energy needs is very important for various reasons. This includes the irresistible depletion of fossil fuels over the next 50 years and environmental damage from burning large quantities of fossil fuels. Thus, it converts solar energy into its most useful form. Electrical energy is critical to meet growing energy needs. Besides, solar panels are environmentally friendly devices, and the Sun is a possible source of energy. Using light energy that reaches the Earth in an environment friendly way, can solve most of the issues affecting energy demand and the environment worldwide. Research work on conversion of solar energy have received greater focus in the last 10 years.[13]

Photovoltaic systems convert solar energy into electric energy. So far, solar power has gone through three generations. Silicon-based solar cells is the first generation, and the second generation is based on thin-film semi-conductors. Third generation photovoltaic devices are DSSC and organic semiconductor solar cells. Compared with first and second generation solar cells, the efficiency of power conversion and stability of 3rd generation solar cells are low. However, the 3rd generation solar cell has many advantages over the 1st and 2nd generations, so it is of interest to researchers around the world. For example, DSSCs are flexible in design and shape, inexpensive, simple to manufacture, colorful, transparent, workable in ambient light conditions, and material cost is low.[13]

Most of the DSSC problems regarding instability are related to the liquid electrolyte of this device. The liquid electrolyte in the DSSC can be replaced with a gel polyelectrolyte (GPE) and thus the problem can be solved. DSSCs based on volatile liquid electrolytes provide higher efficiency of energy conversion, but are associated with disadvantages such as the risk of solvent evaporation, instability, flammability at high temperatures, and dye degradation and desorption. Polyelectrolytes such as semi-solids or gels can solve some of the instability problems associated with liquid electrolytes. For these cells, a gel-like membrane is obtained by trapping a mixture of solvent and liquid salt in a polymer matrix. The polymers commonly used in this system are polymethyl methacrylate (PMMA), polyacrylonitrile (PAN), polyethylene oxide (PEO), phthaloyl chitosan (PhCh), and polyvinylidene fluoride (PVdF). Several attempts have been made to increase the conductivity of polymer-based semi-solid electrolytes. A common method is the use of plasticizers and/or inorganic fillers such as titanium, silicon dioxide, or alumina. The concentration and type of ionic particles in the electrolyte strongly affects the conductivity of these polyelectrolytes. However, in this work, a stable and highly conductive HE for DSSC was obtained by mixing PhCh and PEO.[13]

Dye-sensitive solar cells (DSCs) have earned broad research interest due to their high power conversion efficiency and low price. Improved light collection between 600 and 900 nm range of wavelength, where the state-of-the-art ruthenium complex weak absorption sitizers, which is an assuring path to increase efficiency of power conversion by more than 15%. Versatile research efforts have been conducted to enhance the absorption of light of DSCs by developing highly absorbent dyes and using energy relays.[14]

The plasmonic back reflectors, consisting of 2D arrays of silver nanodomains, have been embedded in a dye-sensitive solid state solar cells (ss-DSC) by nanoimprint lithography. Reflective plates improve absorption by means of excitation of plasmonic modes and the rise of light scattering. The ss-DSCs with plasmonic back reflection exhibit higher extrinsic quantum efficiency, especially in the high wavelength region of the dye absorption band. Therefore, ss-DSC is generated with a sensitizer for ruthenium complexes and high-absorption organic sensitizer had 16 and 12% higher short-circuit optical currents, respectively. They achieved a power conversion efficiency of 3.9 and 5.9%, that is the unexpected results with the same dye.[14]

SS-DSC uses solid state hole transport materials (HTM) to take over the place of traditional liquid electrolytes and HTM highest occupied molecular orbital (HOMO) to tune open circuit voltage to provide a viable high efficiency path. The leakage issues are solved by using of solid materials associated with the corrosiveness and volatility of liquid electrolytes. The mesoporous TiO_2 in the Ss-DSCs consists of photoanode sensitized with a monolayer dye, filled with HTM and coated with a reflective metal contact deposited on the active layer. Recently, optimized ss-DSC is still limited due to recombination of electron-hole and incomplete filling of HTM pores, so the optimized active layer thickness is only 2 μm, which is much lower than the thickness required to achieve adequate light absorption. Efforts to improve light absorption in ss-DSC have mainly focused on the development of highly absorbent dyes and new TiO_2 nanostructures with a large interior surface area for adsorption of dye. Several studies have been undertaken to strengthen the optical characteristics of ss-DSC structures in order to provide better control over photons.[14]

The use of the plasmon effect has been proposed as an effective method for increasing absorption of light in the active layer of solar cells. It has been shown in various thin film solar cell materials such as gallium arsenide, amorphous silicon, polymer, and TiO dye monolayers. Most claimed plasmonic solar cells have much lower efficiencies than advanced devices that are made from the same material, since the active layer used is much thinner than that typically found in cells. Studies of these thin cells showed that the spectral increase in the photocurrent density is due to: (1) an attractive local surface plasmon resonance of metal nanoparticles; (2) light scattering by metal nanoparticles in the dielectric waveguide mode of solar cells; (3) relationship with the regime of surface plasmon polariton (SPP). A significant distinction is made between localized surface plasmon excitation (SP) resonance, which occurs when conduction electrons oscillate in particles of finite size, and SSP, which is a surface electromagnetic wave propagating over a metal surface. The effective excitation of confined SP resonances depends on the shape, geometry, size, and dielectric surrounding of the metal particles, and the electric field amplification is generated only in the immediate vicinity of the metal. Propagating SPP waves can be excited most effectively by creating a periodic grating structure that allows light waves in free space to capture momentum in a plane sufficient to couple with shorter wavelength SPP waves (higher propagation constants).[14]

SPP has the largest field strength at the interface of metal/dielectric, but also has a significant penetration depth (100 nm–1 μm) into the dielectric medium by the side of the metal. For this reason, SPP excitation can increase the absorption of thicker solar cell active layers. It should also be noted that the coupling with the modes of a dielectric waveguide can be enhanced using the plasmon effect. Near the metal particle surface plasmon frequency, the scattering power (i.e., cross section) increases resonantly, and scattering can be enhanced in the dielectric and SPP modes.[14]

Plasmonic DSSCs (PDSSCs) consist of four main components: nano-structured semiconductor, electrolyte, dye molecule, and counter electrode.

For maximum conductivity, the counter electrode is usually covered with transparent conductive glass. The TiO_2 surface is adsorbed with colored dyes for maximum absorption. TiO_2 nanocrystals were used as a continuous film to increase the surface area for enhancing light absorption. Electrolytes are also utilized and are an important element located between the electrodes to complete the path. The photoanode is designed in such a way to capture the maximum possible photon in the visible range through a combination of various metal nanostructures.[15]

In PDSSC, the photon interactions cause the excitement of dye molecules, from the HOMO to the second-lowest unoccupied molecular orbital (LUMO) phases. The conduction band of semiconductor nanostructured TiO_2 films is then injected with electrons. The entire electron population moves through an external load to the counter electrode, the electrolyte transports electrons from the counter electrode to the dye. TiO_2 has non-toxicity, good band edge and anti-photo corrosion stability and TiO_2 has attracted a lot of interest. Due to the content of redox ions, the electrolyte acts as an electronic intermediate between the TiO_2 photo electrode and the counter electrode. Thus, an oxidized dye (photosensitizer) can be reproduced by gathering electrons by means of an oxidized ionic redox mediator. It is possible to calculate the performance and the efficiency and efficiency is the converting of light energy into electricity.[15]

PDSSC consists of various kinds of dyes, and dyes play an important part due to their atomic structure. The separation of charge is initiated by the photon absorption by a dye containing titanium dioxide and an electron transport medium. The work of the PDSSC depends on the respective energy band of the sensitizer and the mobility of the electron transfer action of the dye combination on the surface of the semiconductor and the medium of electron transfer. The key role of the dye is to absorb the visible light spectrum and increase the sensitivity of semiconductors with a wide bandgap. Several principles apply to efficient dye synthesis and efficient PDSSC. (1) The dye needs an absorption range to cover the maximum visible range, and has the highest light gathering power in TiO_2 films. (2) For better anodic electron injection (LUMO), the dye should be close to the anchor group and higher than the conduction state of the TiO_2 semiconductor. (3) For efficient dye regeneration, HUMO must be in a redox energy environment. (4) To protect the aqueous desorption of the dye and TiO_2, the surface of the dye should be in a stable electrolyte with hydrophobic anode contact. (5) Paint must not accumulate on the surface. The optical and electrical properties of TiO_2 can be significantly changed by doping with positive and negative ions.

In TiO_2, an anionic impurity replaces O_2, and a cationic impurity replaces Ti_4. In addition, the sensitizer composition is the best choice for improving PDSSC performance through the use of various additives. For example, an acid can be added to the electrolyte to improve cell performance. Therefore, this prevents the dye from mixing with the TiO_2 surface because the electrons are transferred to the TiO_2 faster. The performance of the device under illumination depends on the difference between the Fermi level of TiO_2 and redox potential of the electrolyte. PDSSC's attracts a lot of attention for its environmental friendliness and high performance compared to other existing materials. Efforts are being made to develop hybrid semiconductor devices and improve the efficiency of dye-sensitive solar cells through the commercialization of renewable solar energy. Various types of natural dyes such as chlorophyll are currently being discussed as sensitizers in PDSSC. It has also been suggested that various plant pigments have the same electronic structure as dyes and can capture photons from solar energy. Photoanodes with electrical conductivity and high resistance, which promote electron transfer, and an external band gap, which prevents the recombination of photo generated charge carriers, have a higher photo catalytic efficiency.[15]

The PDSSC has performed greater activity with the aid of their excessive conversion efficiency as nicely as low cost of manufacturing. Various efforts have been achieved to amplify optimized parameters and machine performance. The critical parameter is to maximize the price era by using the plasmonic consequences and the plasmonic DSSCs have completed an effectivity of around 10.8% with 30% enhancement as in contrast to the reference device. The DSSC with plasmonic-based nanostructures must improve the photosensitizers to soak up most photons. Wen et al investigated the plasmonic impact in DSSC with the aid of the usage of Ag nanoparticles. The efficiency, as properly as dye absorption in dye-sensitized photo voltaic cells, might also favor to be improved via the use of Ag nanoparticles.[15]

The plasmonic effect has received a lot of interest in photovoltaic cells due to its flexibility. Plasmonic nanostructures capture the maximum of incident light and enhance efficiency through various mechanisms. In addition, a number of restrictions have been introduced on the use of plasmonic nanostructures, the most significant of which are their absorbing properties and physical stability.[15]

The stability of plasmonic nanostructures is important in the field of solar cells. Electrical and chemical stability is necessary to obtain organic materials that can be used as passivating layers with standard PDSSC, titanium dioxide, and plasmonic nanostructures to eliminate failed plasmonic effects. The inclusion of these passive layers plays an important role. (1) Plasmonic nanostructures can act as recombination centers for complete current collection and charge separation. These types of defects reduce the active characteristics of plasmonic nanostructures.[15] (2) If the manufacturing process involves heat treatment, noble metal nanostructures lose their optical properties due to deformation. High-temperature annealing processes are sufficient to develop nano composites based on PDSSC, SiO_2 or TiO_2, which retain their structure during annealing and exhibit thermal stability due to the protective layer.[15] 3) Conventional DSSCs using iodine-based electrolytes can cause corrosion and damage to metal-based cells. Wu et al. Corrosion resistance is complemented by a protective layer of TiO2 and a simple iodine-based redox couple. Corrosion is largely prevented by a protective coating that increases chemical resistance. Problems of concern can be minimized by using non-corrosive polymer-based electrolytes. Another important issue is the thickness of the protective film needed to ensure chemical and electrical stability as the near optical distance increases.[15]

LHEs have been developed for various types of solar cells by combining ordered metal nanostructures and can be improved in two ways. One of them is the efficient transfer of energy between light absorbers and metal nanoparticles, which significantly increases the LHE. Spectral alignment of the LSPR band and the absorption spectrum of the photoactive substrate is important to achieve good resonant energy transfer. Secondly, it develops the plasmon response of the solar spectrum and increases the absorption of light in the day time absorption region. Increases light collection. In a narrow spectral range, metal nanostructures resonate with light, reducing the accumulation of plasmons to certain wavelengths. Another factor in the creation of experimental plasmonic devices is the LHE amplification and adjustment of the LSP band position by determining the shape and size of the nanoparticles. Isotropic metal nanoparticles such as Cubic composite nanoparticles with active characteristics exhibiting a pronounced plasmonic field associated with the NIR viewing angle. Plasmon resonance coupling is also a good choice for a wide solar spectrum. The Au-core/Agshell coupled structure exhibits a higher active plasma band than individual nanoparticles due to the maximum energy interaction.[15]

The PDSSCs are attractive and extremely optimistic choice appropriate to ecofriendly and greater than ever conversion efficiency in the energy sector. They are undemanding to make and effective under highest hotness conditions as compared to other various technologies. The great conversion efficiencies of up to 13% arrange been maintained with merely single-junction-based

DSSC, separately from efficiency, the issues counting device stability as perfectly as minimization of expenditure low-price up to the highest limit. The developments of dyes steal to a great extent consideration over the 20 years to use panchromatic dye sensitizers and further boost the efficiency of copied ruthenium-based dyes. Its efficiency is about (~4%) for the reason that of the passing ending of electrolytes.[15]

Finally, Etgar et al. discussed a $CH_3NH_3PbI_3/TiO_2$ heterojunction device based on nanocomposite, where $CH_3NH_3PbI_3$ can act as a light harvester and a hole transportation. Efficiency of recent PSCs from 3 to 22%. Another important question is improved cell stability for maximum usage time in future work. Continuous leakage of liquid electrolyte has also failed Pt catalyst due to instability in DSSC. To prevent this, it is necessary to use a solid-state hole carrier. Likewise, the Pt catalyst changes how the graphene materials have maximum stability under extended potential cycling conditions with the removal of these problems, the life of DSSC has extended to 20 years old, estimated by Gratzel. The interaction of environmental humidity with sunlight is one of the main reasons for the degradation of perovskite crystals. It is also necessary to use a sealing medium suitable for perovskite cells under maximum temperature conditions to avoid environmental influences. Reducing costs and material consumption are essential to continuously improve efficiency. Dyes are also expandable parts that use toxic substances and require further processing. Therefore, natural sensitizers are needed for use in DSSCs derived from flowers and leaves. When selecting optimal natural dyes and processes, several stabilizers are used to increase efficiency. PDSSC contains other expensive components such as FTO and ITO. Greiner et al. discuss and compare CO_2 emissions from DSSC systems and natural gas.[15]

DSSCs have a much lower environmental impact, and the use of vegetable dyes greatly increases their effectiveness. The commercialization of plasmonic DSSCs will be problematic because they do not have a long and short lifetime, in contrast to the long lifetime of single-crystal and polycrystalline Si-based solar cells. The durability of DSSC cells is questionable until solid electrolytes are optimized for good performance. Plasmonic DSSCs can also be commercialized in the future after achieving excellent performance. Plasmonic PSCs are also approaching higher efficiency as an emerging field, but the only problem is their high cost, which can be dealt with by the introduction of new functional materials in the future.[15]

To turn PDSSC into a commercial technology and make it highly efficient and competitive, it is important to reduce material and manufacturing costs to improve reliability and productivity. To this end, model development is the primary task of quantitatively classifying promising semiconductors, electrolytes, dyes, and manufacturing processes. You can use advanced models to streamline your development process and save experimentation and resources. Practical techniques are needed to improve light output and reduce electron loss. In addition, structure-modified solid state DSSCs are featured due to their excellent stability and high efficiency. PDSSC and PSC have developed two major interesting features. They will provide the next generation of meso-superstructured solar cell (MSSCs). Therefore, devices of the mentioned type can have optimal performance. These advanced solar cells can maximize productivity in the future while reducing the use of conventional fuels.[15]

7.3.3 PLASMONIC ORGANIC SOLAR CELL

Organic photovoltaics and plasmonic are utilized to increase sunlight absorption and the coupling of photons and electrons, which in turn enhances device performance. Organic solar cells have two electrodes.[16] One of the electrodes is transparent. Between them, there is an active layer. In this area, free charge carriers are developed. The active layer and each electrode are separated by a buffer zone. This layer ensures a charge selective transport, avoiding impacts of charge recombination that would affect the capability of the device. Conducting polymers absorb photons when light strikes an organic solar cell; as a result, a bound electron hole pair is created with an electrostatic Coulomb force.[16] Exciton separation occurs at the heterojunction's interface. The built-in electric field at the heterojunction's interface transmits the electron to the acceptor material and the hole to the donor material, causing the formation of a photocurrent. With recombination distances ranging from 4 to 20 nm, the excitons produced by light typically have very short lives. The best alternative in terms of performance is the bulk heterojunction (BHJ) solar cell, because there is a greater volume of contact between the acceptor material and donor material, effective charge separation which provides a greater volume of contact between the donor and acceptor, more effective charge separation, and different paths for the movement of free carriers.[16]

Due to their advantages, such as being inexpensive, lightweight, and mechanically flexible, organic photovoltaic cells (OPVs) have attracted a lot of research attention. These cells provide intriguing candidates as a new type of renewable energy source. At present, the power conversion efficiency (PCE) of OPVs has reached a peak of 13–14%.[17] The photoactive layer in the majority of organic semiconductor materials usually quite thin, or around 100 nm or less, to allow diffusion of charge carriers and extraction. Significant amounts of photon energy that could have been converted to electrical energy are lost as a result of insufficient absorption of the incident light. As a result, numerous studies have been encouraged to enhance OPVs' ability to absorb light. Without affecting the photoactive layer's thickness, metal nanostructures can be employed to enhance the surface plasmon resonance (SPR) effect, which will enhance the photoactive layer's ability to absorb light and boost photovoltaic performance. In order to control the plasmonic amplification of the optical absorption, metallic nanoparticles' (NPs) forms, sizes, and chemical compositions can be chemically adjusted. By adding metallic NPs, organic solar cells' ability to trap light can be increased. The electrons of the nanometals can be collectively excited at the metallic-dielectric interfaces in a variety of plasmonic solar cell topologies. Due to this, the electromagnetic field in photoactive layer of the device is enhanced. The Mie theory helps to explain this phenomenon, known as the SPR effect.[17] The far-field scattering effect can be used to boost optical absorption by increasing the incident photons' optical path and decreasing reflection.[17] The metallic nanostructures has ability to restrict electromagnetic waves at the interface of metal dielectric and creates a strong near field which increases the number of absorption events in the near-field SPR effect.[17] The incident sunlight's optical path is prolonged by the far-field scattering, which may aid in photon absorption.[17] Typically, metallic nanoparticles between 30 and 60 nm in size are the most suitable for the forward scattering effect. Plasmonic nanostructures positioned in the front buffer layer require forward scattering (HTL in conventional solar cells or ETL in inverted solar cells). To reduce reflection at the surfaces, the plasmonic nanostructures' geometry has been carefully designed. Increased optical pathways and interaction times between incident sunlight and the absorbing material lead to improved photoactive layer absorption efficiency.[17] Backward scattering must be controlled for the NPs in the back buffer layers because they are distant from the input photons in order to effectively absorb light. The scattering direction may be precisely tuned by changing the size and geometry. Around 50-nm diameter NPs scatter photons in both the forward and reverse directions. Radiation scattered increases in the backward orientation as particle size increases.[17]

When LSPR takes place, the metallic NPs' electrons collectively oscillate in response to the incident radiation, considerably enhancing the electric fields in the vicinity of the NPs' surfaces. Visible and near-infrared wavelengths are frequently where noble metal NP resonance occurs. Typically, the metal nanoparticles' size should be less than the incident light's wavelength in order to excite plasmons. The absorption enhancement in these conditions is mostly caused by the LSPR effect instead of the scattering effect. Smaller than 30 nm metallic NPs behave more like local field enhancers than like scattering centers. The plasmonic near-field of the photoactive layer increases the absorption cross-section, which significantly boosts photon absorption. The increased electric field also be connected to the absorbing layer and thereby increasing its effective absorption cross section, if these kinds of NPs are positioned outside the active layer, such as in the buffer layer.[17]

In contrast to LSPR, SPP typically develops on or near planar metal sheets. SPP is an electromagnetic wave that travels along the planar metal-dielectric contact at infrared or visible frequencies.[17] SPP's charge motions are influenced by both the metal and the dielectric. SPP waves provide a higher magnetic field because they have shorter wavelengths than incident irradiation. The SPP magnetic wave amplitude exponentially decreases as distance increases from the interface into each medium. In the subwavelength region perpendicular to the contact, SPP is spatially constrained.[17]

Using a special self-assembled mono layer of Ag nanoparticles made from a colloidal solution, Yoon et al. demonstrate increased optical absorbance and photocurrent for polythiophene-fullerene bulk hetero junction photovoltaic systems.[18] The distance between the particles can be changed by adding the right nanoparticle to organic capping groups. The self-assembled Ag nanospheres, which have a regulated particle-to-particle spacing and an average diameter of about 4 nm, have a remarkable degree of uniformity, as demonstrated by transmission electron microscopy.[18]

The localized surface plasmon resonance's peak is 465 nm broad at half maximum (95 nm). In the spectral range of 350–650 nm, where the organic bulk hetero junction photo active film absorbs, there is an enhanced optical absorption due to the greater electric field in the photo active layer induced by excited localized surface plasmons in the Ag nano-spheres.[18] In the experimental devices with the self-assembled layer of Ag nanoparticles, the highest induced photo-current efficiency (IPCE) increased to 51.6% at 500 nm under the short-circuit condition, while the IPCE of the reference devices without the plasmon-active Ag nanoparticles is 45.7% at 480 nm.[18] For the experimental devices, short-circuit current density is enhanced when air mass 1.5 global filter illuminations are used with 100 mW/cm^2 incidence intensities. This is due to the photo-generation of excitons being boosted close to the Ag nanoparticles' Plasmon resonance.[18] In bulk hetero junction solar cells that were solution-processed, Morpha et al. employed silver nanoparticle layers that are plasmon active. An indium tin oxide electrode was coated with layers of nanoparticles via vapor-phase deposition.

Due to the rise in the optical electrical field within the photoactive layer, the presence of these particle coatings leads to improved optical absorption and, as a result, increased photo conversion efficiency.[18] The efficiency of conversion of solar energy of a bulk hetero junction photovoltaic device built of poly (3-hexylthiophene)/(6, 6)-phenyl C61butyric acid methyl ester rose from 1.3% ± 0.2% to 2.2% ± 0.1% for devices using thin Plasmon-active layers. Six assessments showed that the improvement factor of 1.7 was statistically significant. Rand et al. examined the optical properties of silver nanoparticles used in conjunction with ultrathin-film organic photovoltaic cells. An array of nanoparticles with an average diameter of 5 nm can be shown to have an increased incident optical field that can penetrate an organic dielectric up to 10 nm from the center.[18] The energy required for the excitation of nanoparticle surface plasmons is also high due to the distance from the improvement. They investigated how plasmon enhancement is affected by cluster spacing, form, and an embedding dielectric medium with a complex dielectric constant, and they proposed a model to account for this long-range enhancement. Due to this phenomenon, tandem organic solar cells are shown to be more effective, and the likelihood of additional gains in solar cell efficiency is investigated.[18] Increasing the cost and efficiency of conversion of photovoltaic DSSCs by taking advantage of the sizable optical cross sections of localized (nanoparticle) surface plasmon resonances is an exciting strategy (LSPRs). Hagglund et al. have investigated this potential for dye-sensitive solar cells. Photoconductivity studies were performed on flat TiO_2 sheets that had been sensitized with a mixture of dye molecules and arrays of nanofabricated elliptical gold discs. It was established that the resonant frequency in the anisotropic, aligned gold discs that is polarization dependent results from the LSPR contribution in order to boost dye charge carrier generation rate.[18]

KEYWORDS

- **photovoltaic**
- **solar cell**
- **nanomaterials**
- **plasmonic**
- **DSSC**

REFERENCES

1. Enrichi, F.; Quandt, A.; Righini, G. C. Plasmonic Enhanced Solar Cells: Summary of Possible Strategies and Recent Results.
2. Wang, B.; Kerr, L. L. Stability of CdS-Coated TiO2 Solar Cells.
3. Atwater, H. A.; Polman, A. Plasmonics for Improved Photovoltaic Devices.
4. Cushing, S. K.; Wu, N. Plasmon-Enhanced Solar Energy Harvesting.
5. Aliberti, P.; Feng, Y.; Takeda, Y. Investigation of Theoretical Efficiency Limit of Hot Carriers Solar Cells with a Bulk Indium Nitride Absorber.
6. Catchpole, K. R.; Polman, A. Plasmonic Solar Cells
7. Voltaics, N.-S. Plasmonic Solar Cells.
8. Silicon Solar Cells—Greenmatch.
9. Spinelli, P.; Ferry, V. E.; van de Groep, J.; van Lare, M.; Verschuuren, M. A.; Schropp, R. E. I.; Atwater, H. A.; Polman, A. Plasmonic Light Trapping in Thin-Film Si Solar Cells.
10. Foslia, C. H.; Thøgersena, A.; Karazhanova, S.; Marsteina, E. S. *Plasmonics for Light Trapping in Silicon Solar Cells*. Department of Solar Energy, Institute for Energy Technology, Instituttveien 18, Kjeller, Norway, 2007.
11. Ali, A.; El-Mellouhi, F.; ... Aïssa, B. Research Progress of Plasmonic Nanostructure-Enhanced Photovoltaic Solar Cells.
12. Dye _ Sensitized Solar Cell – en.m.wikipedia.org
13. Shah, S.; Noor, I. M.; Pitawala, J.; Albinson, I.; Bandara, T. M. W. J.; Mellander, B.-E.; Arof, A. K. Plasmonic Effects of Quantum Size Metal Nanoparticles on Dye-sensitized Solar Cell.
14. Ding, I.-K.; Zhu, J.; Cai, W.; Moon, S.-J.; Cai, N.; Wang, P.; Zakeeruddin, S. M.; Grätzel, M.; Brongersma, M. L.; Cui, Y.; McGehee, M. D. Plasmonic Dye-Sensitized Solar Cells.
15. Asif Javed, H. M.; Sarfaraz, M.; Nisar, M. Z.; Qureshi, A. A.; Fakhar e Alam, M.; Que, W.; Yin, X.; Abd-Rabboh, H. S. M.; Shahid, A.; Ahmad, M. I.; Ullah, S. Plasmonic Dye-Sensitized Solar Cells: Fundamentals, Recent Developments, and Future Perspectives.
16. Notarianni, M.; Vernon, K.; Chou, A.; Aljada, M.; Liu, J.; Motta, N. Plasmonic Effect of Gold Nanoparticles in Organic Solar Cells
17. Feng, L.; Niu, M.; Wen, Z.; Hao, X. Recent Advances of Plasmonic Organic Solar Cells: Photophysical Investigations.
18. Ghosh, B.; Espinoza-González, R. For Improved Photovoltaic Devices.

CHAPTER 8

Applications of Carbon Nanotubes in Energy Storage Materials

K. B. AKHILA and RONY RAJAN PAUL

Department of Chemistry, CMS College, Kottayam, Kerala, India

ABSTRACT

Nanostructered materials are of great interest in energy storage and conversion field due to their favorable mechanical and electrical properties.[1,2] Due to their unique morphology and electrochemical properties energy storage systems have been using carbon nanotubes either as an active anode component or as an additive to improve the electronic conductivity of cathode materials. Among the various allotropes of carbon CNTs is one of the most promising and widely studied allotropes with simplest chemical composition and atomic bonding configuration.[3] CNTs because of their hollow structure provide an extraordinary thermal, mechanical, electrical properties like low density, high tensile strength, enhanced thermal conductivity, high electrical conductivity, high ductility, high thermal and chemical stability.[4] Structure of CNT with slight modifications including defects creation, controlling the distribution of pore size, doping of heteroatoms may lead to complementary properties that enable excellent electrochemical performance. Carbon nanotubes having the potential to improve energy conversion and energy storage applications have been effectively utilized in the storage devices such as Li-ion batteries, supercapacitors, fuel cells and are also being evaluated for applications in renewable energy sources including solar cells and hydrogen storage.[3] In this chapter, we discuss the application of carbon nanotubes in different energy storage materials.

Technological Advancement in Clean Energy Production: Constraints and Solutions for Energy and Electricity Development. Amritanshu Shukla, Kian Hariri Asli, Neha Kanwar Rawat, Ann Rose Abraham, & A. K. Haghi (Eds.)

8.1 INTRODUCTION

Carbon nanotubes are long, thin cylinders of carbon discovered in 1991 by Sumio Iijima.[5] Due to their favorable mechanical and electrical properties, nanostructured materials are of great interest in energy storage and conversion field.[1] Carbon nanotubes which are less than 100 nanometers in diameter and can be as thin as 1 or 2 nm are one of the best examples of true nanotechnology. CNTs are formed up of single or multiple graphene sheets with open or closed ends, which possess large surface area, good electrical properties and is a good conductor and absorber. CNTs are large macromolecules that are unique for their size, shape, and remarkable physical properties. CNTs are tube-shaped substances entirely consisting of carbon and are produced by rolling graphene sheets into cylinders.[6]

FIGURE 8.1 Carbon nanotube.

Depending upon their atomic bonding and dimensionality, CNTs can be classified into various types such as long or short based upon their length, single walled or double walled or multiwalled depending on the no of concentric cylindrical layers in their nanostructure, open (cylindrical tube shape with open ends) or closed type (cylindrical tube shaped capped by a half fullerene at ends), and a spinal structure.[7] CNTs can be further classified based upon their crystallographic configurations such as zig–zag, arm chair, and chiral depending on how the graphene sheet is rolled up. CNTs because of their hollow structure provide extraordinary thermal, mechanical, electrical properties like low density, high tensile strength, enhanced thermal conductivity, high electrical conductivity, high ductility, high thermal and mechanical stability.[3] Depending upon the orientation of graphene lattice with respect to the tube axis, which is called chirality, individual CNTs can be metallic or semiconducting.[8] CNTs which can transmit electrons over long distance are called 1D ballistic conductors.[4,9] The 1D confinement of electrons leads to the ballistic conduction which causes efficient and fast conduction of electrons and carry current with zero resistance.

CNTs have been used in energy storage and conversion systems like alkali metal ion batteries, fuel cells, nano-electronic devices, supercapacitors, and hydrogen storage devices. The extraordinarily high electronic conductivity of CNTs enables CNT and graphite as an additive to composite electrodes and to facilitate activation of poorly conducting electrode materials making them electrochemically active.[5]

8.2 TYPES OF CNTS AND THEIR PROPERTIES

SWNTs are sheets of graphene that have been rolled up to form a long hollow tube, with wall thickness of a single atom. Because of their small diameter and large aspect ratio, SWCNTs are considered 1D materials. Single-walled carbon nanotubes are allotrope of sp^2 hybridized carbon similar to that of fullerenes. The structure of SWCNT is a cylindrical tube including six-membered carbon rings similar to that of graphite.

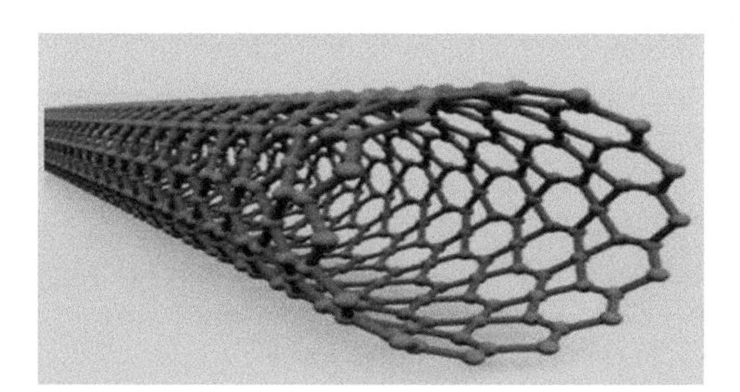

FIGURE 8.2 Single-walled carbon nanotube.

Multiwalled carbon nanotubes consist of multiple carbon nanotubes nested within one another. The number of nanotubes within a MWCNT can vary between 3 and 20. SWCNTs which are structural similar to a single sheet of graphite may be either metallic or semiconducting, depending on the sheet direction about which the graphite sheet is rolled to form a nanotube cylinder.[4] Just like single-walled nanotubes, multiwalled carbon nanotubes also exhibit exceptional electrical, thermal, and mechanical properties. However due to the increased no of walls MWCNTs have a high probability of defects occurring.

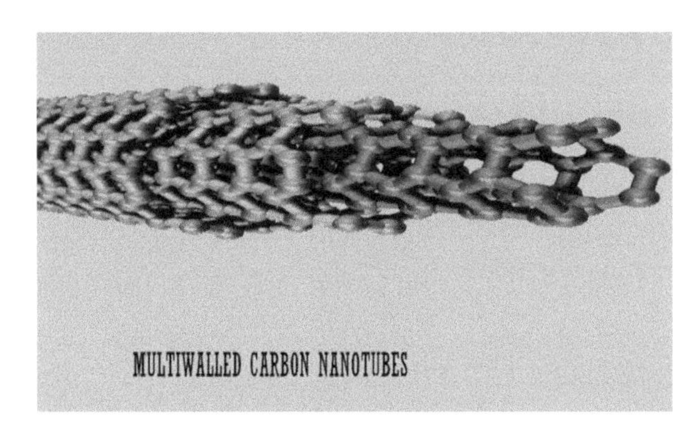

FIGURE 8.3 Multiwalled carbon nanotubes.

The electrical properties of carbon nanotubes are dependent on the orientation of lattice and the lattice orientation is obtainable from a pair of integers n and m that denote the nanotube type(s). Typically, there are three types of nanotubes that include armchair ($n = m$), zig–zag ($n = x$, $m = 0$), and chiral ($n = x$, $m = y$). In addition to the electrical properties CNTs may exhibit either metallic or semiconducting properties depending upon the orientation of lattice. Zig–zag and armchair carbon nanotubes show metallic properties, while chiral nanotubes can be either metallic or semiconducting depending upon the difference between n and m units.[4] Due to the combination of delocalization of electrons across the lattice and the small dimensions in the radial axis constraining movement of charge carriers along the longitudinal axis of tubes, CNTs exhibit also exceptional charge carrier mobilities.

8.3 CNTs AS ENERGY STORAGE MATERIALS

8.3.1 RENEWABLE ENERGY SOURCES

8.3.1.1 HYDROGEN STORAGE

Carbon nanotubes and carbon nanofibers show high hydrogen capacity despite their relatively small surface area and pore volume. CNTs are good hydrogen absorbents with hollow structures and lightweight and can hold hydrogen at a higher density than liquid or solid hydrogen. The hydrogen which is stored inside the carbon nanotube can be gradually released and used as an energy by thermal control.[10] The packing geometry of SWCNTs plays an important role in hydrogen absorption.[5] Hydrogen is stored in the pores formed by the inner tube cavities and inter tube space of single-walled carbon nanotubes and storage density is possibly higher than that of planar graphene surface. The simulations of William et al suggest that the inner tube cavity of SWCNT has a high absorption potential for hydrogen, compared with planar surface and slit pores of similar size. SWCNTs have a very narrow diameter distribution and have dimensions in order of range of carbon attraction interaction.[11] Ritschel et al conducted studies on the hydrogen storage capacity of different carbon nanostructures SWNTs, MWNTs, and CNFs and found out that purified SWNTs showed a reversible storage capacity of 0.63 wt% at room temperature and 45 bar, higher than that of MWNTs and CNFs.

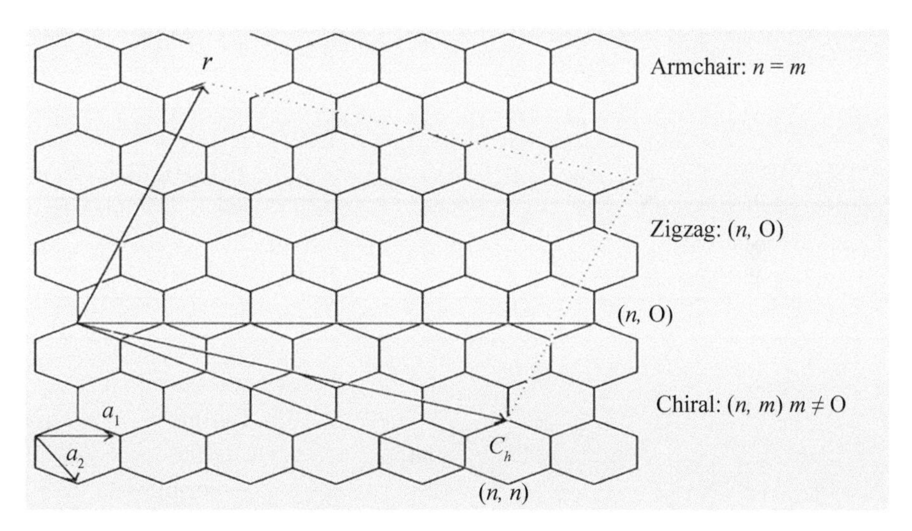

FIGURE 8.4 Structure of carbon nanotube.

8.3.1.2 SOLAR CELL

Solar cell or photovoltaic cell is an electronic device that converts the light energy directly into electrical energy by the photovoltaic effect. Researchers have studied upon solar cells that use graphene as an electrode and bulky balls and carbon nanotubes to absorb light and generate electrons. The unique properties of carbon nanotubes have been explored in light-harvesting and photovoltaic devices. CNTs are one of the most widely explored nanomaterials for photochemical cells because of their extraordinary optical and electronic properties arising from quantum confinement of unique zero band gap graphene structure.[12] CNTs are embedding into solar cells so as to combine them with a related polymer that donates electrons as a solution.[5] Single-walled carbon nanotubes are ideal thin-film photovoltaic materials which are lighter, more flexible, and cheaper to make. For the generation of electric current electron and hole which are formed by the absorption of photons from sunlight must be rapidly separated before the two particles have a chance to come back together and be reabsorbed into the material. SWCNTs which are capable of absorbing light across a wide range of wavelengths from visible to near infrared possess charge carriers (electrons and holes) that move quickly.[12]

TABLE 8.1 Properties of Single-Walled and Multiwalled Nanotubes.

Properties	Units	SWCNTs	MWCNTs
Specific gravity (bulk)	g/cm^3	0.8–1.3	1.8–2.6
Specific area	m^2/g	400-900	200–400
Young's modulus	Pa	~1,000	~1,000
Tensile strength	Pa	(3–50) × 10^{10}	(1–15) × 10^{10}
Thermal conductivity	W/m/K	3,000–6,000	2,000–3,000
Electrical conductivity	S/cm	10^2–10^6	10^3–10^5
Thermal stability temperature in air	°C	550–650	550–650

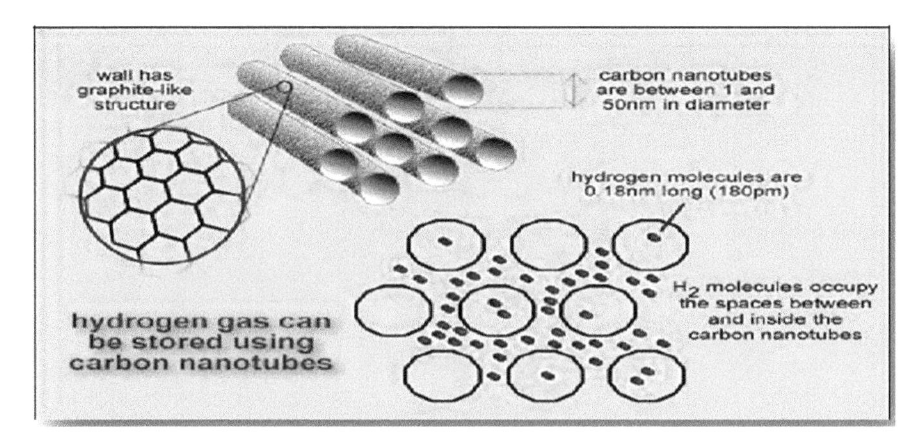

FIGURE 8.5 Mechanism of CNT for hydrogen storage

Source: Reprinted from Ref. [96]. Open access. journal.library.iisc.ernet.in

8.3.2 ENERGY STORAGE DEVICES

8.3.2.1 LITHIUM-ION BATTERIES

Lithium-ion battery is a type of rechargeable battery composed of cells in which lithium ions move from negative electrode to positive electrode through an electrolyte during discharging and back when charging. Lithium-ion batteries which possess high energy density and high open-circuit voltage consist of a negative electrode, a positive electrode, and a conducting electrolyte and store electrical energy in two electrodes in the form of Li- intercalation compounds.[13] The high surface area, quick movement of mobile species, enhanced ionic and electronic conductivity, mechanical resilience, and improved surface reactivity possessed by the nanostructured materials make it suitable as electrodes in LiBs. Graphite has been commonly used as an anode material for LIBs due to its high electronic (in plane) conductivity. However, the capacity of LiBs based on graphite is theoretically limited because of the intercalation of Li into graphite which involves one atom lithium per six carbon atoms. CNTs being an allotrope of graphite due to their unique structures and properties show improved lithium capacity compared to graphite.[5] The unique structural, mechanical, and electrical properties exhibited by carbon nanotubes make it the most suitable anode material for lithium-ion batteries. The enriched chirality of CNTs and the opened structure of CNTs will improve the capacity and electrical transport in CNT-based LiBs.[13]

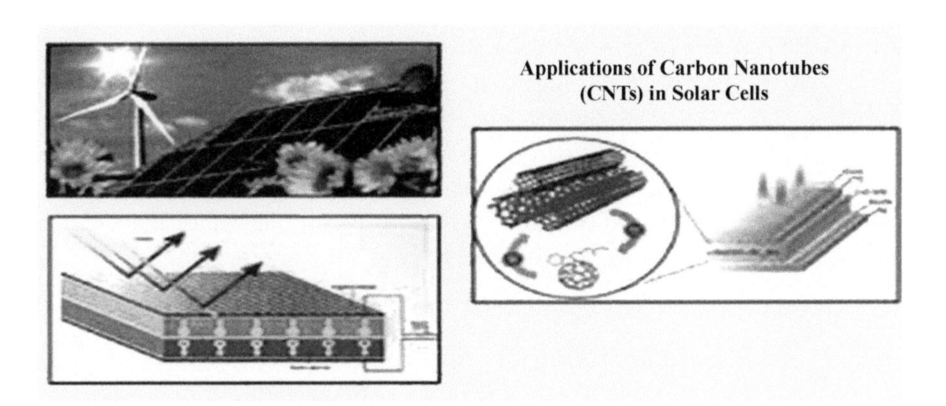

FIGURE 8.6 Application of CNTs in solar cells.

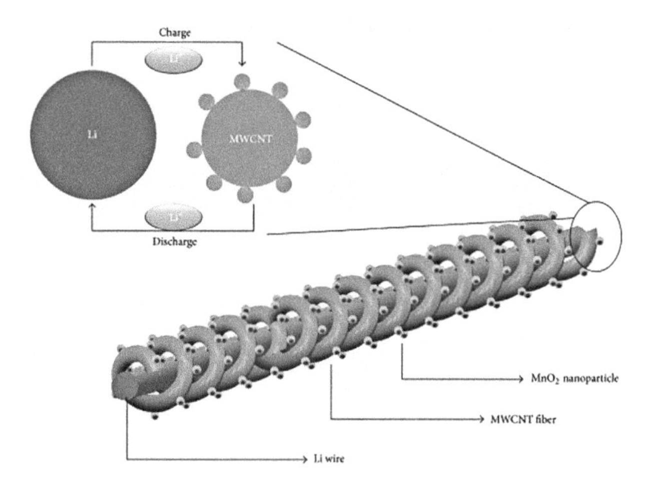

FIGURE 8.7 Application of CNTs in Li-ion batteries.

8.3.2.2 SODIUM-ION BATTERIES

Sodium with its large radius (1.02 A^0) cannot be intercalated comfortably into the layered structure of CNTs; hence, CNTs cannot be used as anode for Na-ion batteries. For enhancing the sodium storage as an anode for sodium-ion batteries, the defect-rich and disordered carbon nanostructures have been synthesized. The doping of heteroatoms such as nitrogen will yield defective CNTs which also enhances the electrical conductivity of carbon nanotubes. A coaxial heterostructure composed of MoS_2 nanosheets coated on carbon nanotubes (CNTs) is reported, where the MoS_2 layers are less stacked at the surface of the CNTs substrate with an enhanced interlayer spacing, which is prepared through a simple one-pot hydrothermal preparation[14] (Figure 8.8).

Due to the pulverization during the charge–discharge cycle most of the alloying and conversion anode materials lose their electron-conducting paths. In those cases CNTs can be used as conductive additive as well as an electrode integrity protector. The battery research community has been encapsulated metal-based anode with CNTs to accommodate the volume expansion during Na insertion to avoid the pulverization.[5]

8.3.2.3 METAL AIR BATTERIES

Metal air batteries have high demand on battery community because of their high energy efficiency, high theoretical density, and lower cost. Metal air batteries are based on the electrochemical coupling of a reactive anode to an electrode open to the air, which is the cathode active material.[5] Carbon nanotubes simultaneously display high electrical conductivities, high specific surface area, and good stability with lattice volume expansion during the charge discharge process. CNTs which possess porous structures with high specific surface area enable rapid electrolyte diffusions and charge transfers which are beneficial for fast charge and discharge. The high electrical conductivity of carbon nanotubes facilitates charge transfer and high specific surface area provides channel for oxygen diffusion and electrolyte. CNTs have been mostly used as a conductive supporting material for metal and metal oxide catalyst particles in metal air batteries. The functionalized CNTs are also employed as electrodes. The most promising metal air batteries include lithium air and zinc air batteries. Rechargeable lithium air batteries are having fairly high energy density comparable to gasoline and much higher than secondary lithium-ion batteries. Zinc air batteries are very safe for electrical vehicles which are fabricated by nonflammable and nonexplosive materials.[5]

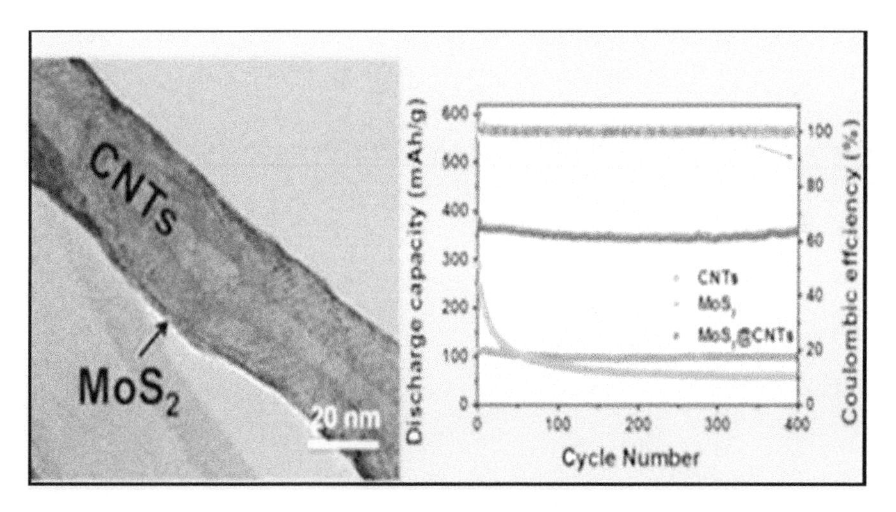

FIGURE 8.8 MoS_2 nanosheets coated on carbon nanotubes by simple one-pot hydrothermal method.

8.3.2.4 SUPERCAPACITOR

Better reversibility, higher power density, and longer life cycle exhibited by the supercapacitors made them attentive and promising for energy-storage devices. Performance of a supercapacitor is highly dependent on the fabrication and morphology of electrode materials. The electrode surface area and porosity of supercapacitor decides the capacitance value of supercapacitor.[5] CNTs have displayed great potential as electrode materials for developing high-performance supercapacitors owing to the properties of carbon nanotubes such as high electrical conductivity, high specific surface area, high charge transport capability, high mesoporosity, and high electrolyte accessibility. Carbon nanotubes with high capacitance are utilized to improve the performance of supercapacitors.[15] Excellent electrical conductivity, high mesoporosity, and high electrolyte accessibility of CNTs result in a high charge transport capability. The formation of defects on surface and open ends by alkaline solution activation increases the surface area of CNTs and exhibits a better capacitance value.[5] The large surface area of single-walled carbon nanotubes is the reason for their enhanced specific capacitance over those of multiwalled carbon nanotubes. However, MWCNTs due to the presence of mesopores and entangled tube structure, facilitating the transport of the ions, could generate capacitance twice as high in comparison to SWCNTs. SWCNTs with better electrical conductivity and high surface area are beneficial for the capacitor applications. Since contact resistance may reduce the performance of supercapacitor polished metal foils are used as current collectors to grow the carbon nanotubes.[5]

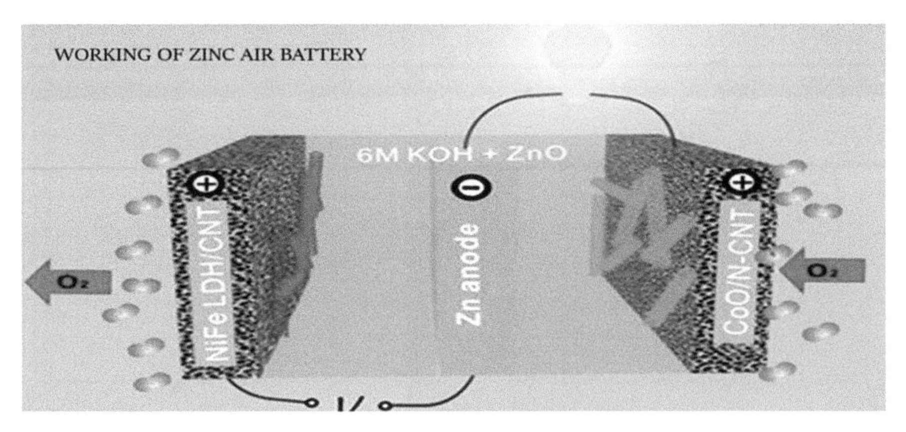

FIGURE 8.9 Schematic of working principle of zinc air battery.

8.3.2.5 FUEL CELLS

Fuel cell is a device that directly transforms chemical energy into electrical energy. The most effective catalyst employed to set out the electrochemical reaction in a fuel cell is Platinum. The effectiveness of platinum catalyst highly depends upon the size of the platinum particles and its distribution pattern over the support structures.[17] For improving the performance of catalyst in the fuel cell, carbon nanotubes have been added to the platinum/carbon catalyst mixture at the anode. Carbon nanotubes can decrease the needs of noble metals which are used as catalyst and improve the performance of fuel cells. Platinum can be fixed in the inner and outer walls of CNTs which constitutes the Pt/CNT structure with good electrocatalytic properties. In addition to the enhancement of catalyst performance, CNTs are capable of enhancing the corrosion resistance and catalyst steadiness.[17]

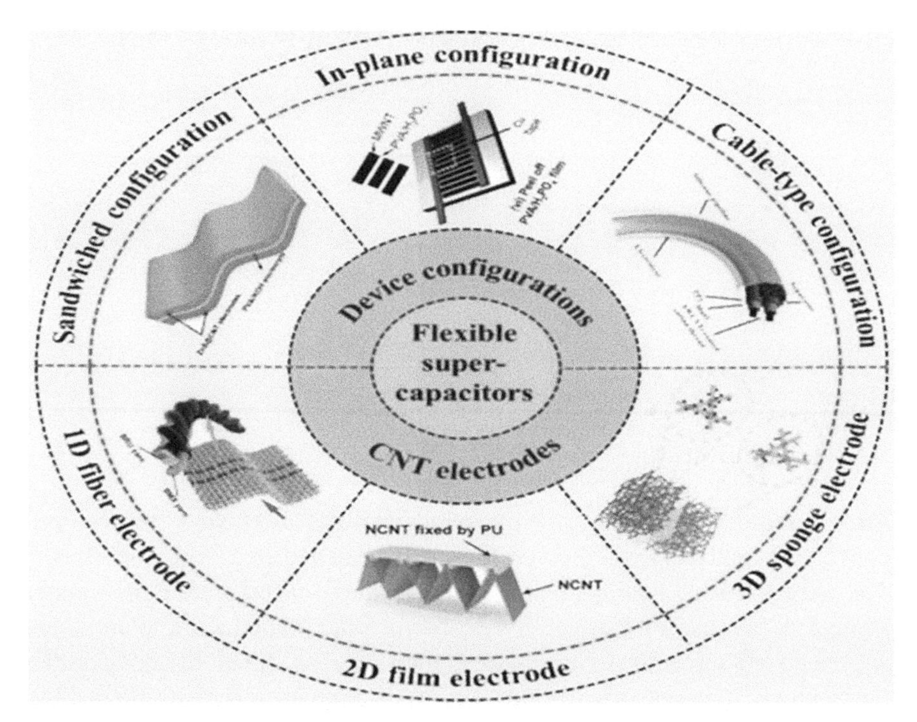

FIGURE 8.10 Applications of supercapacitors.

CNTs can effectively reduce the use of Pt and improve the performance of catalysts. In Figure 8.11 curves a and b correspond to two catalysts in hydrogen fuel cells. Pt/MWCNTs have a better power density and hence a higher Pt loading will yield a better power density. Curves c and d show the performance of Pt/C black and Pt/CNT respectively. The power density of the Pt/CNT electrode is much higher when the current is not too high. Since the conversion efficiencies of different fuel cells are different, we obtain different current densities.[17] Therefore, use of catalyst support has an important significance in fuel cell applications.

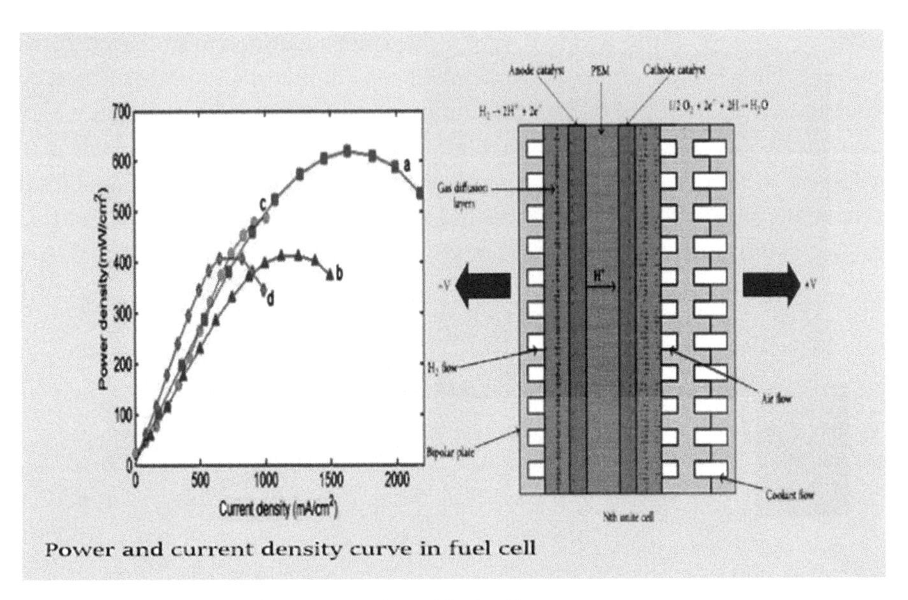

FIGURE 8.11 Schematic representation of fuel cells.

8.4 CONCLUSION

Nanostructured materials are of great interest in energy storage and conversion field due to their favorable mechanical and electrical properties.[1,2] The major reason behind using CNTs as ideal nanomaterial for energy storage is because of their ability to perform as a conductive component and electrode material. CNTs possess their own excellent advantages like lightweight, high electrical conductivity, good physical robustness, and superior electrochemical performance compared with amorphous carbon, carbon black, hard carbon, and other carbon materials that are used in traditional LiBs and SCs.[16] The architecture and quality of CNTs play a vital role in the electrochemical performance exhibited by both batteries and supercapacitors. CNTs can be used as an electrochemically active and inactive electrode component in energy storage devices. CNTs with flexible supporting materials can enable next-generation flexible energy storage devices. Lack of commercially feasible synthesis and purification methods are the main reasons that carbon nanotubes are not still widely used.[3] Many researchers are working for developing new and improved synthesis and treatment methods to modify the performance of CNT-based electrodes in supercapacitors and LIBs. CNTs combined with lithium storage compounds such as metals and inorganic materials enhance the performance of LIBs in terms of specific capacity, stability, and long cycle life. The application of CNTs in the emerging high energy density rechargeable batteries like lithium sulfur and lithium air batteries contributes a major area of research.

KEYWORDS

- **carbon nanotube**
- **lithium-ion battery**
- **sodium-ion battery**
- **hydrogen storage**
- **fuel cell**
- **supercapacitor**
- **solar cell**

REFERENCES

1. Antiohos, D.; Romano, M.; Chen, J.; Razal, J. M. *Carbon Nanotubes for Energy Applications*, 22; pp 495–537.
2. Gupta, N.; Gupta, S. M.; Sharma, S. K. Carbon Nanotubes: Synthesis, Properties and Engineering Applications. Korean Carbon Society 2019. *Carbon Lett.* 2233–4998.
3. Akbari, E.; Buntat, Z. Benefits of Using Carbon Nanotubes in Fuel Cells. A Review. *Int. J. Energy Res.* **2016**. DOI: 10.1002/er.3600
4. Baughman, R. H.; et al. Carbon Nantubes—the Route Toward Application. *Science* **2002,** 297, 297–787.
5. Amin, R.; Kumar, P. R.; Belharouak, I. Crbon Nanotubes: *Applications to Energy Storage Devices*. IntechOpen. DOI: http://dx.doi.org/10.5772/intechopen.94155.
6. Scoville, C.; Cole, R.; Hogg, J.; Farooque, O.; Russell, A. *Carbon Nanotubes.*
7. Kumar, S.; et al. Carbon Nanotubes: A Potential Material for Energy Conversion and Storage. *Prog. Energy Combust. Sci.* **2017,** 1–35.
8. De Volder, M. F.; Tawfick, S. H.; Baughman, R. H.; Hart, A. J. Carbon Nanotubes: Present and Future Commercial Applications. *Science 339* (6119), 535–539.
9. Lan, Y.; Wang, Y.; Ren, Z. F. Physics and Applications of Aligned Carbon Nanotubes. *Adv. Phys.* **2011,** *60* (4), 553–678.

10. Kachhi, B.; Singh, K. K.; Singh, A.; Sharma, D. K.. Role of Carbon Nanotubes as Energy Storage Materials. *Int. J. New Chem.* **2022**. DOI: 10.22034/ijnc.2022.3.4
11. Cheng, H.-M.; Yang, Q.-H.; Liu, C. Hydrogen Storage In Carbon Nanotubes. *Carbon* **2001,** *39*, 1447–1454.
12. Baker, B. A.; Zhang, H.; Cha, T.-G.; Choi, J. H. *Carbon Nanotube Solar Cells*. Woodhead Publishing Limited, 2013.
13. Xiong, Z.; Yun, Y. S.; Jin, H.-J. Application of Carbon nanotubes for Lithium Ion Battery Anodes. *Materials* **2013,** *6,* 1138–1158.
14. Li, W.; Bashir, T.; Wang, J.; Zhou, S.; Yang, S.; Zhao, J.; Gao, L. *Enhanced Sodium Ion Storage Performance of a 2D MoS_2 Anode Material Coated on Carbon Nanotubes* [Online]. https://doi.org/10.1002/celc.202001486.
15. Lu, W. *Carbon Nanotube Supercapacitors* [Online]. Available at http://www.intechopen.com/books/carbon-nanotubes/carbon-nanotube- supercapacitors
16. Sun, L.; et al. Role of Carbon Nanotubes inNovel Energy Storage Devices. *Carbon* **2017,** *122,* 462–474.
17. Akbari, E.; Buntat, Z. Benefits of Using Carbon Nanotubes in Fuel cells: A Review. *Int. J. Energy Res.* DOI: 10.1002/er.3600
18. Iijima, S. Helical Microtubules of Graphitic Carbon. *Nature* **1991,** *354,* 56–58.
19. Arico, A. S.; et al. Nanostructured Materials for Advanced Energy Conversion and Storage Devices. *Nat. Mater.* **2005,** *4* (5), 366–377.
20. Belin, T.; Epron, F. Characterization Methods of Carbon Nanotubes: A Review. *Mater. Sci. Eng. B* **2005,** *119,* 105–118.
21. Varshney, K. Carbon Nanotubes: A Review on Synthesis, Properties, and Applications. *Int. J. Eng. Res. General sci.* **2014,** *2,* 660–677.
22. Kaushik, B. K.; Majumder, M. K. Carbon Nanotube-Based VLSI Interconnects. In *Springer Briefs in Applied Sciences and Technology*, 2015. DOI: 10.1007/978-81-322-2047-3
23. Kierzek, K.; Frackowiak, E.; Lota, G.; Gryglewicz, G.; Machnikowski, J. Electrochemical Capacitors Based on Highly Porous Carbons Prepared by KOH Activation. *Electrochimica Acta* **2004,** *49,* 515–523.
24. Banks, C. E.; Compton, R. G. New Electrodes for Old: From Carbon Nanotubes to Edge Plane Pyrolytic Graphite. *Analyst* **2006,** *131,* 15–21.
25. Thostenson, E. T.; Ren, Z.; Chou, T.-W. Advances in the Science and Technology of Carbon Nanotubes and their Composites: A Review. *Compos.Sci. Technol.* **2001**. https://www.sciencedirect.com/science/article/abs/pii/S0266353801000 94X.
26. Tserpes, K.; Papanikos, P. Finite Element Modeling of Single-Walled Carbon Nanotubes. *Compos. B: Eng.* **2005,** *36,* 468–477.
27. Xiong, Z.; Yun, Y. S.; Jin, H.-J. Applications of Carbon Nanotubes forLithium Ion Battery Anodes. *Materials* **2013,** *6* (3).
28. de las Casas, C.; Li, W. A Review of Application of Carbon Nanotubes for Lithium Ion Battery Anode Material. *J. Power Sources* **2012,** *208,* 74–5.
29. Shimoda, H.; Gao, B.; Tang, X. P.; Kleinhammes, A.; Fleming, L.; Wu, Y.; et al. Lithium Intercalation into Opened Single-Wall Carbon Nanotubes: Storage Capacity and Electronic Properties. *Phys. Rev. Lett.* **2001,** *88* (1), 015502.
30. Nishidate, K.; Hasegawa, M. Energetics of Lithium Ion Adsorption on Defective Carbon Nanotubes. *Phys. Rev. B* **2005,** *71* (24), 245418.
31. Mi, C. H.; Cao, G. S.; Zhao, X. B. A Non-GIC Mechanism of Lithium Storagein Chemical Etched MWNTs. *J. Electroanal. Chem.* **2004,** *562* (2), 217-221.

32. Yang, S.; Huo, J.; Song, H.; Chen, X. A Comparative Study of Electrochemical Properties of Two Kinds of Carbon Nanotubes as Anode Materials for Lithium Ion Batteries. *Electrochim Acta.* **2008,** *53* (5), 2238–2244.
33. Meunier, V.; Kephart, J.; Roland, C.; Bernholc, J. *Ab Initio* Investigations of Lithium Diffusion in Carbon Nanotube Systems. *Phys. Rev. Lett.* **2002,** *88* (7), 075506.
34. Prem, K. T.; Ramesh, R.; Lin, Y. Y.; Fey, G. T.-K. Tin-Filled Carbon Nanotubesas Insertion Anode Materials for Lithiumionbatteries. *Electrochem. Commun.* **2004,** *6* (6), 520–525.
35. Wang, X. X.; Wang, J. N.; Chang, H.; Zhang, Y. F. Preparation of Short Carbon Nanotubes and Application as an Electrode Material in Li-Ion Batteries. *Adv. Funct. Mater.* **2007,** *17* (17), 3613–3618.
36. Garau, C.; Frontera, A.; Quiñoncro, D.; Costa, A.; Ballester, P.; Deyà, P. M. *Ab initio* Investigations of Lithium Diffusion in Single-Walled Carbon Nanotubes. *Chem. Phys.* **2004,** *297* (1), 85–91.
37. Eom, J. –Y.; Kwon, H. -S. Effects of the Chemical Etching of Single-Walled Carbon Nanotubes on their Lithium Storage Properties. *Mater. Chem. Phys.* **2011,** *126* (1), 108–113.
38. Wang, L.; Guo, W.; Lu, P.; Zhang, T.; Hou, F.; Liang, J. A Flexible and Boron-Doped Carbon Nanotube Film for High-Performance Li Storage. *Front. Chem.* **2019,** *7* (832).
39. Akbari, E.; Buntat, Z. Benefits of Using Carbon Nanotubes in Fuel Cells: A Review. *Int. J. Energy Res.* **2017,** *41* (1), 92–102.
40. Landi, B. J.; Ganter, M. J.; Cress, C. D.; DiLeo, R. A.; Raffaelle, R. P. Carbon Nanotubes for Lithium Ion Batteries. *Energy Environ. Sci.* **2009,** *2* (6), 638–654.
41. Pan, H.; Li, J.; Feng, Y.; Carbon Nanotubes for Supercapacitor. *Nanoscaleresearch Lett.* **2010,** *5* (3), 654–668.
42. Cheng, H.-M.; Yang, Q.-H.; Liu, C. Hydrogen Storage in Carbon Nanotubes. *Carbon* **2001,** *39* (10), 1447–1454.
43. Dresselhaus, M. S.; Dresselhaus, G.; Saito, R. Physics of Carbon Nanotubes. *Carbon* **1995,** *33* (7), 883–891.
44. Monea, B. F.; Ionete, E. I.; Spiridon, S. I.; Ion-Ebrasu, D.; Petre, E. Carbon Nanotubes and Carbon Nanotube Structures Used for Temperature Measurement. *Sensors* **2019,** *19* (11).
45. Amelinckx, S.; Lucas, A.; Lambin, P. Electron Diffraction and Microscopy of Nanotubes. *Rep. Prog. Phys.* **1999,** *62* (11), 1471–1524.
46. Belin, T.; Epron, F. Characterization Methods of Carbon Nanotubes: A Review. *Mater. Sci. Eng. B* **2005,** *119* (2), 105–118.
47. Iijima, S.; Ichihashi, T. Single-Shell Carbon Nanotubes of 1-nm Diameter. *Nature* **1993,** *363* (6430), 603–605.
48. Yu, M.-F.; Lourie, O.; Dyer, M. J.; Moloni, K.; Kelly, T. F.; Ruoff, R. S. Strength and Breaking Mechanism of Multiwalled Carbon Nanotubes Under Tensile Load. *Science* **2000,** *287* (5453), 637.
49. Ando, Y.; Zhao, X.; Shimoyama, H.; Sakai, G.; Kaneto, K. Physical Properties of Multiwalled Carbon Nanotubes. *Int. J. Inorg. Mater.* **1999,** *1* (1), 77–82.
50. Zhang, Y.; Bunes, B. R.; Wu, N.; Ansari, A.; Rajabali, S.; Zang, L. Sensing Methamphetamine with Chemiresistivesensors based on Polythiopheneblended Single-Walled Carbon Nanotubes. *Sens. Actuators B: Chem.* **2018,** *255,* 1814–1818.
51. Shobin, L. R.; Manivannan, S. Silver Nanowires-Single Walled Carbon Nanotubes Heterostructurechemiresistors. *Sens. Actuators B: Chem.* **2018,** *256,* 7–17.

52. Rakhi, R. B.; Sethupathi, K.; Ramaprabhu, S. Electron Field Emission Properties of Conducting Polymer Coated Multi Walled Carbon Nanotubes. *Appl. Surf. Sci.* **2008,** *254* (21), 6770–6774.
53. Yan, Y.; Miao, J.; Yang, Z.; Xiao, F.-X.; Yang, H. B.; Liu, B.; et al. Carbon Nanotube Catalysts: Recent Advances in Synthesis, Characterization and Applications. *Chem. Soc. Rev.* **2015,** *44* (10), 3295–3346.
54. Saito, S. Carbon Nanotubes for Next-Generation Electronics Devices. *Science* **1997,** *278* (5335), 77.
55. Hodge, S. A.; Bayazit, M. K.; Coleman, K. S.; Shaffer, M. S. P. Unweaving the Rainbow: A Review of the Relationship Between Single-Walled Carbon Nanotube Molecular Structures and their Chemical Reactivity. *Chem. Soc. Rev.* **2012,** *41* (12), 4409–4429.
56. Zandiatashbar, A.; Lee, G.-H.; An, S. J.; Lee, S.; Mathew, N.; Terrones, M.; et al. Effect of Defects on the Intrinsic Strength and Stiffness of Graphene. *Nat. Comm.* **2014,** *5* (1), 3186.
57. Terrones, H.; Lv, R.; Terrones, M.; Dresselhaus, M. S. The Role of Defects and Doping in 2D Graphene Sheets and 1D Nanoribbons. *Rep. Prog. Phys.* **2012,** *75* (6), 062501.
58. Wei, B. Q.; Vajtai, R.; Ajayan, P. M. Reliability and Current Carrying Capacity of Carbon Nanotubes. *Appl. Phys. Lett.* **2001,** *79* (8), 1172–1174.
59. Pugno, N. M. The Role of Defects in the Design of Space Elevator Cable: From nanotube to Megatube. *Acta Mater.* **2007,** *55* (15), 5269–5279.
60. Robertson, D. H.; Brenner, D. W.; Mintmire, J. W. Energetics of Nanoscale Graphitic Tubules. *Phys. Rev. B.* **1992,** *45* (21), 12592–12595.
61. Hone, J.; Batlogg, B.; Benes, Z.; Johnson, A. T.; Fischer, J. E. Quantized Phonon Spectrum of Single-Wall Carbon Nanotubes. *Science* 2000**,** *289* (5485), 1730.
62. Yamashita, S. Nonlinear Optics in Carbon Nanotube, Graphene, and Related 2D Materials. *APL Photon.* **2018,** *4* (3), 034301.
63. Matsuda, K. 1 - Fundamental Optical Properties of Carbon Nanotubes and Graphene. In *Carbon Nanotubes and Graphene for Photonic Applications*; Yamashita. S., Saito, Y., Choi, J. H., Eds.; Woodhead Publishing, 2013; pp 3–25.
64. Eder, D. Carbon Nanotube–Inorganic Hybrids. *Chem Rev.* **2010,** *110* (3), 1348–1385.
65. Xiong, Z.; Yun, Y. S.; Jin, H.-J. Applications of Carbon Nanotubes for Lithium Ion Battery Anodes. *Materials.* **2013,** *6* (3).
66. de las Casas, C.; Li, W. A Review of Application of Carbon Nanotubes for Lithium Ion Battery Anode Material. *J. Power Sources* **2012,** *208,* 74–85.
67. Shimoda, H.; Gao, B.; Tang, X. P.; Kleinhammes, A.; Fleming, L.; Wu, Y.; et al. Lithium Intercalation into Opened Single-Wall Carbon Nanotubes: Storage Capacity and Electronic Properties. *Phys. Rev. Lett.* 2001, *88* (1), 015502.
68. Nishidate, K.; Hasegawa, M. Energetics of Lithium Ion Adsorption on Defective Carbon Nanotubes. *Phys. Rev. B* **2005,** *71* (24), 245418.
69. Mi, C. H.; Cao, G. S.; Zhao, X. B. A Non-GIC Mechanism of Lithium Storage in Chemical Etched MWNTs. *J. Electroanal. Chem.* **2004,** *562* (2), 217–221.
70. Yang, S.; Huo, J.; Song, H.; Chen, X. A Comparative Study of Electrochemical Properties of Two Kinds of Carbon Nanotubes as Anode Materials for Lithium Ion Batteries. *Electrochim. Acta* **2008,** *53* (5), 2238–2244.
71. Meunier, V.; Kephart, J.; Roland, C.; Bernholc, J. *Ab Initio* Investigation of Lithium Diffusion in Carbon Nanotube Systems. *Phys. Rev. Lett.* **2002,** *88* (7), 075506.

72. Prem Kumar, T.; Ramesh, R.; Lin, Y. Y..; Fey, G. T.-K. Tin-Filled Carbon Nanotubes as Insertion Anode Materials for Lithiumion Batteries. *Electrochem. Commun.* **2004,** *6* (6), 520–525.
73. Sun, J.; Huang, Y.; Fu, C.; Wang, Z.; Huang, Y.; Zhu, M.; et al. Highperformance Stretchable Yarn Supercapacitor based on PPy@CNTs@ Urethane Elastic Fiber Core Spun Yarn. *Nano Energy* **2016,** *27,* 230–237.
74. Song, C.; Yun, J.; Keum, K.; Jeong, Y. R.; Park, H.; Lee, H.; et al. High Performance Wire-Type Supercapacitor with Ppy/CNT-Ionic Liquid/AuNP/Carbon Fiber Electrode and Ionic Liquid based Electrolyte. *Carbon* **2019,** *144,* 639–648.
75. Tan, L.; Liu, Z.-Q.; Li, N.; Zhang, J.-Y.; Zhang, L.; Chen, S. CuSe Decorated Carbon Nanotubes as a High Performance Cathode Catalyst for Microbial Fuel Cells. *Electrochim. Acta* **2016,** *213,* 283–290.
76. Deng, D.; Yu, L.; Chen, X.; Wang, G.; Jin, L.; Pan, X.; et al. Iron Encapsulated within Pod-like Carbon Nanotubes for Oxygen Reduction Reaction. *Angew. Chem. Int. Ed.* **2013,** *52* (1), 371–375.
77. Wen, L.; Liming, D. Carbon Nanotube Supercapacitors. 2010.
78. Yun, Y. S.; Yoon, G.; Kang, K.; Jin, H.-J. High-Performance Supercapacitors Based on Defect-Engineered Carbon Nanotubes. *Carbon* **2014,** *80,* 246–254.
79. Frackowiak, E.; Jurewicz, K.; Delpeux, S.; Béguin, F. Nanotubular Materials for Supercapacitors. *J. Power Sour.* **2001,** *97–98,* 822–825.
80. Futaba, D. N.; Hata, K.; Yamada, T.; Hiraoka, T.; Hayamizu, Y.; Kakudate, Y.; et al. Shape-Engineerable and Highly Densely Packed Single-Walled Carbon Nanotubes and Their Application as Super-Capacitor Electrodes. *Nature materials* **2006,** *5* (12), 987–994.
81. An, K. H.; Kim, W. S.; Park, Y. S.; Choi, Y. C.; Lee, S. M.; Chung, D. C.; et al. Supercapacitors Using Single-Walled Carbon Nanotube Electrodes. *Adv. Mater.* **2001,** *13* (7), 497–500.
82. Du, C.; Yeh, J.; Pan, N. High Power Density Supercapacitors Using Locally Aligned Carbon Nanotube Electrodes. *Nanotechnology* **2005,** *16* (4), 350–353.
83. Belin, T.; Epron, F. Characterization Methods of Carbon Nanotubes: A Review. *Mater. Sci. Eng. B* **2005,** *119,* 105–118.
84. Rahmandoust, M.; Öchsner, A. Buckling Behaviour and Natural Frequency of Zigzag and Armchair Single-Walled Carbon Nanotubes. *Nano Res.* **2012,** ***16,*** 153–160.
85. Science and Technology of Carbon Nanotubes and their Composites: A Review. *Compos. Sci. Technol.* **2001**. https://www.sciencedirect.com/science/article/abs/pii/S026635380100094 X
86. Wong, K. V.; Bachelier, B. Carbon Nanotubes Used for Renewable Energy Applications and Environmental Protection/Remediation: A Review. *J. Energy Resour. Technol.* **2013,** *136.*
87. Sgobba, V.; Guldi, D. M. Carbon Nanotubes as Integrative Materials for Organic Photovoltaic Devices. *J. Mater. Chem.* **2008,** *18,* 153–157.
88. Cataldo, S.; Salice, P.; Menna, E.; Pignataro, B. Carbon Nanotubes and Organic Solar Cells. *Energy Environ. Sci.* 2011. https://pubs.rsc.org/en/content/articlelanding/2012/EE/C1EE02276H
89. Scharber, M. C.; et al. *Design Rules for Donors in Bulk-Heterojunction Solar Cells-Towards 10 % Energy-Conversion Efficiency*. Wiley Online Library. 2006. https://onlinelibrary.wiley.com/doi/10.1002/adma.200501717

90. Fan, W.; Zhang, L.; Liu, T. *Graphene-Carbon Nanotube Hybrids for Energy and Environmental Applications*; Ghent University Library, 1970. https://lib.ugent.be/catalog/ebk01:3710000000943925.
91. Cheng, H.; Shapter, J. G.; Li, Y.; Gao, G. Recent Progress of Advanced Anode Materials of Lithium-Ion Batteries. *J. Energy Chem.* 2020.https://www.sciencedirect.com/science/article/abs/pii/S2095495620306197
92. Wilson, I. A. G.; Hall, P.; Rennie, A. Energy Storage in Electrochemical Capacitors: Designing Functional Materials to Improve Performance. *Energy Environ. Sci.* **2016**. https://www.academia.edu/5480796/Energy_storage_in_electrochemical_capacitors_designing_functional_materials_to_improve_performance
93. Frackowiak, E.; Metenier, K.; Bertagna, V.; Beguin, F. Supercapacitor Electrodes from Multiwalled Carbon Nanotubes. *Appl. Phys. Lett.* **2000,** *77,* 2421–2423.
94. Li Li, Hui Yang, Dongxiang Zhou, Yingyue Zhou, "Progress in Application of CNTs in Lithium-Ion Batteries", Journal of Nanomaterials, vol. 2014, Article ID 187891, 8 pages, 2014. https://doi.org/10.1155/2014/187891.
95. Li W, Bashir T, Wang J, Zhou S, Yang S, Zhao J, Gao L. Enhanced Sodium-Ion Storage Performance of a 2D MoS2 Anode Material Coated on Carbon Nanotubes. ChemElectroChem. 2021 Mar 1;8(5):903-10.
96. Ranveer, A. C. 2015. Carbon Nanotubes and Its Environmental Applications. Journal of Environmental Science Computer Science and Engineering & Technology 4(2):304-311. Open access.

CHAPTER 9

Hydrogen Energy and Its Storage in 2D Nanomaterials: Insights from Density Functional Theory Simulations

BRINTI MONDAL[1] and BRAHMANANDA CHAKRABORTY[2,3]

[1]*Indian Institute of Technology (IIT), Bombay, Department of Energy Science and Engineering (DESE), India*

[2]*High Pressure and Synchrotron Radiation Physics Division, Bhabha Atomic Research Centre, Trombay, Mumbai, India*

[3]*Homi Bhabha National Institute, Mumbai, India*

ABSTRACT

One of the most difficult issues of the twenty-first century is to develop alternative fuel for automotive vehicles that can replace commonly utilized fossil fuels. Because of its pure energy and abundance in nature, hydrogen is one of the best options. Furthermore, when compared to liquid hydrocarbons (47 MJ kg^-1), it has the highest energy value per unit weight (142 MJ kg^-1). Because of their huge size and greater energy costs for liquefaction, standard hydrogen storage solutions, such as high-pressure tanks and liquid state storage, are inapplicable. Solid-state storage may become a feasible solution if the storage media can absorb a substantial amount of hydrogen (6.5 wt%) and release it readily (which can correspond to the binding energy range of -0.2 eV/H_2 to -0.7 eV/H_2, as advised by the United States Department of Energy (DoE)). Hence compared to storing hydrogen on metal or metal alloy surfaces via chemisorption, storing hydrogen on porous

Technological Advancement in Clean Energy Production: Constraints and Solutions for Energy and Electricity Development. Amritanshu Shukla, Kian Hariri Asli, Neha Kanwar Rawat, Ann Rose Abraham, & A. K. Haghi (Eds.)

carbon-based materials or on their analogous structures via physisorption is beneficial. Especially, 2D materials (graphene, graphyne, graphitic carbon, zeolite templated carbon, etc) that can utilize both surfaces simultaneously result in higher gravimetric hydrogen storage. But often these materials suffer from the issue of low desorption temperature. Research suggests high gravimetric wt% can be obtained using 2D materials if the surface of those materials can be decorated with either alkali or transition metals which store hydrogen via an interaction weaker than chemisorption, but stronger than physisorption (Kubas interaction), suitable for fuel cell applications as targeted.

In this chapter, we have discussed the recent advancement of metal-doped 2D nanomaterial-based hydrogen storage research using density functional theory (DFT) calculation. The benefit of using DFT is saving experimental costs and understanding the mechanism in a more detailed way. For this chapter, we have focused on the calculation of adsorption energy, calculation of gravimetric wt%, density of states analysis, charge transfer analysis, and often metal–metal clustering possibilities.

9.1 INTRODUCTION

9.1.1 NEED FOR ALTERNATE ENERGY

Fossil fuels have long dominated the world's energy sources since the industrial revolution. In the modern world, fossil fuels also continue to rule the roughly 1.5 trillion dollars global energy market as the primary supplier of day-to-day energy sources. According to the World Energy Outlook 2007, the energy produced from fossil fuels will still continue to be the main source and will still be needed to meet 84% of global energy demand in 2030.[1] However, there are two main problems associated with these conventional energy sources: they are finite, which means they could run out, and they release carbon dioxide, which is having a negative effect on the environment, and also their volatile prices which are influenced by political factors. If we take a look at the past 40 years of energy scenario, we can see the huge demand for fast and smart life that eventually increased the demand for energy leading to the depletion of natural resources, faster than interpreted. Kober et al.[2] showed in their study (Figure 9.1) that from 18.0 billion tonnes in 1978 to 33.7 billion tonnes in 2018, global energy-related carbon dioxide emissions increased by 87%. The increase in the use of nuclear and renewable energy led to a reduction in the carbon intensity of the global energy supply by 13%.

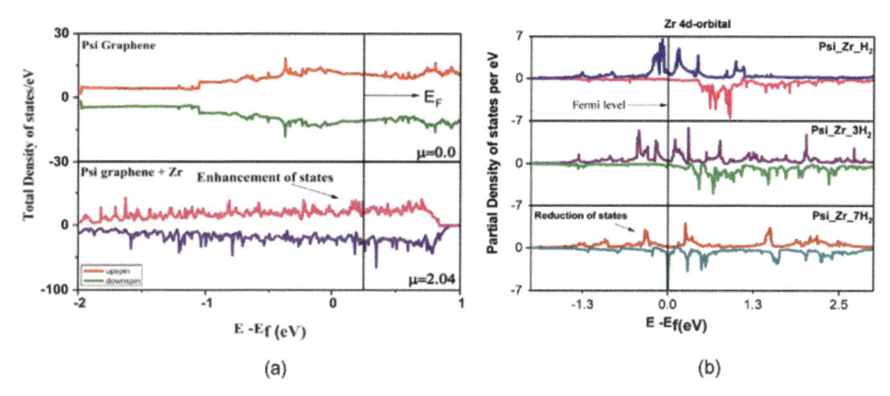

FIGURE 9.1 (a) TDOS of ψ-graphene and Zr+ ψ-graphene b) PDOS of Zr+ψ-graphene+H_2.
Source: Reproduced with permission from Ref. [77]

In essence, that encourages people to investigate alternative fuels for sustaining advancement, such as biodiesel, bio alcohol, hydrogen, etc. The primary drivers behind the development of alternative fuel technologies are specifically the increase in cost and harm to the environment caused by the depletion of fossil fuels.

9.1.2 ALTERNATE SOURCE OF ENERGY

From hydropower in the 19th century to nuclear energy in the 20th, there has long been a supply of alternative energy.[3] But today's alternative energy has a much wider range and is more widely applicable than ever before. We are in a transition period when alternative, clean energy is being developed and used more and more, and this will eventually change the definition of what constitutes conventional energy and the predominant energy used worldwide. There are multiple sources of alternating energy that also include renewables. Renewable energy sources like hydro, solar, wind, geothermal, biomass, etc are among the top considerations when discussing alternating energy sources. But the problems associated with this kind of energy source are, they are very occasional. But sources like nuclear and hydrogen do not suffer from these. Though the nuclear source is one of the cleanest modes of energy generation, there are severe risks associated with it, which have declined its use in reality.[3] So, the option that remains here is "Hydrogen" to be discussed further.

9.1.3 HYDROGEN ENERGY: MERIT AND DEMERIT

According to the IEA report of 2019, the demand for H_2 fuel has gone up 3-folds than in 1975. Its unique properties and versatile applicability have gained tremendous business and political attraction apart from the attraction of researchers and environmentalists, throughout the world (Figure 9.2).

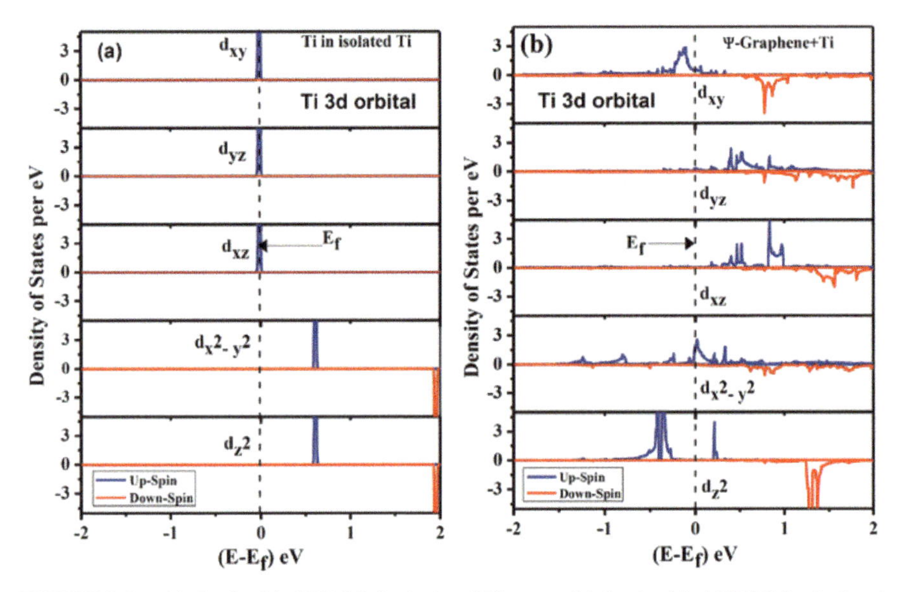

FIGURE 9.2 (a) d suborbital PDOS for isolated Ti atom. (b) d suborbital PDOS for isolated Ti+ ψ-graphene.

Source: Reproduced with permission from Ref. [78]

Like any other technology, it also suffers from several bottlenecks and drawbacks. In this section, we will discuss the merits and demerits of hydrogen energy.

The most important characteristics of hydrogen fuel to be used are:

- Combustion of this fuel produces water and not polluting carbon dioxide or carbon monoxide along with other polluting gaseous products.
- It has the highest energy density of 143 MJ/Kg compared to any other fuels.[4]

- One of the most notable characteristics of hydrogen fuel is its ability to regenerate, and it can be produced in practically infinite quantities using renewable energy sources.[5]
- Another distinctive characteristic of hydrogen is that it is not only a clean source of energy when used as a carrier, but also that it is produced in a clean way.[6]

Among the very few difficulties of replacing conventional fossil fuel with hydrogen is, there are sadly no practical innate reservoirs for molecular hydrogen on earth.[7] A large portion of the hydrogen present on the Earth is bonded to oxygen in water and to carbon in living, dead, and/or fossilized biomass. Thus, it is necessary to separate hydrogen from any of its compounds using a primary energy source, such as solar energy, biomass, electricity, gas, or hydrocarbons. Another associated issue is, due its lightweight, it is very difficult to store. Hence, there has been intense research focusing on these two criteria, which we will be discussing in the following sections.

9.1.4 HYDROGEN PRODUCTION

Finding a proper source for hydrogen which can meet the requirements of a fully developed hydrogen economy, and which does not require fossil fuels as feedstocks, is the principal problem associated with hydrogen production, due to its nonavailability in molecular form. Major amount of hydrogen is produced by reforming natural gas, which is not a very preferable option due to continuous use of limited natural resources and production of carbon dioxide—exactly the issues that we are trying to be saved from. Hence, a more mature technology is required as the demand is increasing.

Production of hydrogen by water splitting is a potentially desirable and renewable method to dominate hydrogen production in coming days.[8,9] Water-hydrogen is not only a cleaner transformation, but it is also reversible. But again, this technology has the problem of energy efficiency. More energy is required to split the water than the final recovery. So, another area of research is born from this section, which focuses only on energy efficient water splitting to produce hydrogen.[10] The roadmap of hydrogen production is shown in Figure 9.3, and the current hydrogen production technologies have been categorized in Figure 9.4.

According to the findings of the environmental performance rating (Figure 9.5) of the renewable resources of hydrogen production, the wind is the source with the lowest GHG emissions, while biomass has the highest

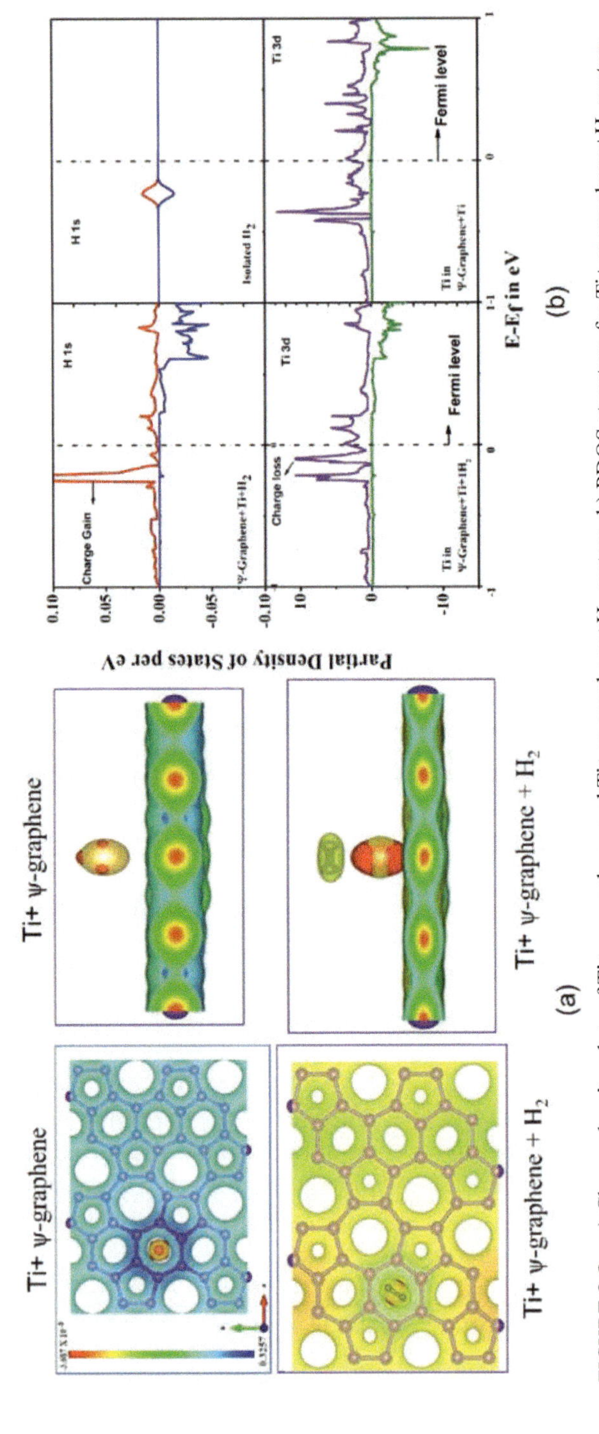

FIGURE 9.3 a) Charge density plots of Ti+ ψ-graphene and Ti+ ψ-graphene+H_2 system. b) PDOS structure for Ti+ ψ-graphene+H_2 system.

Source: Reproduced with permission from Ref. [78]

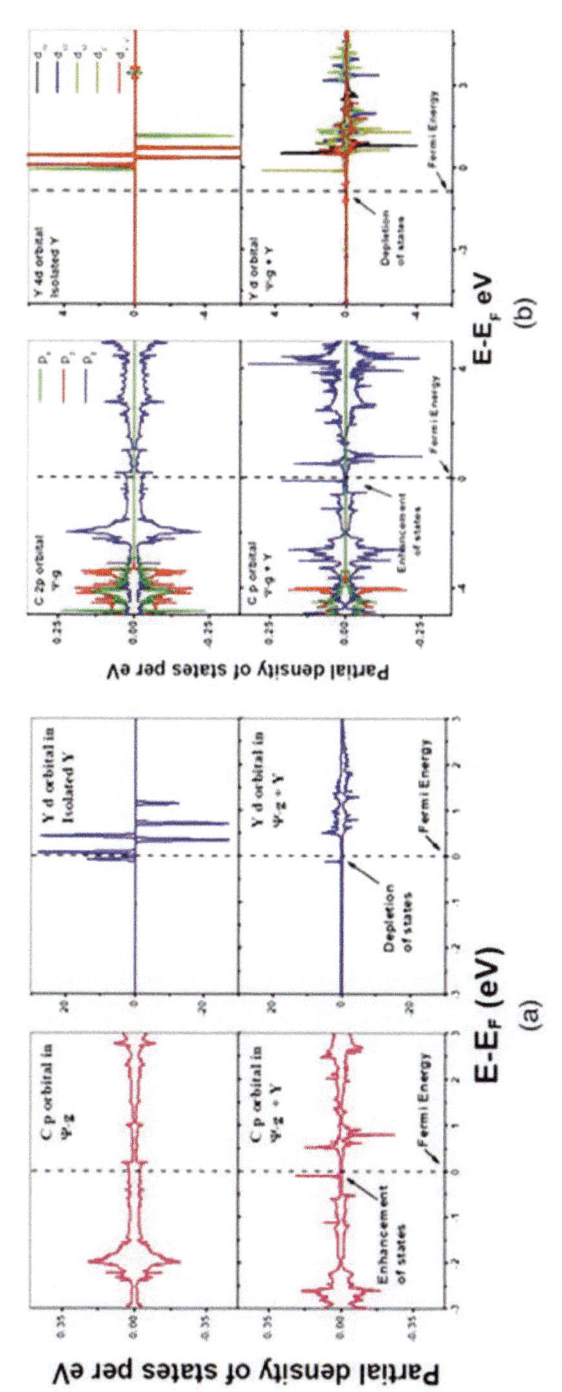

FIGURE 9.4 a) Interaction of Y and ψ-graphene described via PDOS. b) Suborbital analysis of interaction of Y and ψ-graphene described via PDOS.

Source: Reproduced with permission from Ref. [79]

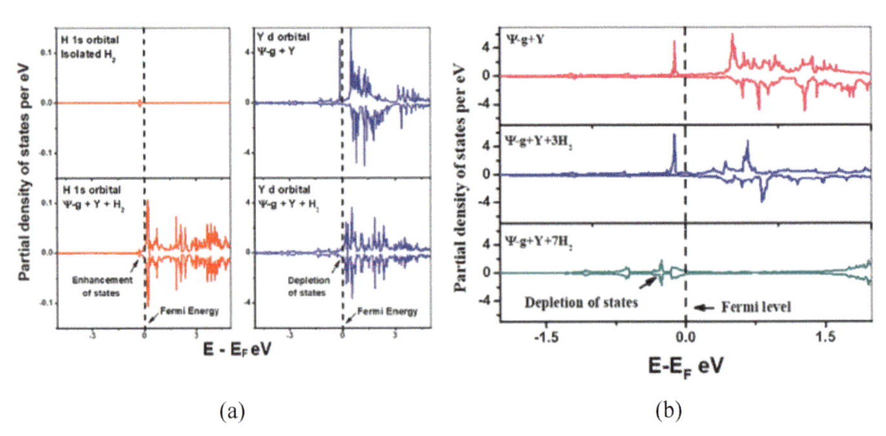

FIGURE 9.5 Interaction of hydrogen with Y+ψ-graphene system.

Source: Reproduced with permission from Ref. [79]

emissions. The least land is needed for solar energy, whereas the most land is needed for nuclear energy. In terms of water discharge quality, solar and wind are the best sources, but geothermal and nuclear tend to be the worst. Nuclear power produces the most solid waste, whereas solar power produces the least. The findings indicate that, as a source for producing hydrogen, solar has the greatest average performance (74%), followed by hydro and wind (60%), biomass (58%), nuclear (46%), and geothermal (46%).[11] The majority of these procedures undoubtedly have great social performances, such as the low negative influence on public health, are highly dependable and have very little bad impact on the land, air, and water supplies. Lowering initial and ongoing expenses will improve the economic performance of solar-based hydrogen generation technologies, making them competitive with conventional systems. Both geothermal and hydro have some concerns related to influence on the quality of the water discharged and the natural habitat, and these problems lower their performance score.

9.1.5 HYDROGEN STORAGE, CHALLENGES AND METHODS

Depending on temperature, hydrogen can exist in various states including gaseous, liquid, and solid. The available forms are shown in the phase diagram (Figure 9.6.a).

At room temperature and atmospheric pressure, 1 kilogram of hydrogen gas requires a volume of 11 m^3; therefore, hydrogen storage essentially entails reducing the vast volume of the hydrogen gas. Compressing hydrogen must be done in order to increase its density in a storage system, and the temperature must be lowered below the critical temperature, or the repulsion must be lessened by hydrogen interacting with another substance. The ability to absorb and release hydrogen in a reversible manner is the second crucial requirement for a hydrogen storage system. The target hydrogen storage has been shown in Figure 9.6.b, and the conventional hydrogen storage methods have been shown in Figure 9.7.[12]

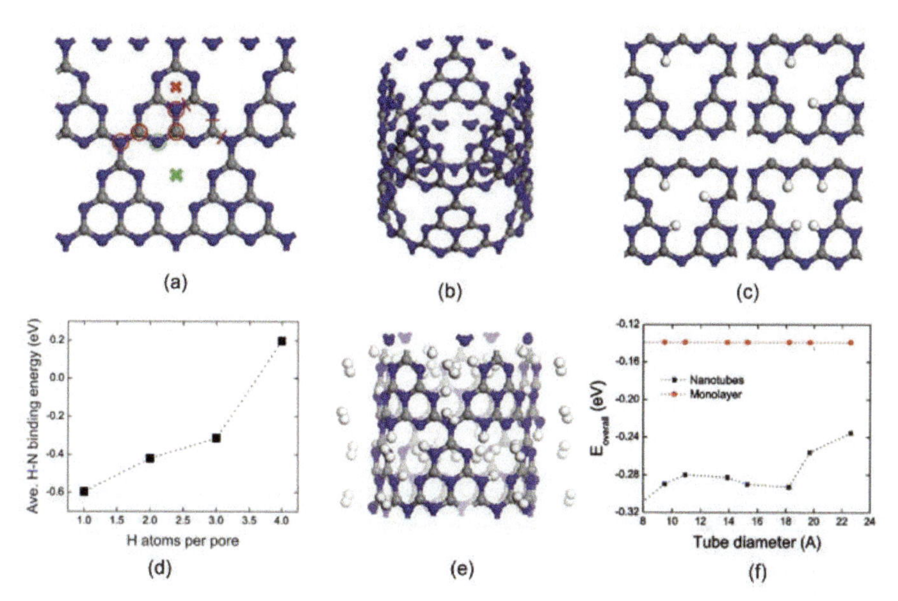

FIGURE 9.6 (a) optimized structure of *g-C3N4*. (b) *g-C3N4 nanotube* illustration. (c) Simultaneous hydrogen addition over one pore of *g-C3N4 nanotube*. (d) Average H-N bonding energy with several H atoms per pore. (e) illustration of hydrogen adsorption on *g-C3N4 nanotube*. (f) Evolution of adsorption energy over tube.

Source: Reproduced with permission from Ref. [80]

The storage procedure of hydrogen can also be divided into physical and chemical methods. Physical methods include storing the gas in either gaseous or liquid form in pressurized containers or at very low temperature. Chemical method includes the storing of hydrogen onto some suitable materials (like metal hydride).

Hydrogen may be adsorbed on a substance primarily in three basic ways (Figure 9.8). In the case of physisorption, hydrogen is still a molecule and forms a weak surface bond with an energy of about meV.[13] As a result, it desorbs at extremely low temperatures. When chemisorption occurs, the H2 molecule splits into its component atoms, diffuses into the medium, and

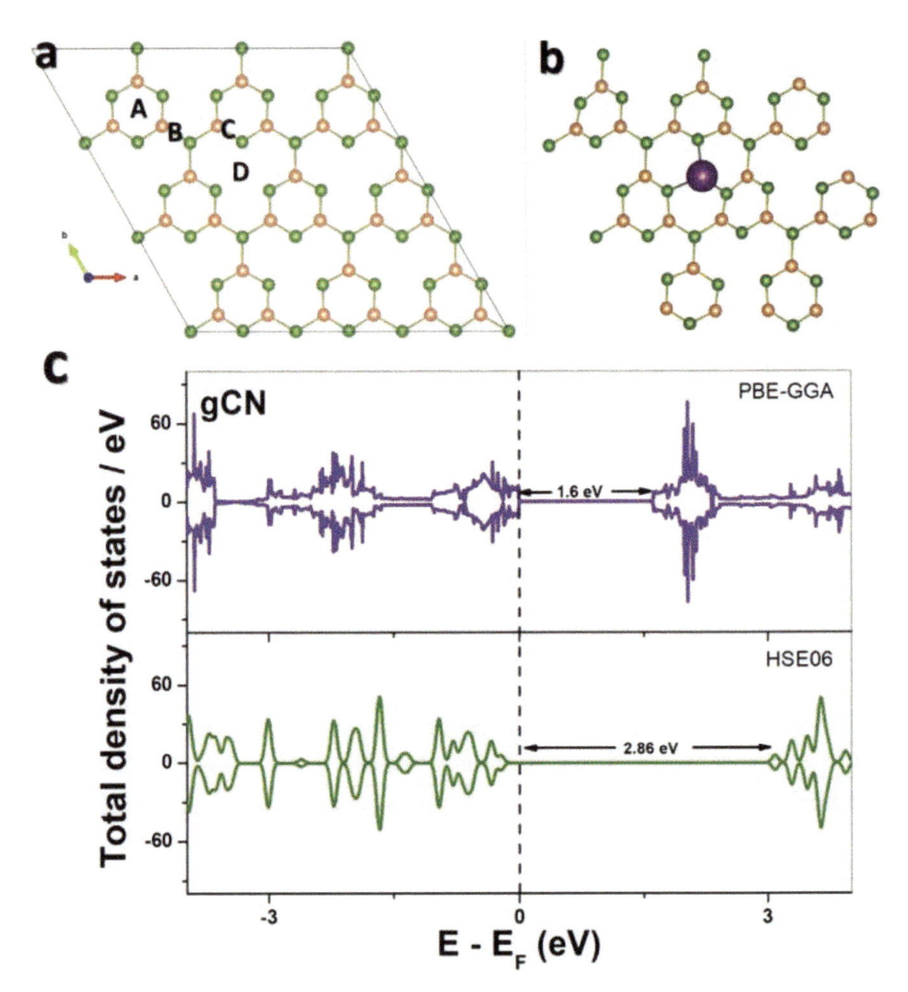

FIGURE 9.7 a) g-C3N4 supercell. b) Y-doped g-C3N4 supercell. c) TDOS of g-C3N4 and Y-doped g-C3N4.

Source: Reproduced with permission from Ref. [81]

attaches chemically with a binding energy in the 2–4 eV range.[13] Strong bonding results in desorption at higher temperatures. The link between the H atoms in an H2 molecule is weakened but not broken in the third kind of binding. Under normal pressure and temperature conditions, the strength of binding, which is halfway between physisorption and chemisorption (binding energy in the 0.1–0.8 eV range), is perfect for storing hydrogen. There are two distinct ancestors of this kind of quasi-molecular binding. According to Kubas,[14] this quasi-molecular bonding is caused by charge donation from the H2 molecule to the unoccupied d orbitals of transition-metal atoms and back-donation from the transition-metal atom to the H2 molecule's anti-bonding orbital. On the other hand, Niu et al.[15] have demonstrated that the polarization of the H2 molecule by a positively charged metal ion allows it to bond to the metal cation in a quasi-molecular form. Multiple hydrogen atoms can attach to a single metal in both scenarios.

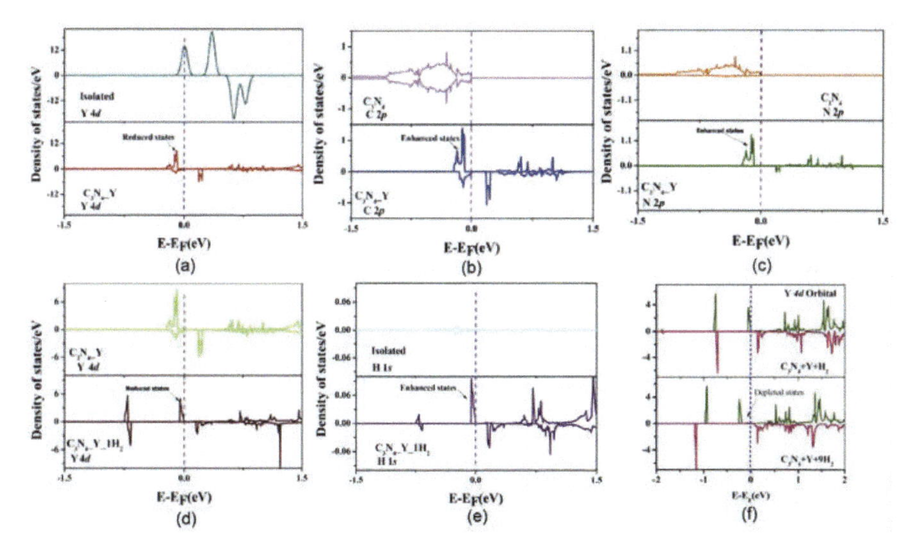

FIGURE 9.8 PDOS analysis of (a) Y 4d orbital of isolated Y and Y doped on g-C3N4 supercell. (b) C 2p orbital. (c) N 2p orbital. (d) Y 4d orbital in g-C3N4 supercell before and after H_2 addition. (e) H 1s orbital. f) Y 4d orbital with simultaneous H_2 addition.

Source: Reproduced with permission from Ref. [81]

The large size and weight of the storage tank, the excessive need for compression energy, the high cost of liquefaction, the boiling-off of liquid, etc. are problems with physical hydrogen storage systems.[16] Together, these factors render the physical process of storing hydrogen inefficient from an energy and financial standpoint.[17] While chemically storing hydrogen in a solid form is a far more practical alternative, it must meet the US Department of Energy's requirements of holding a minimum of 6.5 weight percent of H_2. The hydrogen should also be released while keeping the host's original structure, and the hydrogen's binding energy (BE) should range from −0.2 to −0.7 eV/H_2.[18] Chemisorption is used to store hydrogen in solid form in metal hydrides and complex hydrides, which makes the process irreversible and prevents desorption.[16,19]

A comparative study has been shown in Figure 9.9 for the available hydrogen storage systems. The summary of Figure 9.9 is listed in Table 9.1.

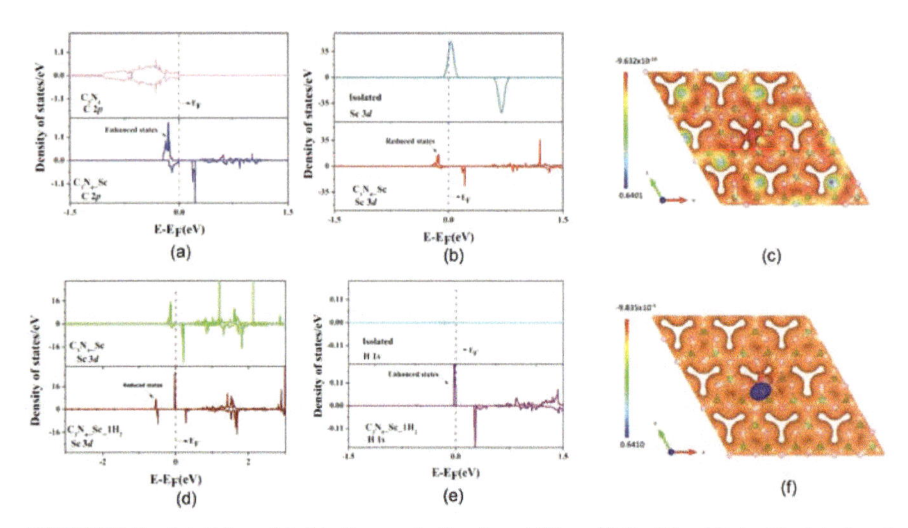

FIGURE 9.9 (a) C 2p orbital before and after Sc addition. b) Sc 3d orbital of isolated and after addition on *g*-C_3N_4. c) Charge density plot for Sc+*g*-C_3N_4. d) Sc 3d orbital in Sc+*g*-C_3N_4 and Sc+*g*-C_3N_4+H_2. (e, f) Charge density plots for Sc+*g*-C_3N_4+H_2.

Source: Reproduced with permission from Ref. [84]

TABLE 9.1 Merits and Demerits of Conventional Hydrogen Storage Mediums.[20]

Type of storage	Pros	Cons
Storage under high pressure	- Well-established - Commercial	- Cost of high-density storage is high
Liquid H_2	- Highest density - Commercial	- Energy-intensive, low-temperature storage - Constant leakage
Chemical storage as hydrides (organic and inorganic)	- High volume density - Extensively studied mechanisms	- Management of generated waste remains a concern
Nanomaterials	- Studies show high energy density	- Currently in early phases of research

From the discussion above, we can summarize the fact that solid-state hydrogen storage, specially, nanomaterials, have enough potential to dominate the hydrogen storage area, but as mentioned in Table 9.1, this is still in the phase of early research. So, in this chapter we will be focusing on solid-state storage of hydrogen to aware the readers about recent advantages and future possibilities of this field.

9.2 CONVENTIONAL SOLID-STATE HYDROGEN STORAGE MATERIALS

9.2.1 METAL HYDRIDES AND METAL ALLOYS

Metal hydrides and metal alloys have been explored for a very long time,[21] and those based on the intermetallic complexes A2B, AB, AB2, and AB5 are particularly interesting for hydrogen storage. A is an element that may create stable hydrides, such as a rare earth metal or a transition metal from the left side of the periodic table. As opposed to this, B elements create unstable hydrides (BHy), such as transition metals from the right side of the periodic table. This will provide the resultant hydride AmBnHz intermediate qualities between AHxand BHy.[22] Table 9.2 lists some examples of intermetallic compounds that might be used to store hydrogen.

TABLE 9.2 Different Types of Metal Alloys and Hydrides Along with their Hydrogen Storage Performance.[23–25]

Type	Metal	Hydride	H2 wt %	T (@1 bar) (K)
A2B	Mg2Ni	Mg_2NiH_4	3.59	528
AB	FeTi	$FeTiH_2$	1.89	265
AB2	ZrMn2	$ZrMn_2H_2$	1.77	713
AB3	LaCo3			
AB5	LaNi5	$LaNi_5H_6$	1.37	285
	Mg	MgH_2	7.60	552

So, overall, we see a low wt% in case of metal alloys, whereas Mg-hydride shows higher wt%. But in practical the hydrogen is very strongly bonded in case of metal hydrides that causes poor kinetics and irreversibility. Complex hydrides like [AlH4]- and [BH4-] have also been investigated for hydrogen storage applications. Though these have moderate storage capacity, still the chemisorption makes the reversibility difficult.

9.2.2 METAL ORGANIC FRAMEWORKS

In recent years, a family of physical storage materials called Metal Organic Frameworks (MOFs) has undergone substantial development for the storage of hydrogen. MOFs are made up of inorganic clusters joined by organic linkers to form strong connections. An organized network of channels or pores is made possible by the linked frameworks of MOFs. Because MOFs are porous, they can store comparatively smaller molecules like CO_2, CH_4, and H_2 as gas. These structures' pores have the capacity to both catch and release gas, and this action is reversible. The pore size and surface area of MOFs can be customized during synthesis in order to provide the specific function that is required of that MOF. Due to their high porosity, high surface areas, high crystallinity, low density, relative affordability, and ease of synthesis, MOFs are ideal for gas adsorption applications. They provide excellent reversibility and quick kinetics for H_2 storage.[26] The recent research on MOF as hydrogen storage system has been listed in Table 9.3.

TABLE 9.3 Literature Review of MOFs for Hydrogen Storage.

Materials	BET area (m^2)	wt %	Reference
MOF-5	3510	7.8	27
IRMOF-20	4070	9.1	27
NU-1101	4340	9.1	28
NU-1102	3720	9.6	28
NU-1103	6245	12.6	28
SNU-70	4940	10.6	29
UMCM-9	5040	11.3	29
NU- 100/PCN- 610	6050	13.9	29
NPF-200	5830	11.4	30
NU-1500-Al	3560	8.2	31
NU-1501-Fe	7140	13.2	31
NU-1501-Al	7310	14.0	31
MFU-41-Li	4070	9.4	32

In terms of actual performance, system-level hydrogen capabilities are more important than material-level values. For crystalline materials like MOFs, the volumetric capacity is frequently calculated using the crystallographic density. Due to the intrinsic mechanical stability of materials and packing-related voids, the actual packing density is typically lower than the ideal crystallographic density. As a cumulative effect the hydrogen storage capacity of MOF at room temperatures is very low and the process ability difficulty of the material limits its practical application.

9.2.3 ZEOLITES

Zeolites are one of several porous, aluminosilicate minerals that come in a variety of crystalline forms, chemical compositions, and physical characteristics. Zeolites are attracting attention for their capacity to decrease ecologically hazardous species like, for example, nitrogen oxides and carbon dioxide, generated by the burning of fossil fuels, in addition to possible uses in hydrogen storage. The existence of micro- and/or mesopores inside zeolites' structures is crucial since it dictates their selective characteristics. The transport of guest species in zeolites is governed by the configuration of these pores. Zeolites were among the first inorganic porous minerals to be studied for gas storage, and as a result, they provide well-recognized examples of how hydrogen may be held in nanoscale channels.[26] Under cryogenic circumstances, hydrogen is held in zeolites by physisorption and encapsulation (trapping). In this second scenario, molecules are driven at extreme pressure into the typically inaccessible zeolitic cages pressure and temperature. After reaching room temperature, in the pores, hydrogen is entrapped and can be released by increasing the temperature or exerting pressure. Zeolites' storage capacity is unfortunately limited by their high mass density of the structure (containing Si, Al, O, and heavy cations).[26]

The size, efficiency, and secure on board gas storage of hydrogen fuel cells (HFC) are the three characteristics that restrict their utilization in addition to cost, operability, and durability difficulties. As an alternative to the conventional hydrogen storage systems, the use of nanomaterials has been suggested to address these problems. The DoE's driving range limit of 300 miles could be increased by using nanomaterials since they could give rise to a system with a higher density. Researchers' primary attention in this field at the moment is on carbon-based materials that include graphite, carbon nanotubes (CNTs), and metal hydrides. The efficient use of carbon-based materials as hydrogen storage media was initially described by Dillon et al.[33] Due to their light weight, large specific surface area, improved stability, etc., carbon nanomaterials are favored. At room temperature, however, these materials do not work well for practical applications because the binding energy of hydrogen molecules with them is too low due to weak Van der Waals interaction.[34,35] These, however, are fantastic hosts for holding hydrogen at lower temperatures. For instance, highly pure SWCNT has a desorption temperature of 80 K, which is substantially lower than atmospheric temperature, and a hydrogen storage capacity of about 8.25 wt%.[36] Doping of the carbon nanostructures with metals like Sc, Ti, Ni, Ca, Li etc. can overcome the issue of sluggish desorption of hydrogen at room temperature.[37]

Metal-doped fullerenes have first been researched in order to examine the impact of doping on carbon-based materials. Li and K-doped fullerene displayed storage capacities that were far below DoE standards at 4.5 weight percent and 4.0 weight percent, respectively.[38,39,40] Similar investigations have been conducted on CNTs after fullerene. In contrast to pure CNTs, which have only been reported to store 0.53 wt% of H_2, Pd and V doped CNTs store 0.66 wt% and 0.69 wt% of H_2, respectively.[41] Ti-doped SWCNT has an 8 wt% hydrogen storage capacity.[42] In a DFT research on Y-doped SWCNT, Chakraborty et al. reported the H2 binding energy of −0.41 eV/H_2 and the hydrogen storage capacity of 6.1 wt%.[34]

However, because several sides of the structure can be employed, it has been found that carbon's 2D allotropes are more efficient H_2 storage systems than CNTs and fullerenes.[35] According to the experimental data from Gu et al.,[43] Ni/Al-doped graphene has an H_2 intake of about 5.7 Wt%. The combination of Ni nanoparticles and MgH_2, doped on a material resembling graphene, increases the system's H_2 intake to a level greater than 6.5 weight percent, according to experimental evidence presented by Tarasov et al.[44] Thus, in this chapter we will be discussing about the recent advancements in 2D materials for hydrogen storage materials, but from the DFT perspective. "Why DFT?"- This answer is given in the next section.

9.3 IMPORTANCE OF THEORETICAL SIMULATIONS

In order to forecast how hydrogen molecules would interact with nanomaterials and, in the case of functionalized nanomaterials, with functionalized atoms or groups, theoretical simulations are crucial. Extensive computer simulations are now quite possible because of the development of supercomputers with high processing speeds and plenty of memory. The computation of theoretical data on hydrogen storage is also possible because of the abundance of well-established computer programs, both open access and commercial. As a result, it is now possible to create hydrogen storage databases for a large number of systems and suggest viable systems for experimentation to experimentalists thanks to the availability of reliable software packages and supercomputers.

In this section, we will be discussing about the advantage of theoretical simulations and the applicability of it in terms of hydrogen storage research in particular.

9.3.1 ADVANTAGE OF THEORETICAL SIMULATIONS

As any other field, the theoretical research has recently been increased exponentially for analysing suitable materials for hydrogen storage. Though without experimental validation, the practical application is tense to zero, still the theoretical analysis provides a background of the area and saves time, money, and resources as well. Along with this benefit, important data that is difficult to obtain via experimentation, such as optimal adsorption configuration, adsorption energy, charge transfer, etc., can be obtained by simulation approaches. It might be challenging or perhaps impossible to obtain an accurate image of orbital interactions and charge transfer from experimental observations, but theoretical simulations can offer a great picture in this regard. It is easy to determine from the simulation data if a particular H2 is experiencing physisorption, chemisorption, or Kubas sorts of interactions.[14,45] Theoretical simulations may also be used to determine the desorption temperature and the stability of the structures. The metal–metal cluster is one of the significant problems for hydrogen storage in metal-doped carbon nanomaterials or polymers. In order to prevent metal–metal clustering, we may acquire a sense of the potential metal loading pattern from the computation of the diffusion energy barrier of metals. However, estimating the hydrogen weight percentage from theoretical simulations has its difficulties. The simulation methodologies and configuration settings have an impact on the simulation outcomes. Compared to GGA[46] exchange correlation, for instance, LDA[46] exchange correlation over binds the H_2 molecules. Therefore, we must exercise caution when selecting simulation methodologies and making predictions based on simulation data. Weak Van der Waals interactions exist for the adsorption of hydrogen molecules on carbon nanostructures or polymers. So, while calculating the binding energy of hydrogen molecules on the host, we need to carefully take such interactions into account. The following subsections will highlight significant parameters obtained from theoretical data, the significance of dispersion corrections, and the sensitivity of theoretical data while keeping in mind the role of theoretical simulations for forecasting hydrogen storage parameters and the limitations of simulation methods.

9.3.2 THEORETICAL PARAMETERS FOR HYDROGEN STORAGE

Adsorption configuration, adsorption energy, charge transfer, and orbital interactions are all difficult to obtain in experiments, but these are important parameters to understand the systems behavior, can be clearly depicted in theoretical simulations.

9.3.2.1 ADSORPTION CONFIGURATION

For hydrogen storage simulations, identifying the adsorption configuration that uses the least amount of energy is a crucial step. It will take longer for us to optimize the shape until the forces between the atoms are very low (usually −0.01 eV/H_2). As weak Van der Waal forces are involved in the hydrogen adsorption on carbon nanomaterials and polymers, the convergence process may take longer. There is also a possibility that the gas molecules may occasionally become trapped at local minima, causing us to miss the precise arrangement that uses the least amount of energy. To prevent this, it is advised to create 2–3 distinct initial setups, let them all rest on their own, and then select the stable configuration with the least amount of energy. If the system comprises a transition metal with a magnetic signature, spin polarization should be taken into account when calculating the relaxation and total energy. The stable configuration also depends on simulations' settings for variables like the plane wave's energy cut-off and the Brillouin zone's K-point sampling. Therefore, it is necessary to examine the convergence with regard to the plane wave's energy cut-off and the quantity of K-points. The strength of the bonding may be inferred from the distance between the hydrogen molecule and the closest atom of the host or dopant. Stronger adsorption occurs at shorter distances (shorter bond lengths).

9.3.2.2 ADSORPTION ENERGY

A crucial factor in determining a material's capacity to store hydrogen is its H_2 adsorption energy. This energy must fall inside of a window in order for the hydrogen storage device to be useful. It will be challenging to release them if the adsorption energy is too high, yet if the H2 molecules are bound weakly, they may be desorbed from the system before being used. In this regard, DOE has stipulated that for the practical use of hydrogen as fuel, the binding energy of H2 must lie within the range of −0.2 to −0.7 eV.[47] The eq 9.1 may be used to calculate the adsorption energy of H_2 on the host.

$$E_{ads} = E_{Host+H_2} - E_{Host} - E_{H_2} \tag{9.1}$$

where E_{Host} is the energy of nanostructure, doped with desired dopant. According to tradition, a negative adsorption energy value means that binding is energetically advantageous. The three energies in the equation above should all be calculated using the same simulation settings because the calculated energies can change depending on the settings. For various adsorption sites, the adsorption energy could change. A suitable vacuum should be taken into consideration to prevent interactions between periodic pictures if there is no periodic border condition in any direction (for molecules, clusters, etc.). From Figure 9.10 we see that GGA under binds the system and as a consequence, consideration of van der Waals force along with GGA increases the binding energy.

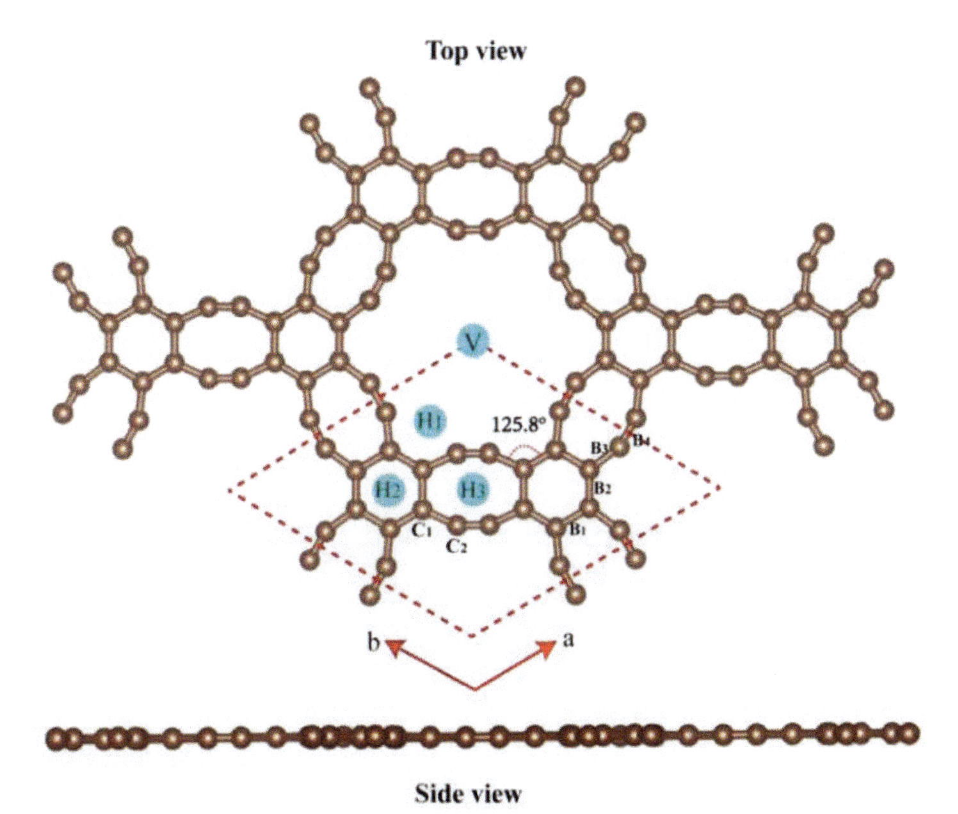

FIGURE 9.10 Optimized structure of holey graphene.

Source: Reproduced with permission from Ref. [86]

9.3.2.3 HYDROGEN WT%

During DFT simulation, we can also have access to determine the hydrogen wt%. The DoE has the criteria of minimum 6.5 %.[47] The wt% can be calculated using eq 9.2. It is mainly dependent on the type of nanostructure and its weight, the type of doping material and its weight and also maximum number of hydrogen that can be adsorbed by 1 doped material. This discussion also tells us that, there are multiple suitable sites where the dopant can be added, and all these sites need to be considered for perfect calculation.

$$wt\% = \frac{Toatal\ weight\ of\ the\ adsorbed\ H_2}{Total\ weight\ of\ the\ adsorbad\ H_2 + Total\ weight\ of XTC} \tag{9.2}$$

In Figure 9.11 we see multiple occupied sites on a nanomaterial by a metal atom, and we can also see that when only 1 side is occupied the wt% is calculated to be 7.08, whereas when both sides of the nanomaterial are decorated with metal atoms, the calculated wt% is 9.34.[48]

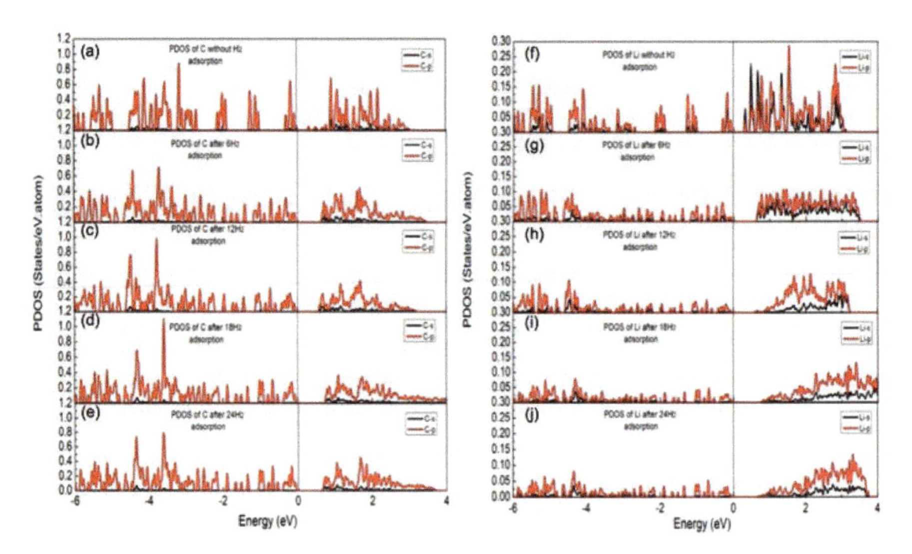

FIGURE 9.11 PDOS analysis of Lithium decorated holey graphyne.

Source: Reproduced with permission from Ref. [86]

9.3.2.4 CHARGE TRANSFER AND ORBITAL INTERACTIONS

Simulation results can be used to gain theoretical understanding of the mechanisms underlying bonding and charge transfer. The quantity of charge being transmitted between the bound species may be quantified. Charge transfer gives us some insight into the bonding process. The H2 molecule may adhere to the host by Kubas-type interactions, physisorption, or chemisorption. For chemisorption, physisorption, and Kubas kinds of contact, the bonding is strong, weak, and intermediate, respectively. Bonding strength increases with charge transfer. For maximum nanomaterials, hydrogen atoms are bonded via physisorption, where charge transfer is less. But in case of chemisorption, as the charge transfer is higher, the bonding is stronger and as a result, the system is irreversible for hydrogen desorption. Hence, both physisorption and chemisorption cannot be used to store hydrogen in a realistic manner. For nanomaterials doped with metal atoms, there is a charge transfer from the metal's d orbitals to the H s orbital, followed by back donation through the Kubas interaction. The binding energy in this instance falls within the DoE-recommended range of 0.2–0.7 eV and is most acceptable for the real-world use of hydrogen storage. Therefore, the charge transfer can reveal whether an interaction is Kubas-type, chemisorption-type, or physisorption-type. One must pay attention to the quantity of valence electrons used in the simulations in order to determine the charge transfer. Two effective methods to determine the amount of charge transfer during orbital contacts are Bader charge analysis[49] and Mulliken's charge analysis.[50] Using visualization tools like XCRYSDEN,[51] VESTA,[52] etc., the graphical plot of charge density may be generated to see the spatial fluctuation of charge density. The plot of partial density of states (PDOS), which describes the contribution of density of states from each orbital of each atom, may be used to qualitatively determine the orbital interactions and hybridization. The charge transfer from the Bader charge analysis, the partial density of state analysis, and the charge density map should all be consistent.

9.4 HYDROGEN STORAGE IN PRISTINE 2D NANOMATERIALS: RECENT WORK

9.4.1 GRAPHENE

Graphene is made up of a single atomic layer of sp^2 hybridized carbon in a hexagonal lattice, with large specific surface area.[53] Since its experimental manifestation in 2004, graphene has been a renowned star for scientists in numerous domains, and as a result, graphene-based materials and technologies have been extensively researched. In terms of materials, graphene has been regarded as a type of adaptable and one-of-a-kind building block for useful materials.[53] It has been widely studied for different applications in energy storage including battery, capacitors, solar cells, and also hydrogen storage.

9.4.1.1 TI-DOPED GRAPHENE

Liu et al.[54] reported Ti-doped graphene material for hydrogen storage using DFT calculation. Their calculation showed each Ti can adsorb up to 4 hydrogen molecules leading to a total uptake of 7.8 wt%. The bonding between Ti and graphene was classified as Dewar interaction, where electrons were donated from 4s and $3d_{z2}$ of Ti to the π orbitals of graphene and a back donation takes place simultaneously from π^* of graphene to $3d_{xz}$ and $3d_{yz}$ of Ti. This phenomenon was well observed in PDOS analysis (Figure 9.12), where we observe distortion in 4s and $3d_{z2}$ and shift of π and π^* band toward lower energy, indicating strong hybridisation. The dispersion in $3d_{xz}$ and $3d_{yz}$ is because of the back donation from the Ti atom. According to the Bader charge analysis the charge transferred amount is 0.86 e from Ti to graphene.

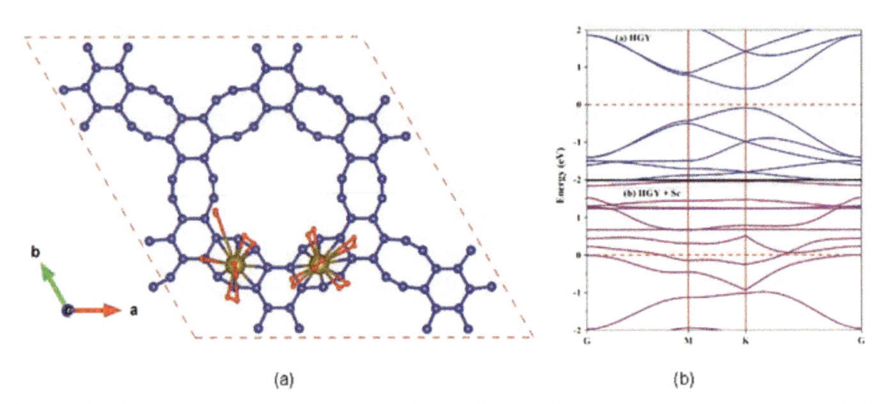

FIGURE 9.12 a) Optimized structure of HGY+Sc+H_2. b) Band structure of pristine HGY and HGY+Sc.[100]

Source: Reproduced with permission from Ref. [100]

The interaction with H2 and Ti is Kubas type of interaction where $3d_{xy}$ and $3d_{x2-y2}$ overlap with H2 σ and σ^*. The amount of transferred charge from Ti to H depends mainly on distance between Ti and H, but, in general, it lies between the range of 1.07–1.16 e. But the charge accepted by the H2 molecules is within the range of 0.13–0.20 e.

9.4.1.2 Y DECORATED GRAPHENE

Y-doped graphene was studied for hydrogen storage material using first-principle DFT calculations by Liu et al. Each Y was able to adsorb maximum of 6 H2 molecule with an average adsorption energy of 0.495 eV and with a total maximum capacity of 5.78 wt%. According to the calculation of Liu et al.[55] the interaction was mainly caused by 5d orbital of Y and H 1s. The interaction between Y and graphene can be understood more clearly from Mulliken charge analysis. The electronic configuration of Y before bonding to graphene is $4d^1 5s^2$. According to Mulliken population analysis, the electronic configurations of Y and C change after Y adsorption to $4d^{1.4165}s^{0.8505}p^{0.112}$ and $2s^{1.2922}p^{2.744,}$ respectively. In comparison to the contributions of other orbitals, the electron change of the C 3d-orbital is insignificant. 1.150 e electrons clearly travel from the Y-5s orbital to the C-2p orbital. As the Y atom approaches the graphene sheet, it becomes positively charged with a charge of 0.522 e, while the carbon atoms nearby become negatively charged. This electron potential difference generates an electric field between the metal atom and graphene, resulting in the donation of some electrons from graphene's occupied orbitals to the Y-4d and 5p orbitals and increasing the electron potential difference.[55]

Desnavi et al.[56] also reported Y-doped graphene with an average binding energy of 0.415 eV where Y decorated graphene may adsorb up to four hydrogen molecules. With an average desorption temperature of 530.44 K, all hydrogen atoms were reported to be physisorbed with a maximum gravimetric hydrogen storage capacity of 6.17 %.[56]

9.4.1.3 ZR-DOPED GRAPHENE

Yadav et al.[18] predicted using DFT that a single Zr atom can adsorb 9 H_2 molecules, when decorated on graphene surface, and the resulting gravimetric wt% of the system was calculated to be 11 wt % with the desorption temperature of 433 K, which is very suitable for fuel cell applications. According to their research, the charge transfer to the hydrogen molecule is greater in a system with a large magnetic moment, resulting in a higher desorption temperature (may be higher than prescribed limit for hydrogen storage by DoE). As the magnetic moment decreases, T_D approaches the optimum range for fuel cell applications. This work suggests that altering the magnetic nature of the system by doping may be an efficient technique to bring T_D inside the required window.

DFT analysis shows that the initial magnetic moment of isolated Zr atom is 2 μ_B and of pristine graphene is 0 μ_B. But when Zr is added on to graphene, the system attains a higher magnetic moment of 3 μ_B. Also, as number of H2 is increased from 1, 5, to 9 the magnetic value is found to decrease (2.0 μ_B, 0.42 μ_B, 0.30 μ_B). From charge density plots (Figure 9.13.a), we can see that the overall area of the iso-surface, which represents a qualitative estimate of the magnetic moment, decreases as the system adsorbs more and more H_2 molecules. According to the Bader charge analysis, the net charge gain by hydrogen molecules decreases as the system absorbs more and more hydrogen molecules and becomes less and less magnetic. As a result, as the magnetic moment of the system increases, more net charges are transferred to the H2 molecules, resulting in higher binding energy. The charge transfer will be reduced when the system becomes less magnetic or nonmagnetic, resulting in a decreased binding energy. As a result, the binding energy of H2 molecules adsorbed on TM-doped carbon nanostructures is affected by the system's magnetic signature. The empirical relation reported is shown in Figure 9.13.b.

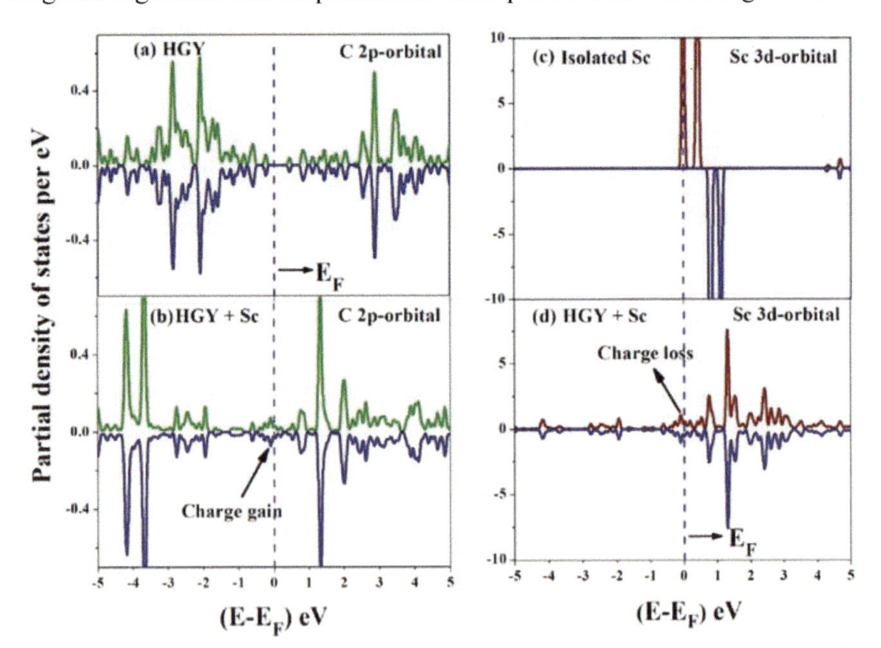

FIGURE 9.13 PDOS analysis of (a) C 2p orbital in pristine HGY. (b) C 2p orbital in HGY+Sc. (c) 3d orbital of isolated Sc. (d) 3d orbital of is HGY+Sc.

Source: Reproduced with permission from Ref. [100]

From PDOS analysis (Figure 9.14) we can see that the charge of an isolated Zr atom remains on the dz^2 and dx^2-y^2 sub-orbitals in addition to the s orbital. When Zr is attached to a graphene surface, the magnitude of the partial density of states in the occupied dz^2 and dx^2-y^2 suborbital decreases significantly, indicating charge transfer from these suborbitals to the C p orbital. As a result of charge redistribution in the graphene + Zr system, several occupied states occur in the d_{xy} suborbital. The intensity of occupied states in the valence band near to the Fermi level decreases for dz^2, indicating charge transfer from dz^2 to the H 1s orbital. Furthermore, the decrease in the intensity of the occupied dx^2-y^2 suborbital toward the Fermi level implies a charge donation from dx^2-y^2 to the H s orbital.

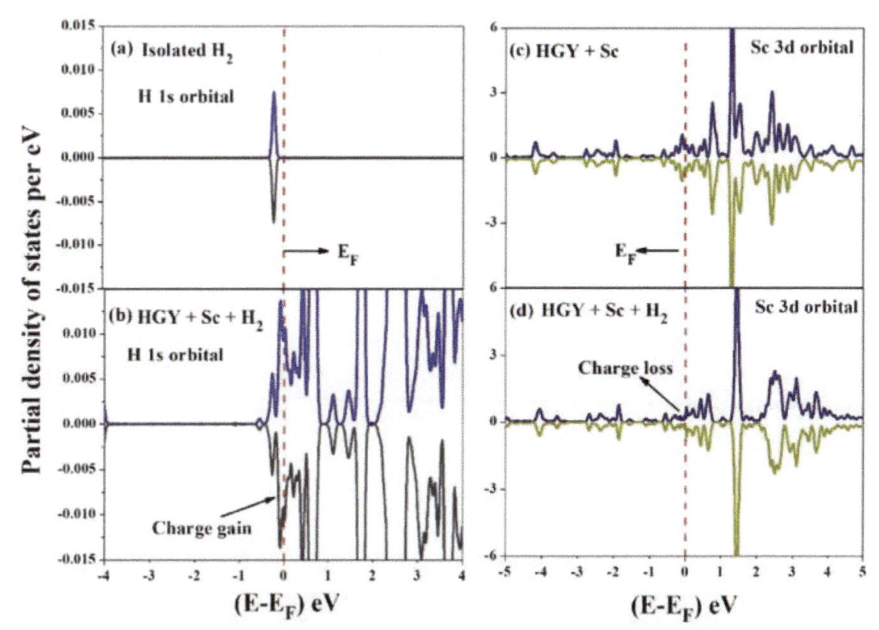

FIGURE 9.14 PDOS analysis of (a) H 1s orbital of isolated hydrogen, (b) H 1s orbital in HGY+Sc+H_2, (c) 3d orbital of HGY+Sc, (d) Sc 3d orbital of is HGY+Sc+H_2.

Source: Reproduced with permission from Ref. [100]

9.4.2 BORON NITRIDE SHEET

Boron nitride has been examined as suitable hydrogen storage materials over past 20 years, through experimental and theoretical advancements, mainly due to its porous structure and unique defects in the nanostructure.[57] Various nanostructures have been investigated for hydrogen storage application of BN family, which includes BN nanotubes, BN nanosheets etc. All these structures have been reported in terms of their performances in Table 9.4.[57] As we see from the table, that remarkable hydrogen storage performance is shown in BN nanosheets and hence in this section we will be focusing our discussion based on recent advancements in BN nanosheets as hydrogen storage systems.

TABLE 9.4 BN Structures and Their Hydrogen Storage Performance [NT: Nanotube, NF: Nanofiber, MWNT: Multiwalled Nanotubes].

System	BET area	T (K)	Pressure (bar)	H2 wt%	Ref
Porous turbostatic structure	960	77	1	1.01	57
Porous microbelts	1144	77	10	2.3	58
Porous spongy structures	1900	77	10	2.57	59
Milled hexagonal structures	–	298	10	2.6	60
MWNTs	150	293	100	1.8	61
Bamboo-like NTs	210	293	100	2.6	61
	230	298	100	3	62
Collapsed NTs	789.1	293	100	4.2	63
BNNTs		298	60 9	2.2 0.87	64 65
BNNFs	260	298	100	2.9	66
Hollow spheres	215	298	100	4.07	67
O-doped NSs	536	298	50	5.7	68
LiBH4-doped hexagonal BN	122	100	1	2.3	69

Using the van der Waals correction parameter in DFT, Chhetri et al.[70] reported the potential of -BN as a viable hydrogen storage material. The most advantageous H adsorption site was found to be the hollow hexagonal site with a binding energy of −0.212 eV, among the four potential sites. The average adsorption energy is seen to decrease as the number of H molecules rises. The low average adsorption for all the H adsorbed systems and the difficulty in reversible adsorption and desorption processes led to the low predicted desorption temperature (Table 9.5). Their research has shown the 6.7 wt% hydrogen absorption capability, with an average adsorption energy of −0.122 eV/H, which is higher than the DOE's weight percent benchmark figure. Low charge transfer from the BN nanosheet to the H molecules, as shown by the Bader charge analysis in Table 9.5, resulted in poor H molecule attachment to the host material.[70]

TABLE 9.5 Detailed Analysis of the Performance of a BN NS for Hydrogen Storage Application.[70]

Number of hydrogen	H2 adsorption energy (eV)	Wt %	Desorption T (K)	Charge transfer (e)
0	–	–	–	–
1	−0.21	0.80	270	0.005
2	−0.17	1.70	212	0.019
3	−0.161	2.60	206	0.020
4	−0.159	3.40	203	0.024
5	−0.153	4.30	196	0.037
6	−0.143	5.10	183	0.050
7	−0.134	5.90	172	0.065
8	−0.128	6.70	164	0.046

9.4.2.1 *PT (PD)-DOPED BORON NITRIDE SHEET*

Ren et al.[71] reported a first-principles investigation of hydrogen adsorption on a sheet of hexagonal boron nitride doped with platinum (Pt) and palladium (Pd). The findings demonstrate both solitary Pt and Pd atoms, with binding energies of −5.028 and −4.113 eV, respectively, which prefer to locate on the top of N atoms compared to other 4 possible adsorption sites as shown in Figure 9.15.a. In accordance with the Mulliken charge analysis of single Pt and Pd atoms adsorbed on BN sheet, Pt and Pd atoms carry 0.25 and 0.17 c, respectively, showing that metal atoms give electrons to the nearby boron and nitrogen atoms on the BN sheet.[71]

As a result of the charge transfer behavior, the metal atoms become cationic and widely form heteropolar bonds with the nearby boron and nitrogen atoms. An increase in the absorption of the H2 molecule results from the formation of more dipole moments as a result. A single Pt/Pd atom can chemically attach up to three hydrogen molecules at a time, with the H-H bonds of H_2 molecules being noticeably longer. Pt-/Pd-doped BN sheets have typical H_2 molecule binding energies between −1.010 and −0.705 eV, which are greater than those of pristine BN sheets.[71] PDOS analysis shows (Figure 9.15.c and 15.d) that the initial hydrogen adsorption is mainly the result of interaction between H-1s orbital and the metal atom. With increase in number of H_2 molecules, we see that there is a partial donation from H-1s orbital to the metal atom near Fermi level accompanied by a sale time back donation from metal to antibonding σ* orbitals of H_2. As a result of this Kubas type interaction we see enlargement in H-H bond. The interaction of H_2 molecules causes the H 1s peak to expand and split into two peaks, as seen in Figure 9.15.c and 9.15.d. The band widening of the H 1s orbitals gets more exceptional when more H2 molecules are adsorbed on metal, demonstrating the importance of the H_2–H_2 interaction.[71]

9.4.3 GRAPHENE

In 1987, Baughman, Eckhardt, and Kertesz projected the structure of the carbon allotrope graphyne.[72] The layer structure of graphyne is seen in Figure 9.16, where it is possible to see the coexistence of sp and sp2 carbon atoms. The layers may be created by substituting acetylenic linkages for one-third of the carbon–carbon bonds in graphite to create the graphyne.[73] The two main categories of graphynes are graphynes-n (Figure 9.16.a), in which the aromatic rings are separated by n acetylenic bonds, and x, y, and z-graphynes (Figure 9.16.a-d), which contain at least some nonaromatic sp2 carbons.[74] But the synthesis of these classes of materials was always a challenge until 2022. In 2022 Desyatkin et al.[74] reported the scalable synthesis of multilayer basic graphyne-n family member with n = 1 orγ-graphyne.

Quite a lot of studies have been performed on metal-doped graphynes/γ-graphynes. In this section, we will try to discuss about the best performed systems in detail.

9.4.3.1 YTTRIUM DECORATED GRAPHYNE

A Yttrium atom attached to the surface of graphene is predicted to be able to adsorb up to nine molecular hydrogen (H2), with a uniform binding energy of about −0.3 eV/H2 and an average desorption temperature of about 400 K (ideal for fuel cell applications), resulting in 10 weight percent of hydrogen, which is significantly more than the DoE's requirement.[35]

The triangular hollow and the hexagonal hollow are two adsorption sites that may be seen when the graphyne structure is taken into account. The sp2 hybridized C atoms with an out-of-the-plane pz (π/π*) orbital surround the hexagonal hollow site, whereas both sp^2 and sp hybridized C atoms, including the C atoms on the acetylene linkage, are present at the triangular hollow site. The in-plane p_x-p_y orbitals that are present in graphyne as a result of the sp hybridized C atoms also contribute to the π and π*states. The

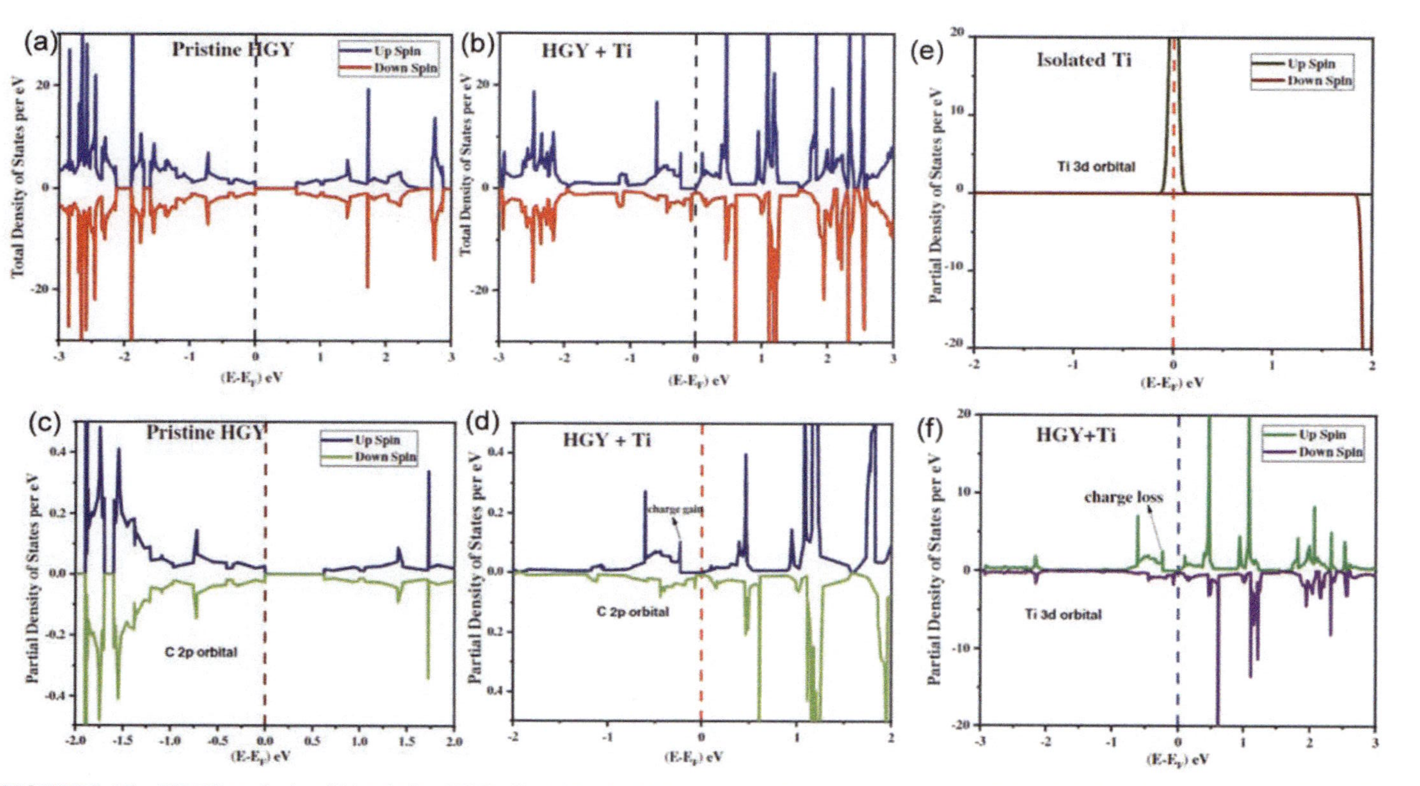

FIGURE 9.15 TDOS analysis of (a) pristine HGY, (b) HGY+Ti; PDOS of (c) C 2p orbital in pristine HGY, (d) C 2p orbital in pristine HGY+Ti, (e) Ti 3d orbital of isolated Ti, (f) 3d orbital of is HGY+Ti.

Source: Reproduced with permission from Ref. [89]

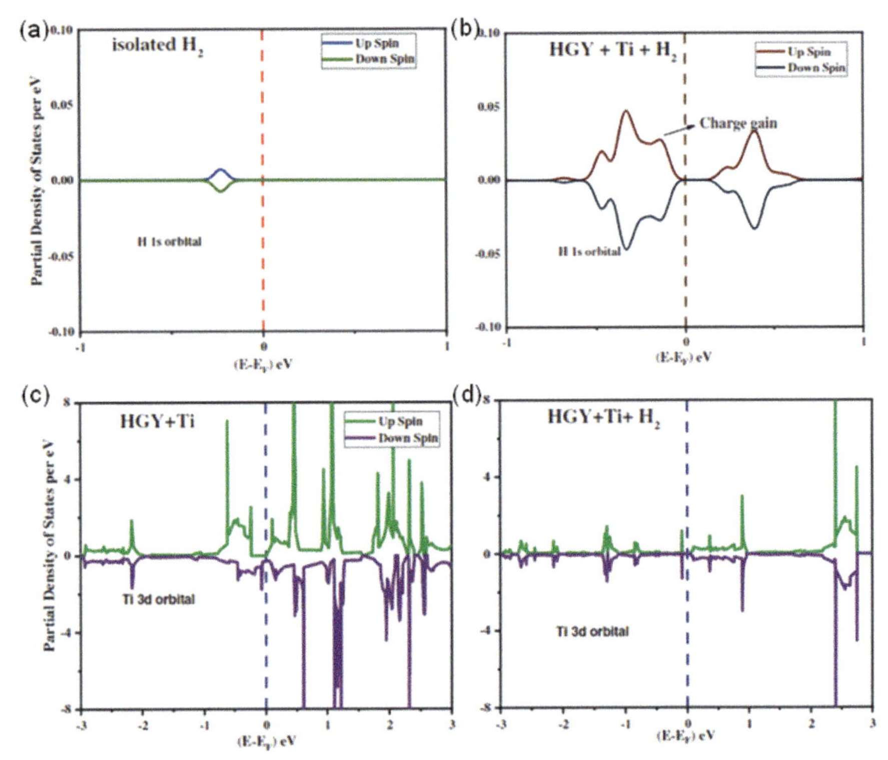

FIGURE 9.16 PDOS analysis of (a) H 1s orbital of isolated hydrogen, (b) H 1s orbital in HGY+Ti+H_2, (c) 3d orbital of HGY+Ti, (d) Sc 3d orbital of is HGY+Ti+H_2.

Source: Reproduced with permission from Ref. [89]

triangular hollow site is preferred as shown in Figure 9.17.a, because of the presence of addition p_x-p_y orbitals that combine with filled $d_{x^2-y^2}$ orbitals of Y atom, accompanied by a charge transfer of 1.72e.[35] But, in the hexagonal site the amount of charge transferred from Y to graphyne was calculated as 1.61 e. At triangular hollow site the Y atoms were able to adsorb maximum of 5 H_2 molecule with binding energy −0.36 eV, whereas at the hexagonal hollow sites, the Y atoms were able to adsorb 9 H_2 molecules with the binding energy of −0.24 eV.[35] The interaction between H2 molecules and Y+graphyne system has been identified as Kubas interaction by measuring the charge depletion from Y atom with addition of H2 molecules. As shown in Figure 9.17.b, we can see the there is a continuous loss of charge in the near Fermi region of Y d orbital upon addition of 9 H2 compared to 5 H2. Also, the diffusion energy barrier for Y was calculated as 0.92 eV.

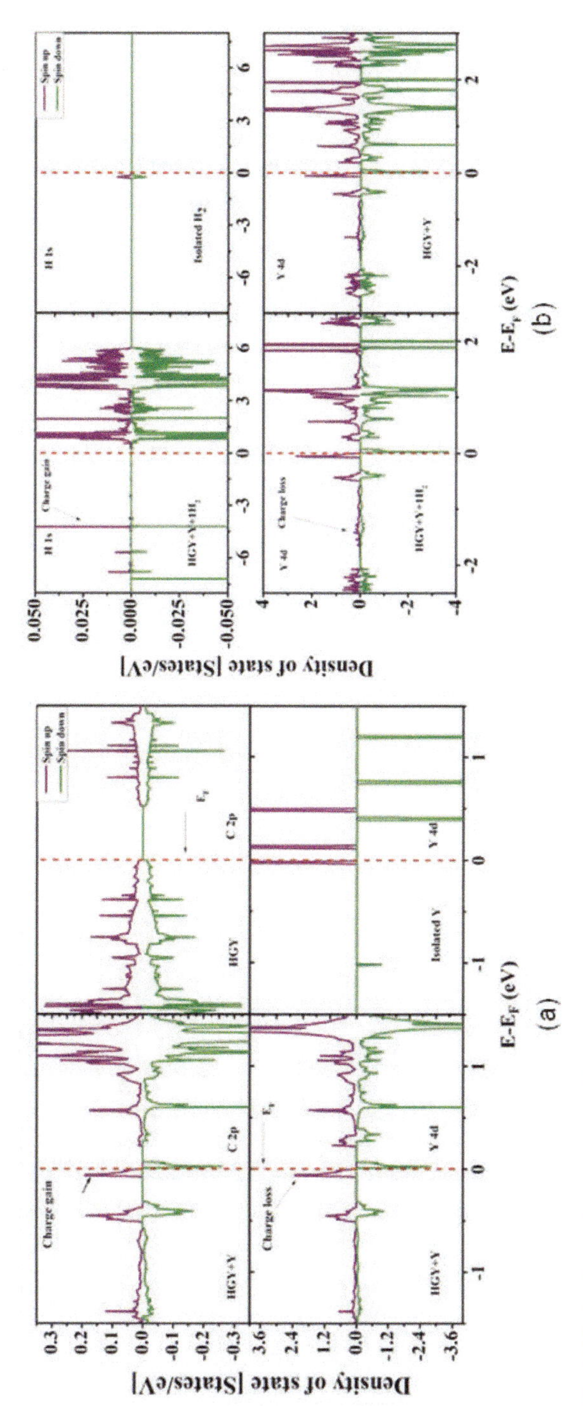

FIGURE 9.17 a) Interaction of Y atom with HGY, b) interaction of H2 with HGY+Y.

Source: Reproduced with permission from Ref. [48]

9.4.3.2 LI/CA/SC/TI DECORATED GRAPHYNE

Guo et al.[73] reported the study of Li, Ca, Sc, and Ti dope graphyne structures for hydrogen storage applications. The detailed DFT calculations have been shown in Table 9.6. The PDOS analysis (Figure 9.18) tells us about the fact that the interaction between the hydrogen and the metal atom embedded graphyne is mainly because of the electric field induced by the difference in charge density around the metal atoms. Guo et al. indicated that Ca, Sc, and Ti have better interaction with graphyne due to the presence of their empty d orbital that leads to stronger hybridization compared to Li. The orbitals of adsorbed hydrogen are positioned considerably below the Fermi level for the Li/graphyne complex and are essentially unaltered. For the Ca, Sc, and Ti/graphyne complex, stronger polarization of the adsorbed hydrogen orbitals causes them to approach the vacant 3d orbitals of the metals. This phenomenon is known as orbital hybridization. The electrostatic field created by the metals on the graphyne has a major role in modulating this kind of hybridization. Hence, it can be proved that the stronger the hybridization, the stronger the electric field's intensity is.[73]

TABLE 9.6 Parameters Obtained from DFT Calculation for Li/Ca/Sc/Ti Decorated Graphyne.[73]

Metal	Binding energy (M)	Adsorption energy	M-H (Å)	H-H (Å)	Wt%
Li	−2.70	−0.26	2	0.793	18.6
Ca	−2.41	−0.34	2.29	0.821	10.5
Sc	−4.85	−0.60	2.12	0.818	9.88
Ti	−5.11	−0.54	1.98	0.825	9.54

9.4.4 PSI-GRAPHENE

One of the most unexplored, yet most stable and one of the most recently reported allotropes is called Psi-graphene (ψ-graphene), with the chemical formula of $C_{12}H_8$.[75] ψ-graphene is made up of pentagonal, hexagonal, and heptagonal carbon rings, is metallic in nature, and belongs to the P_{2mg} space group.[76] In Figure 9.19.a, we can see an optimized structure of ψ-graphene, with s-indacene backbone. The first optimized lattice parameters reported were a = 6.70 Å and b = 4.84 Å, and it consists of 4 different kinds of C atoms, as shown in Figure XX as C1, C2, C3, and C4.[75] In Figure 9.19.a, 1 set of C1–C1 bond has been marked specially with a blue circle, as the bond length of this particular bond is higher (1.51 Å), compared to other similar bonds, that were within the range of 1.41 Å–1.44 Å. It was justified as the intersection of 3 different s-indacene units.[75] Li et al.[75] first reported ψ-graphene for Li storage. But its unique properties like high mechanical and thermal stability make its application interesting toward hydrogen storage. Li et al.[75] also proposed possible synthesis paths for ψ-graphene (as shown in Figure 9.19.b), but there is no practical synthesis reported yet.

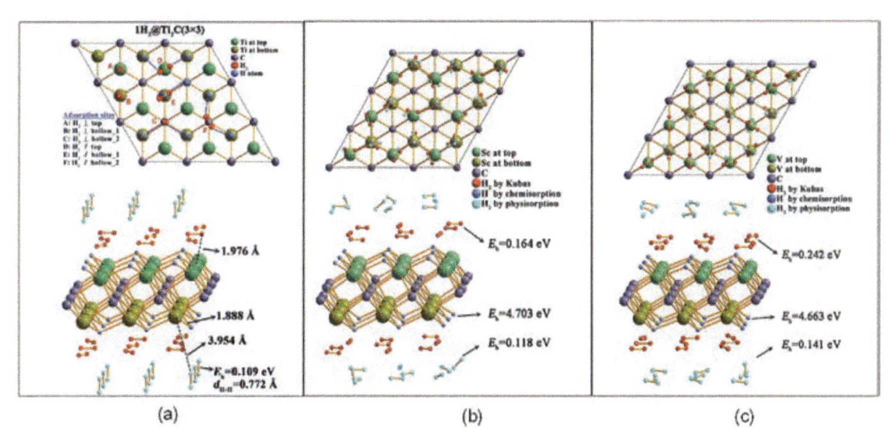

FIGURE 9.18 MXenes for hydrogen storage application (a)Ti2C, (b) Sc_2C, (c) V_2C.
Source: Reproduced with permission from Ref. [93]

There are very few reports till now on ψ-graphene as potential hydrogen storage systems. But the results are promising. We will discuss those research in this section in detail.

9.4.4.1 ZIRCONIUM DECORATED Ψ-GRAPHENE

Using simulations based on Density Functional Theory, Nair et al.[77] investigated the hydrogen storage potential of psi-graphene with Zr decorating. To prevent clustering at both surfaces, they doped Zr atoms at the opposite hexagonal rings. The binding energy of Zr on psi-graphene was calculated to be around −3.54 eV as the result of charge transfer from the Zr 4d orbital to the C 2p orbital. The system was found to be stable at room temperature. 9 H_2 molecules were added to each Zr and the interaction has been demonstrated using TDOS, PDOS, and Bader charge analysis. The system's weight percentage works out to be roughly 11.3 wt%, which is significantly more than the DoE standard.

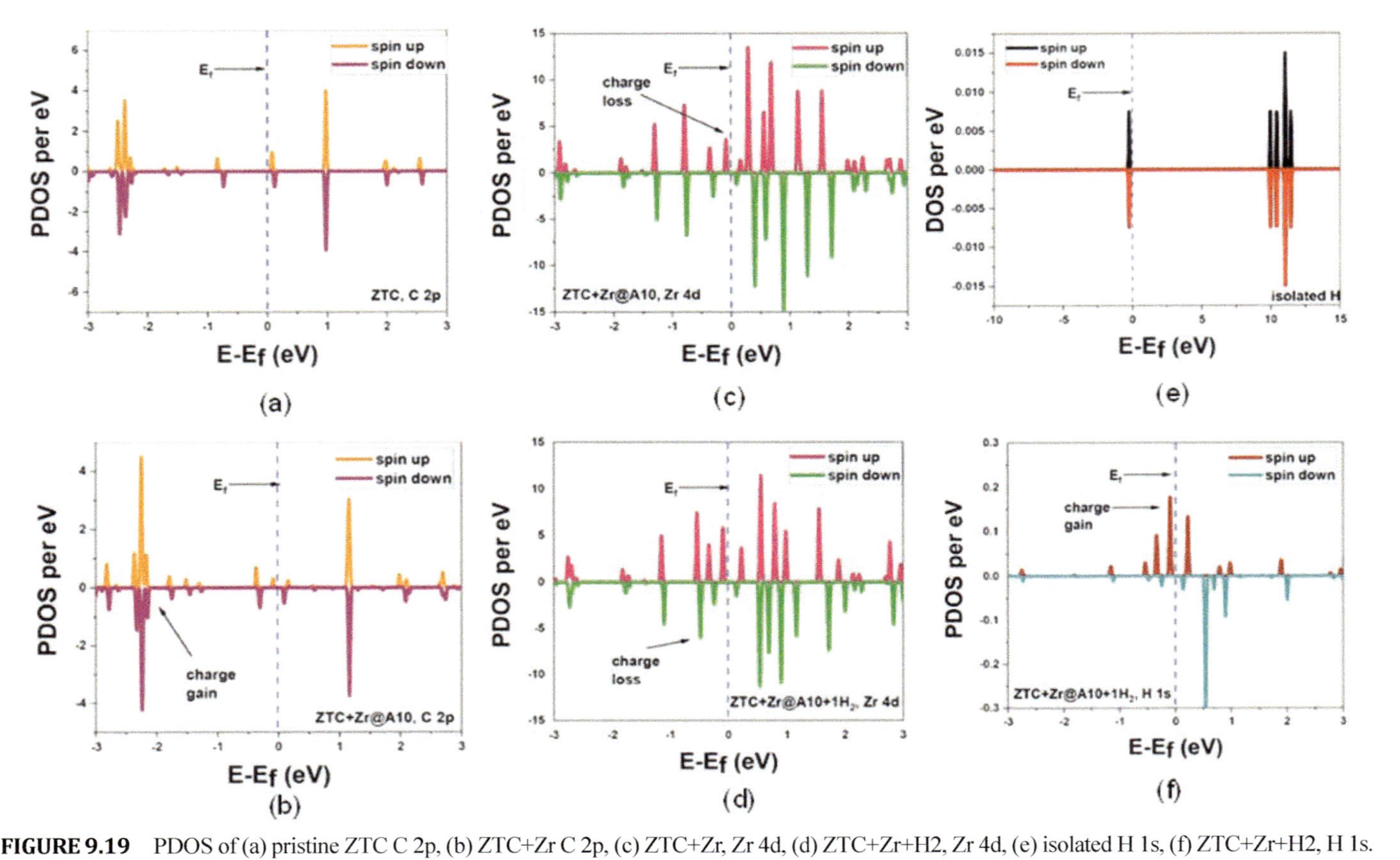

FIGURE 9.19 PDOS of (a) pristine ZTC C 2p, (b) ZTC+Zr C 2p, (c) ZTC+Zr, Zr 4d, (d) ZTC+Zr+H2, Zr 4d, (e) isolated H 1s, (f) ZTC+Zr+H2, H 1s.

Source: Reproduced with permission from Ref. [98]

The TDOS analysis shows (Figure 9.20.a) significant change in density of states and magnetic moment, indicating successful Zr adsorption on ψ-graphene. PDOS analysis shows the charge transfer of Zr-4d to H-1s orbital. Figure 9.20.b shows continuous reduction in density of states upon hydrogen addition showing the charge transfer. Bader charge analysis shows 0.125 e charge gain by hydrogen and 0.16 e charge loss by Zr (for charge donation to ψ-graphene and donation to the adsorbed hydrogen molecules).

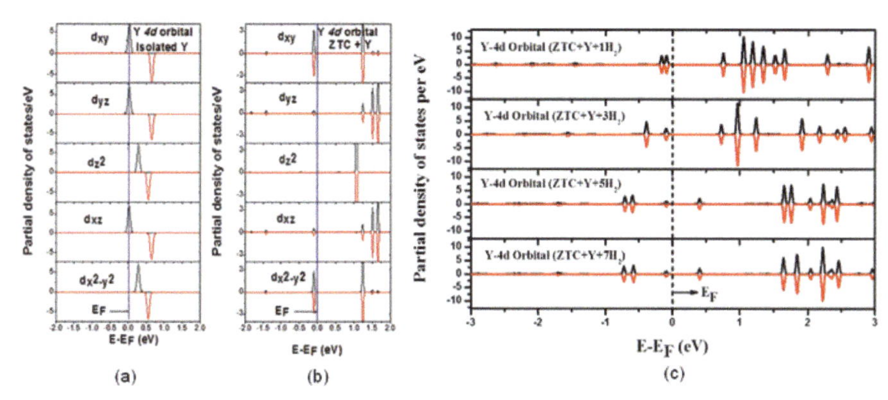

FIGURE 9.20 (a) Suborbital analysis for Y 4d orbital in isolated Y, (b) ZTC+Y, (c) PDOS of ZTC+Y+H_2.

Source: Reproduced with permission from Ref. [99]

9.4.4.2 TITANIUM DECORATED Ψ-GRAPHENE

Chakraborty et al.[78] reported the hydrogen storage capability of Ti-doped ψ-graphene that showed 13.14 wt% of hydrogen storage capacity. Ti was adsorbed on ψ-graphene via the charge transfer from the Ti 3d orbital to the C 2p orbital. And this charge donation caused the adsorption energy of −3.39 eV between Ti and ψ-graphene. This phenomenon can be observed more closely by the analysis of each suborbital PDOS as shown in Figure 9.21. We can clearly notice the change in charge distribution comparing isolated Ti 3d orbital (mainly from d_{xy}, d_{yz}, and d_{xz}) and Ti 3d after adding on to ψ-graphene. According to Bader charge analysis, the transferred charge amount from Ti to C is 1.44 e. Red indicates a charge loss zone, whereas green (reduced gain) and blue suggest an electron gain region (more gain) (Figure 9.22.a).

Further analysis is continued with studying the PDOS and Bader charge analysis between before and after stages of hydrogen adsorption on Ti+ψ-graphene.

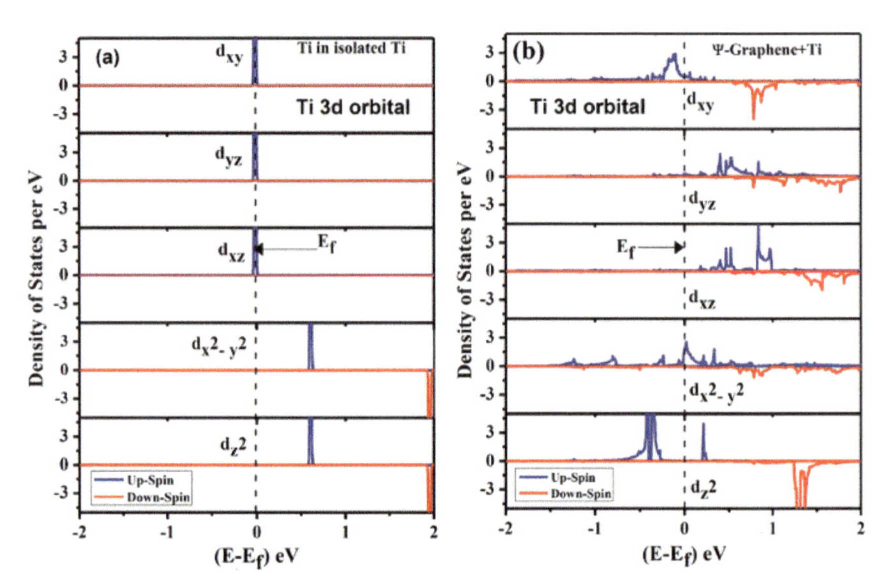

FIGURE 9.21 a) d suborbital PDOS for isolated Ti atom, b) d suborbital PDOS for isolated Ti+ ψ-graphene.

Source: Reproduced with permission from Ref. [78]

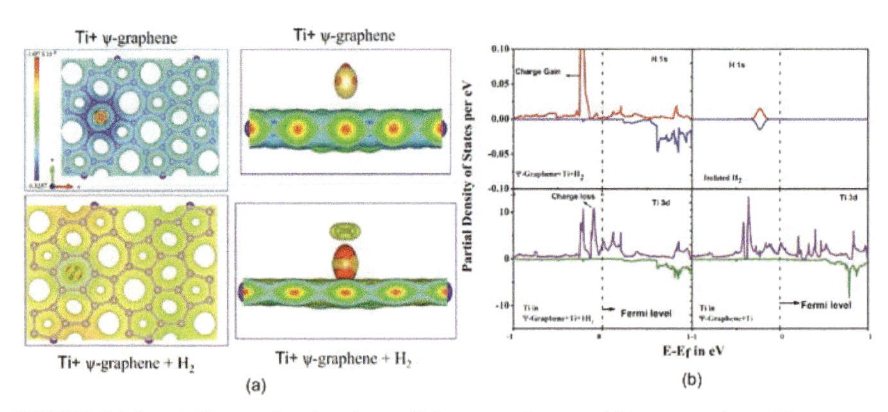

FIGURE 9.22 a) Charge density plots of Ti+ ψ-graphene and Ti+ ψ-graphene+H_2 system, b) PDOS structure for Ti+ ψ-graphene+H_2 system.

Source: Reproduced with permission from Ref. [78]

The enhanced state in H 1s in Figure 9.22.b after the addition on Ti+ ψ-graphene and the depleted state of Ti 3d indicates charge transfer from 3d to 1s. In the Bader charge analysis, we can visualize the same phenomenon as well, and the calculation gives the transferred charge value of 0.074 e.

The diffusion barrier to avoid Ti-Ti clustering was found to be 1.6 eV, which is sufficiently high to prevent metal clustering and making this system feasible for reversible practical applications.

9.4.4.3 YTTRIUM DECORATED Ψ-GRAPHENE

Chakraborty et al.[79] reported Y decorated ψ-graphene as a suitable hydrogen storage medium by DFT investigation. Y was adsorbed on ψ-graphene with a binding energy of −3.06 eV, 2 Y toms were favorably added on the top and bottom of the graphene ring, and each Y atom was able to adsorb 7 H_2 each, resulting in the gravimetric wt% of 8.31.[79]

The suborbital analysis (Figure 9.23) for the density of states of C 2p and Y 4d orbital shows charge transfer from C-2p_z to Y-4d_{xz} and Y-4$d_{x^2-y^2}$. The Bader charge calculation reveals that Y atoms lose 1.2 e charge, while the nearest C atoms of ψ-graphene simultaneously gain 0.92 e. But, addition of 1 H_2 atoms shows transfer of 0.3860 e charge from Yto σ* of H.[79] The back donation from H to Y creates the Kubas interaction, which increases the H–H bonding upon continuous addition, which is shown in Table 9.7. The exact same thing can also be seen in the PDOS analysis of Figure 9.24.a. There is a massive enhancement in the near Fermi region of H-1s when added to Y+ψ-graphene system. Also Figure 9.24b shows continuous depletion of Y 4d orbital depletion, signifying continuous charge transfer.[79]

9.4.5 GRAPHITIC CARBON

Carbon nitride type molecules have interesting properties for various applications including energy storage, electronics, sensor, catalysis, etc.[80] The applications are dependent on the variable properties, which are mainly dependent on C and N atomic ratio and their arrangement in the structure. Among the multiple possible arrangements, one extremely stable heptazine phase is known as graphitic carbon nitride (g-C3N4) which was identified as a suitable hydrogen storage system due to its tubular shape, porous nature, and high surface area.[80]

There are multiple possible hydrogen storage sites which are marked in Figure 9.25.a and these sites exit on both the outer and inner surfaces of the g-C3N4 nanotube (Figure 9.25.b). According to the calculations of Koh et al.[80] the most suitable sites for the H adsorption on the g-C3N4 nanotube are the N around the pores, and up to 3 H_2 molecules were energetically adsorbed on each pore (Figure 9.25.c). The variation in H-N binding energy with number of H_2 addition and the simulated structure of g-C3N4 is shown in Figure 9.25.d.

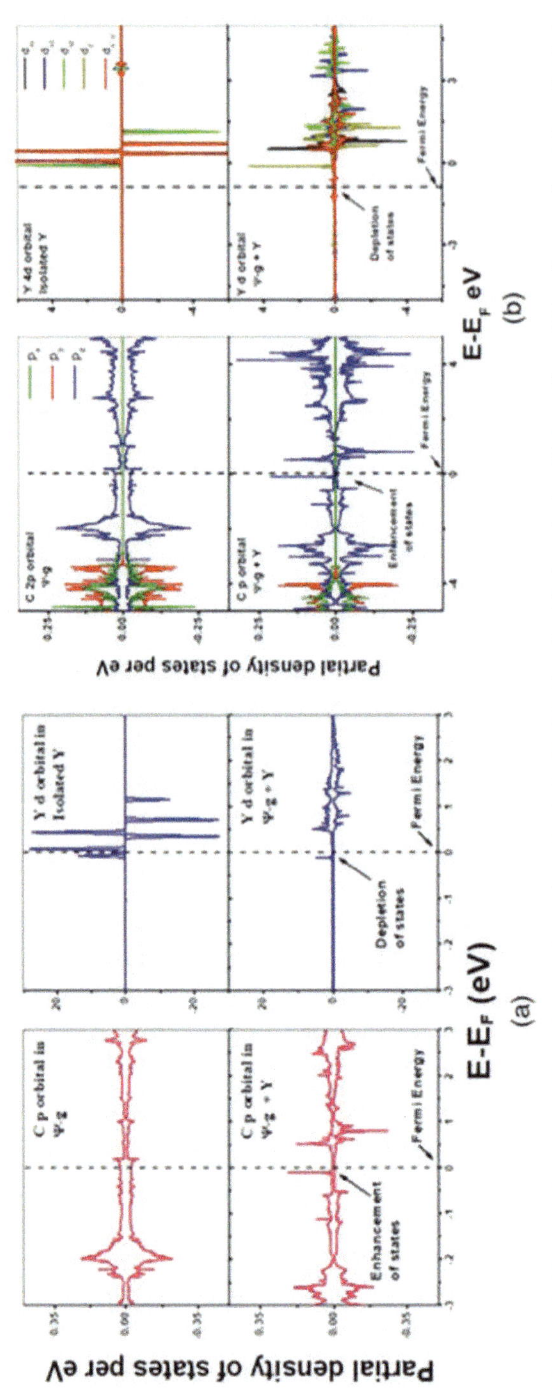

FIGURE 9.23 a) Interaction of Y and ψ-graphene described via PDOS, b) suborbital analysis of interaction of Y and ψ-graphene described via PDOS.

Source: Reproduced with permission from Ref. [79]

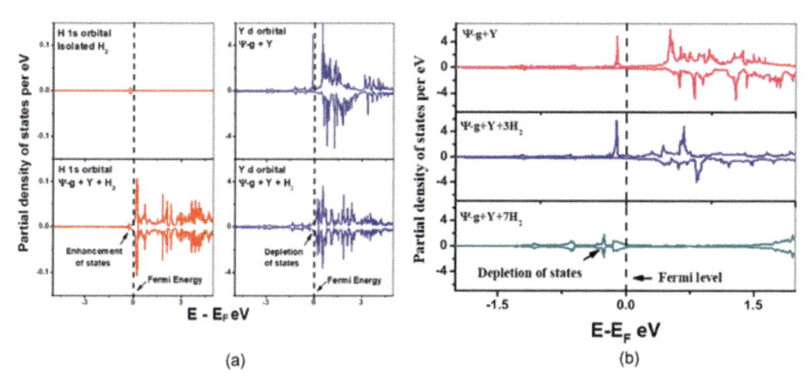

FIGURE 9.24 Interaction of hydrogen with Y+ψ-graphene system.

Source: Reproduced with permission from Ref. [79]

TABLE 9.7 Detail Calculated Parameters for H_2 Interaction on Y+ψ-Graphene System.[79]

Y+ ψ-graphene+ n H_2			
N	**Y-H distance (Å)**	**H-H bond length (Å)**	**Adsorption energy (eV)**
1	2.60	0.77	−0.56
3	2.19	0.77	−0.37
5	2.21	0.80	−0.52
7	4.09	0.80	−0.19

FIGURE 9.25 a) Optimized structure of *g-C3N4*, b) *g-C3N4 nanotube* illustration, c) simultaneous hydrogen addition over one pore of *g-C3N4 nanotube,* d) average H–N bonding energy with several H atoms per pore, e) illustration of hydrogen adsorption on *g-C3N4 nanotube*, f) evolution of adsorption energy over tube.

Source: Reproduced with permission from Ref. [80]

Another important aspect of the study reported by Koh et al.[80] is the variation of binding energy with the tube diameter, as shown in Figure 9.25.f, which illustrates the fact that the average favorable radius of the tube is 9.5 to 18 Å, with a binding energy of −0.29 eV. This graph also gives us the information of favored nanotubular structures over monolayer structures, for hydrogen storage. The maximum wt% reported in this system is 5.45 which includes both physisorbed and chemisorbed hydrogen.[80]

Further studies have been done on this system with metal doping to enhance its performance as a solid-state hydrogen storage medium.

9.4.5.1 Y DECORATED GRAPHITIC CARBON NITRIDE (G-C3N4)

Mane et al.[81] simulated a 3 × 3 supercell of g-C3N4 with lattice parameters of 14.37 Å × 14.37 Å. The stable structure is shown in Figure 9.26.a. This structure has the band gap of 1.6 (PBE-GGA) and 2.86 eV (HSE06), whereas the experimentally reported band gap for this molecule is 2.72 eV. Among 4 prospective adsorption sites, Y atom preferred the hollow site in between 3 triazine rings with a binding energy of −6.852 eV (Figure 9.26.b). From PDOS analysis (Figure 9.27) we see enhancement in the C 2p and N 2p orbitals near Fermi level and depletion of Y 4d orbitals; Bader charge analysis shows charge transfer of 1.653 e from Y to g-C3N4.[81]

Mane et al.[81] reported Y+ g-C3N4 system could successfully adsorb 9 H_2 molecules and as a result, the gravimetric wt% of the system is reported to be 8.55. The investigation of charge transfer phenomenon is as usual done via PDOS and Bader charge analysis. The PDOS graphs (Figure 9.27) show us that upon addition of H_2 on Y+g-C3N4 system H 1s orbital gains charge and Y 4d orbital continuously losses charge with addition of 1-9 H_2 molecules. According to the Bader charge calculation, the Y atom transfers 0.044 e charge to nearest H_2 molecules, whereas each H_2 molecule gains 0.0226 e charge.

Also, the calculation of diffusion energy barrier shows that the possibility of metal–metal clustering is negligible enough as the existing diffusion energy barrier for this system is calculated as 3.07 eV and the calculated desorption temperature is 384.24. All these criteria are well suitable according to DoE requirements.

9.4.5.2 FE, RU, OS DECORATED GRAPHITIC CARBON NITRIDE (G-C3N4)

In a recent study, Habibi et al.[82] used DFT calculations to examine the hydrogen adsorption behavior over the transition metals (TM = Os, Ru, and

Fe) contained in g-C3N4 as well as the binding energies of these elements over the material. The embedded Os, Ru, and Fe elements are observed to have calculated binding energies of −5.697, −5.288, and −5.195 eV, respectively, on the g-C3N4 materials. The results showed that the Os is highly adsorbed on the porous structure of g-C3N4 because of the overlap between the Os 5d orbitals and the N 2p orbitals, than that of the Ru and Fe atoms.[82] Each transition metal atom was reported to adsorb 6 H2 molecules. The adsorption details are provided in Table 9.8. From this data we can understand that the most favorable system in this case is Os+g-C3N4. But the adsorption energies are maximum time out of the DoE criteria. Hence though these systems are good for hydrogen storage application, but they have the drawback of irreversibility to be applied in FCV applications in practical.[82]

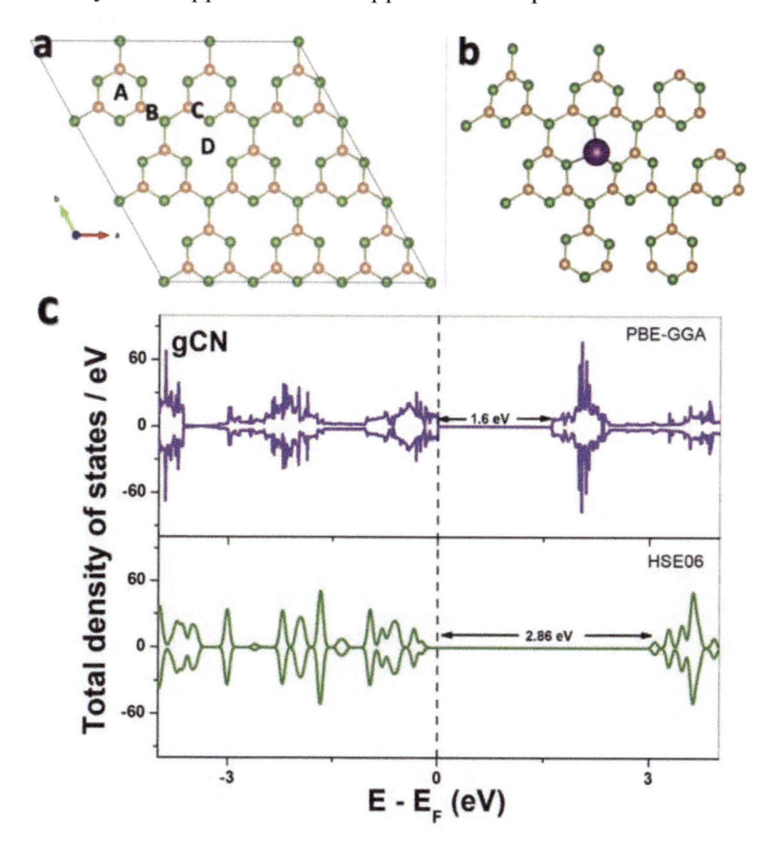

FIGURE 9.26 a) g-C3N4 supercell, b) Y-doped g-C3N4 supercell, c) TDOS of g-C3N4 and Y doped g-C3N4.

Source: Reproduced with permission from Ref. [81]

TABLE 9.8 Adsorption Details of Fe, Ru, Os Decorated Graphitic Carbon Nitride for Hydrogen Storage Application.[82]

Metal+ g-C3N4	Adsorption energy (eV)						$\Delta q\ (H_2)$	Δq (M)	H-H (Å)
	1 H_2	2 H_2	3 H_2	4 H_2	5 H_2	6 H_2			
Fe	−1.305	−1.039	−0.662	−0.146	−0.182	−0.185	0.022	−0.432	0.749
Ru	−2.208	−1.901	−0.682	−0.152	−0.901	−0.362	0.025	−0.542	0.751
Os	−2.452	−2.250	−0.692	−0.663	−1.091	−1.053	0.027	−0.550	0.755

9.4.5.3 SC, TI, NI, V DECORATED GRAPHITIC CARBON NITRIDE (G-C3N4)

Panigrahi et al.[83] explored the structural, electrical, charge transfer, and H_2 adsorption mechanisms of metal functionalized g-C3N4 sheets using spin-polarized DFT-D2 level modeling. Their calculations suggest that g-C3N4 binds to metal dopants like Sc, Ti, Ni, and V rather strongly and even more so to each dopant's specific dimer structure, which is substantially stronger than the observed cohesive energies for the dopant. As a result, these metal dopants on g-C3N4 can produce a uniform coverage effect while preventing metal–metal clustering. The thermal stabilities of all the four metal functionalized g-C3N4 sheets are confirmed by AIMD simulations at 400 K. A considerable charge transfer from the metal dopants to the N atoms of the g-C3N4 sheet is shown by the Bader charge analysis. As a result, the metal dopants polarize the H_2 molecules by acting as cationic entities with strong attraction for them. All the calculation data are listed in Table 9.9. Dimers functionalized to g-C3N4 sheets are found to be more promising for H_2 storage than monomers because the binding energy of each H_2 molecule is between −0.6 and −0.9 eV/H_2.

TABLE 9.9 Parameters for Sc, Ti, Ni, V Decorated Graphitic Carbon Nitride System as Hydrogen Storage [SS: Single sided; DS/ Double sided].[83]

	BE/M (eV)	BE/M2(eV)	BE/H_2(eV)			Desorption temperature (K)		
			SS	DS	Dimer	SS	DS	Dimer
Sc	−7.26	−8.36	−0.13	−0.21	−0.69	166	269	883
Ti	−6.97	−10.67	−0.02	−0.16	−0.93	25.6	205	1190
Ni	−3.01	−4.43	−0.38	−0.27	−0.46	486	345	589
V	−8.27	−6.42	−0.12	−0.05	−0.28	154	64	358

9.4.5.4 SC DECORATED TRIAZINE GRAPHITIC CARBON NITRIDE (G-C3N4)

Foreseeing whether g-C3N4 would work as a practical hydrogen storage technology for onboard fuel applications, systematic DFT simulations have been carried out on Sc-doped triazine-based g-C3N4. This system has been not considered much, whereas heptazine base g-C3N4 has recently been an interesting material for hydrogen research with promising possibilities. Hence, Chakraborty et al.[84] continued their study on this Sc-doped triazine-based g-C3N4. With a storage capacity of 8.55 wt%, which is higher than the DOE's aim of 6.5%, a single Sc atom coupled to the surface of g-C3N4 may reversibly bind 7H2. Furthermore, ab initio MD simulations show that Sc is energetically favorable to be attached to g-C3N4 with a high binding energy of −7.13 eV, and that this bond persists even at higher temperatures of up to 500 K.[84] A significant 2.79 eV diffusion energy barrier prevents Sc from forming clusters over g-C3N4, which reduces the likelihood of cluster formation. An average hydrogen binding energy of around −0.39 eV/H2 and a desorption temperature of 458.28 K are observed in this system.[84] Using PDOS and Bader Charge Analysis, the mechanism of interaction of Sc with g-C3N4 and hydrogen molecules on Sc-g-C3N4 has been clarified. It (Figure 9.28) shows that Sc loses charge while interacting with g-C3N4, where a charge transfer from Sc 3d to C 2p orbital is detected.[84] With a net charge transfer from the hydrogen molecule's Sc 3d orbital to its 1s orbital, the binding of hydrogen may adhere to Kubas' style of binding. According to the Bader charge analysis (Figure 9.28), Sc losses 1.98 e charge to g-C3N4 and 0.11 e charge to H_2.[84]

9.4.6 HOLLEY GRAPHYNE

Liu et al.[85] recently achieved the effective synthesis of holey graphyne (HGY), a *p*-type semiconducting, 2D graphyne like material with high hall mobility. In HGY, the ratio of *sp* to sp^2 carbon is 0.5, and there are connections between the six and eight vertex carbon rings (Figure 9.29). Single-layer HGY adopts a hexagonal lattice with space group P6/mm that resembles graphene. There are two distinct kinds of carbon atoms, totalling 24 in each unit cell of HGY (C1 and C2). C–C bonds come in four different varieties: two sp^2-sp^2 bonds (B1 = 1.463 Å, B2 = 1.397 Å), one sp-sp^2 bond (B3 = 1.414 Å), and one sp-sp bond (B4 = 1.227 Å). Similar to GY, the HGY sheet is composed of hexagonal benzene rings linked together by acetylenic connections (C).[,48,85,86] In GY structures, three acetylenic linkages surround to

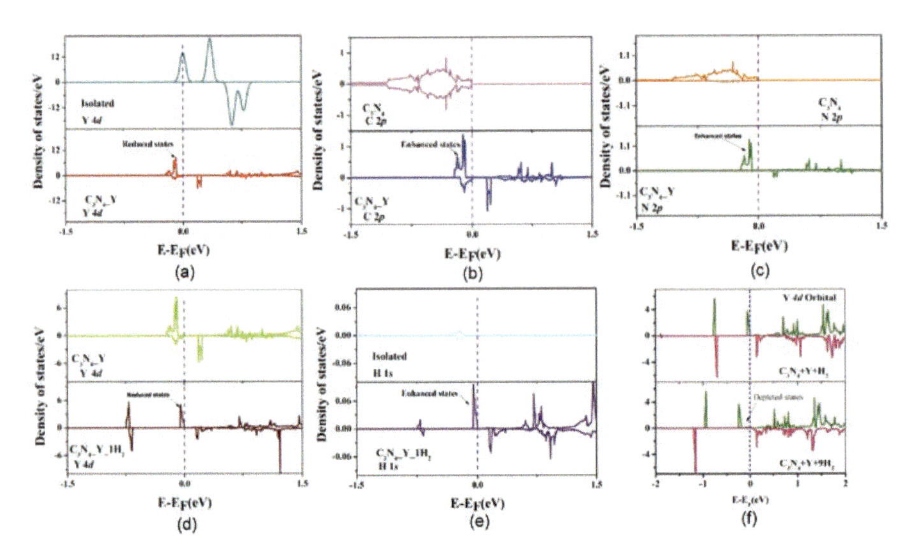

FIGURE 9.27 PDOS analysis of (a) Y 4d orbital of isolated Y and Y doped on g-C3N4 supercell, (b) C 2p orbital, (c) N 2p orbital, (d) Y 4d orbital in g-C3N4 supercell before and after H_2 addition, (e) H 1s orbital, (f) Y 4d orbital with simultaneous H_2 addition; reproduced.

Source: Reproduced with permission from Ref. [81]

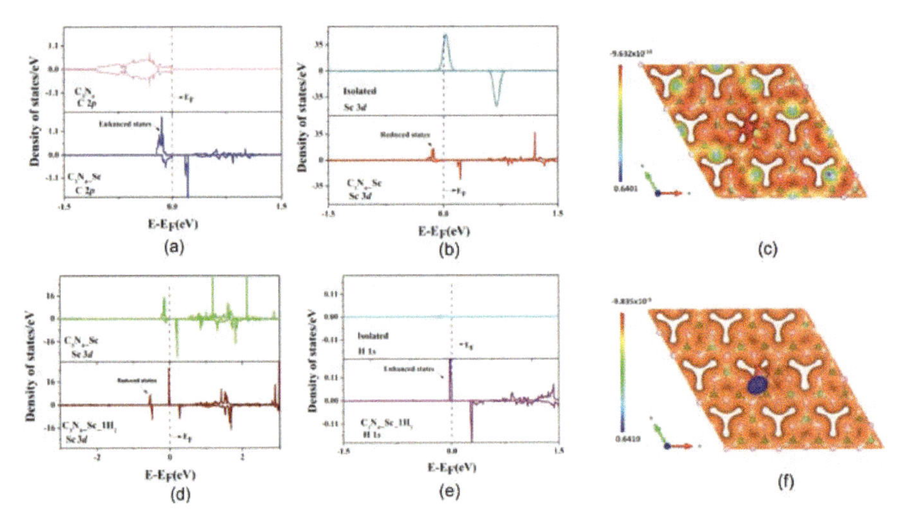

FIGURE 9.28 a) C 2p orbital before and after Sc addition, b) Sc 3d orbital of isolated and after addition on *g-C_3N_4*, c) charge density plot for Sc+*g-C_3N_4*, d) Sc 3d orbital in Sc+*g-C_3N_4 and* Sc+*g-C_3N_4+H_2*, e), f) charge density plots for Sc+*g-C_3N_4+H_2*.

Source: Reproduced with permission from Ref. [84]

form a large triangular pore, and the angles between acetylenic linkages and benzene rings are exactly 120°, whereas in HGY structures, six acetylenic linkages surround to form a large pore, and the angles between acetylenic linkages and benzene rings are approximately 125.8° (Figure 9.29).[86] In addition, two nearby acetylenic connections in HGY simultaneously attach to the nearby benzene rings to form an eight-vertex ring. The consistent holes created by the acetylene linkage connecting the benzene rings in HGY make them appropriate for use in energy storage applications. Different high symmetry adsorption sites are hole sites (V, H1, H2, and H3), two top sites (C1, C2), and four bridge sites (B1, B2, B3, and B4) (Figure 9.29).[86]

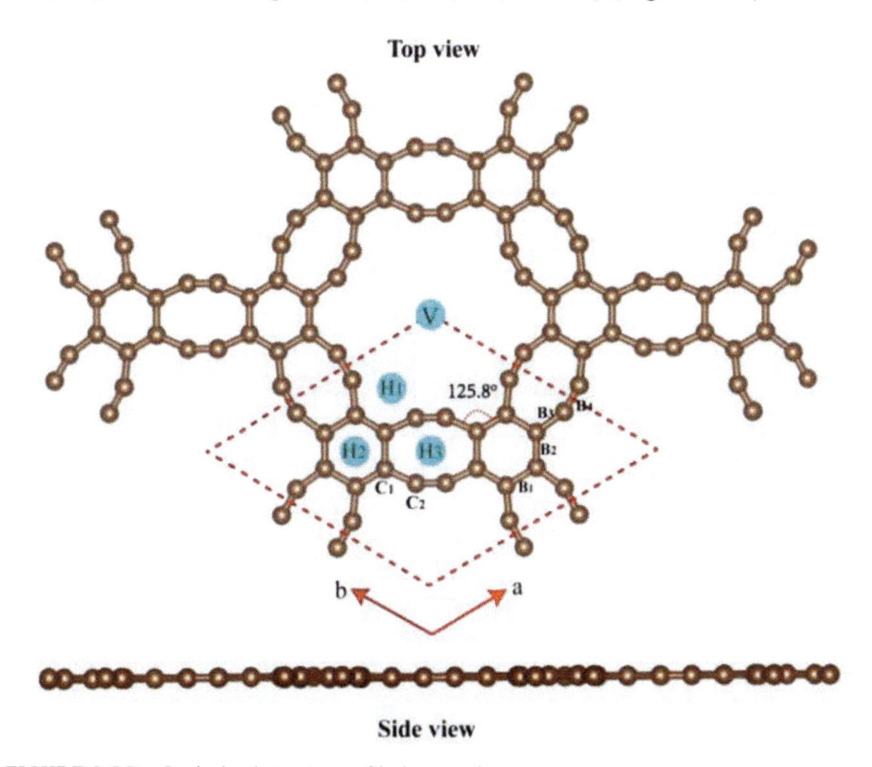

FIGURE 9.29 Optimized structure of holey graphene.

Source: Reproduced with permission from Ref. [86]

Gao et al.[86] investigated the hydrogen storage capability of pristine HGY system, but the binding energy values at each possible adsorption site were between −0.04 eV and −0.06 eV, which is too low for proper application. Hence further importance was given on doped HGY structures.

9.4.6.1 LITHIUM DECORATED HOLEY GRAPHYNE

Gao et al.[86] investigated the Li-decorated HGY'sH$_2$ storage behavior. These complexes can have a H storage capacity of up to 12.8 wt % with 6 Li atoms decorated on HGY and each Li adsorbing 24 H$_2$ molecules, according to numerical studies; this gravimetric density is much greater than that of several common GY nanostructures with alkali–metal decorations. The adsorption energy per H$_2$ molecule, however, was −0.22 eV, which is ideal for reversible H2 adsorption/desorption at temperatures close to ambient temperature. Also, Li–Li clustering possibility has been measured by Gao et al.,[86] and the results were satisfactory, showing 0.44 eV, 0.26 eV, and 2.80 eV as the diffusion barriers of all possible paths; whereas the thermal energy of Li at high temperature of 450 K was only 0.038 eV. Hence the clustering possibility was discarded.[86]

The interaction of the hydrogen and the host has been analyzed using PDOS, charge density plots, and Bader charge analysis. Figure 9.30 demonstrates the PDOS of 6Li-HGY before and after H$_2$ adsorption. Before and after the H2 adsorption, there is a little variation in the PDOS of C, which suggests that there is only a weak contact between the H$_2$ molecules and the host materials, namely the acetylenic links of the HGY sheet. But the PDOS of Li following H$_2$ adsorption shows a sizable variance which clarifies that the H2 molecules mostly interact with the Li atoms.[86]

Depending on the type of rearrangement, two types of adsorbed hydrogen molecules have been observed in the system. Closer to Li atom (Li-H distance 2 Å, H1 Type) and slightly far from the Li atom (Li-H distance 3.77 Å, H2 Type), charge density analysis reveals the polarization between the adsorbed hydrogen and the substrate, where H1 type hydrogen molecules are more polarized than compared to H2 type hydrogen molecules. Bader charge analysis qua notifies this effect showing 0.08 e charge transfer from Li to H1 and 0.05 e charge transfer from Li to H2.

As a result of this week interaction, the desorption temperature of the system is only 282 K, which is lower than room temperature, making this system difficult for the application for reversible hydrogen storage.

9.4.6.2 SCANDIUM DECORATED HOLEY GRAPHYNE

Mahamiya et al.[87] used molecular dynamics modeling and density functional theory to demonstrate the hydrogen adsorption and desorption characteristics of the Sc-doped HGY (Figure 9.31.a). With a binding energy of around 4.56 eV, the Sc atom is fixed to the top of the HGY's octagon. They reported that each adsorbed Sc atom can bind maximum of 5 hydrogen molecules, and each holey graphene can accommodate a total of 6 Sc atoms (3 on top, 3 at bottom) resulting in a gravimetric weight percentage for the system of 9.80 wt%, which is significantly greater than the DoE-US standards. The Sc decorated HGY system's average desorption temperature was calculated to be 464 K, making it ideal for real-world fuel cell applications.

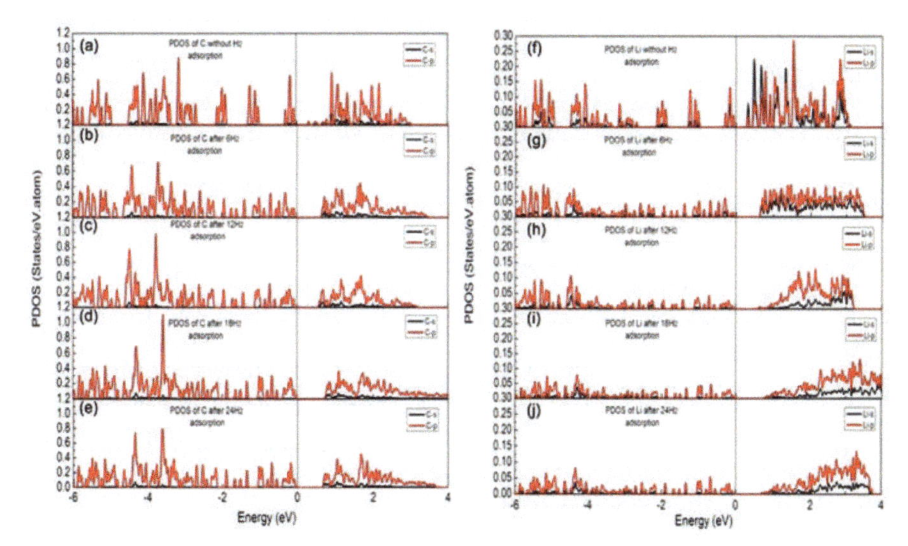

FIGURE 9.30 PDOS analysis of Lithium decorated holey graphyne.

Source: Reproduced with permission from Ref. [86]

The band structure of HGY and HGY+Sc shows (Figure 9.31.b) evolution of metallic character upon doping. According to Modak etal.,[88] compared to semiconducting carbon nanostructures, metallic carbon nanostructures are able to adsorb more hydrogen molecules through Kubas interactions. Using the similar theory, we can explain the high performance of Sc-doped HGY. But for further analysis, Mahamiya et al.[87] analyzed PDOS of pristine HGY with Sc-doped HGY, as shown in Figure XX. The enhancement in C-2p states upon addition of Sc signifies charge gain, whereas the loss of Sc 3d states near Fermi level compared to isolated Sc atom signifies the charge transfer of 3d of Sc to C 2p of HGY.

Further analysis was done upon addition of hydrogen atoms on Sc. Charge transfer from 3d orbital of HGY+ Sc is visible in PDOS (Figure 9.32). The charge is transferred to the 1s orbital of H_2, and as a result, we see enhanced states near the Fermi level of hydrogen, attached to Sc-doped HGY. Bader charge analysis also showed a transfer of 1.38 e charge to HGY and 0.08 e charge to H_2 of the HGY+Sc+H_2 system (Figure 9.33). For practical applications, diffusion energy barrier was calculated using two different possible pathways, and both the paths have higher energy barrier of 3.15 eV and 2.93 eV, which is much higher than the thermal energy of the Sc at desorption temperature and hence, the possibility of metal–metal clustering is very low.

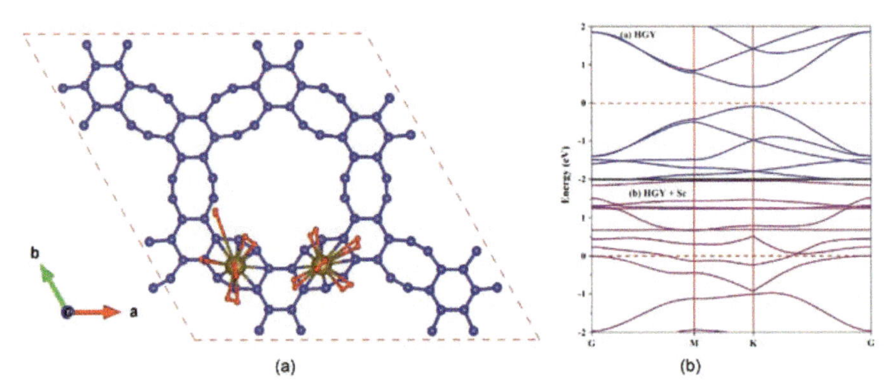

FIGURE 9.31 a) Optimized structure of HGY+Sc+H_2, b) band structure of pristine HGY and HGY+Sc.

Source: Reproduced with permission from Ref. [100]

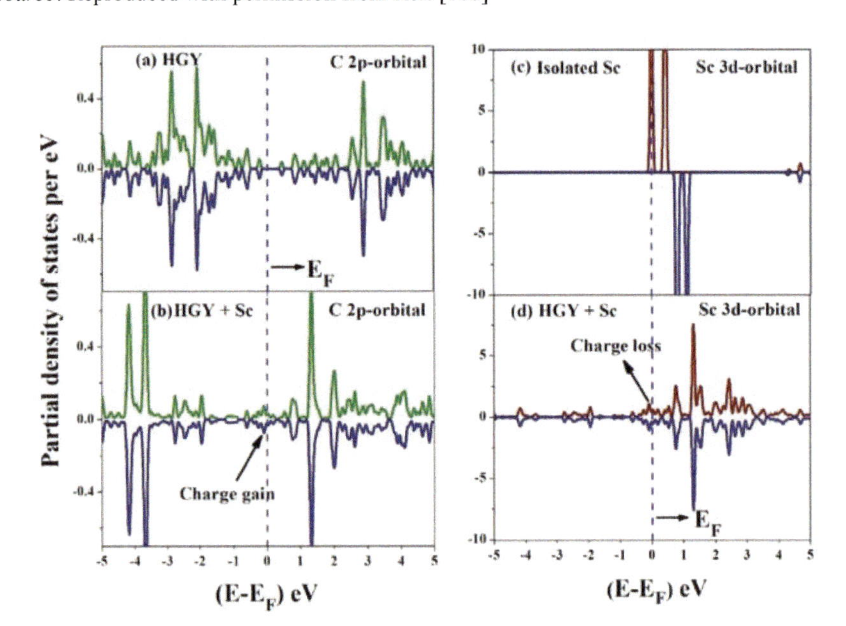

FIGURE 9.32 PDOS analysis of (a) C 2p orbital in pristine HGY, (b) C 2p orbital in HGY+Sc, (c) 3d orbital of isolated Sc, (d) 3d orbital of is HGY+Sc.

Source: Reproduced with permission from Ref. [100]

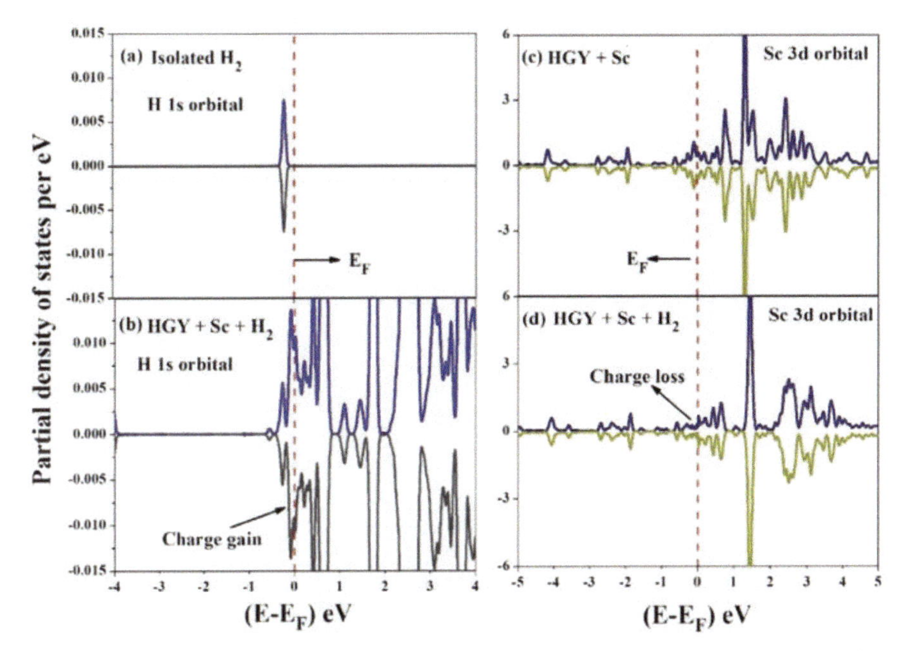

FIGURE 9.33 PDOS analysis of (a) H 1s orbital of isolated hydrogen, (b) H 1s orbital in HGY+Sc+H_2, (c) 3d orbital of HGY+Sc, (d) Sc 3d orbital of is HGY+Sc+H_2.

Source: Reproduced with permission from Ref. [100]

9.4.6.3 TITANIUM DECORATED HOLE GRAPHYNE

Dewangan et al.[89] investigated the H_2 adsorption and desorption characteristics of the Ti-functionalized holey graphyne system using density functional theory simulations. According to the simulation results, the Ti atom has a strong Dewar interaction with the holey graphyne sheet, which results in a binding energy of −4.16 eV. In the Ti-functionalized HGY system, charge is donated from the carbon atom's π orbitals to Ti-d orbitals while at the same time, charge is back donated from full Ti-d orbitals to the carbon atoms' unoccupied π* antibonding orbitals in the HGY surface. As a result, the Dewar contact between the Ti atom and the HGY surface creates a strong connection. Both TDOS (Figure 9.34. a-b) and PDOS (Figure 9.34. c-e) show the enhanced interaction between Ti and HGY.

With an average H_2 adsorption energy of around 0.38 eV/H2, the Ti-functionalized holey graphyne may absorb up to 7H2 molecules, resulting in a hydrogen gravimetric density of 10.52 wt%. Table XX sows the Kubas interaction between the adsorbed hydrogen and Ti+HGY where the H6H

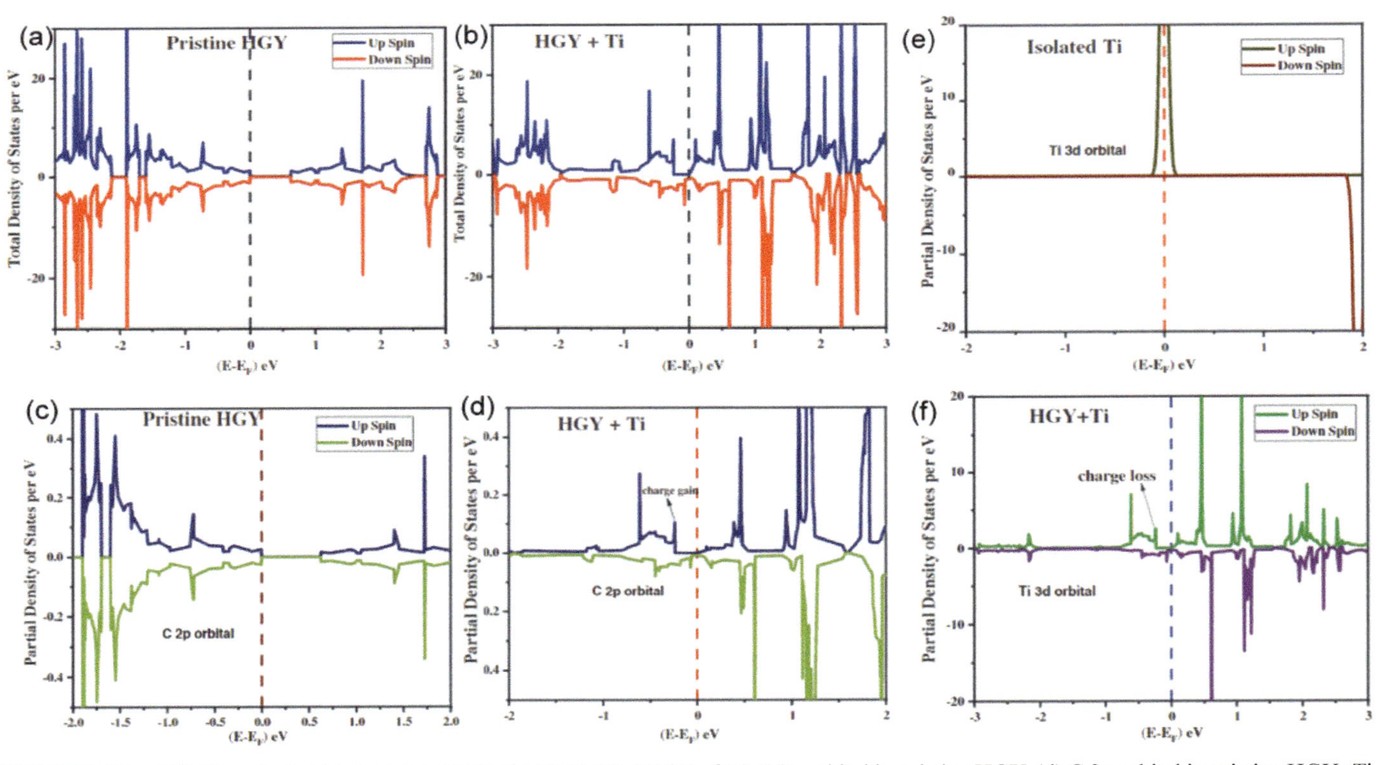

FIGURE 9.34 TDOS analysis of (a) pristine HGY, (b) HGY+Ti; PDOS of (c) C 2p orbital in pristine HGY, (d) C 2p orbital in pristine HGY+Ti, (e) Ti 3d orbital of isolated Ti, (f) 3d orbital of is HGY+Ti.

Source: Reproduced with permission from Ref. [89]

distance is elongated from ideal 0.75 to upto 0.82 Å, without breaking the H-H bonding. The PDOS analysis, as shown in Figure 9.34.e-f and Figure 9.35, shows charge transfer from Ti 3d to H 1s orbital. The Bader charge analysis, as shown in Table 9.10, also quantifies the same argument.

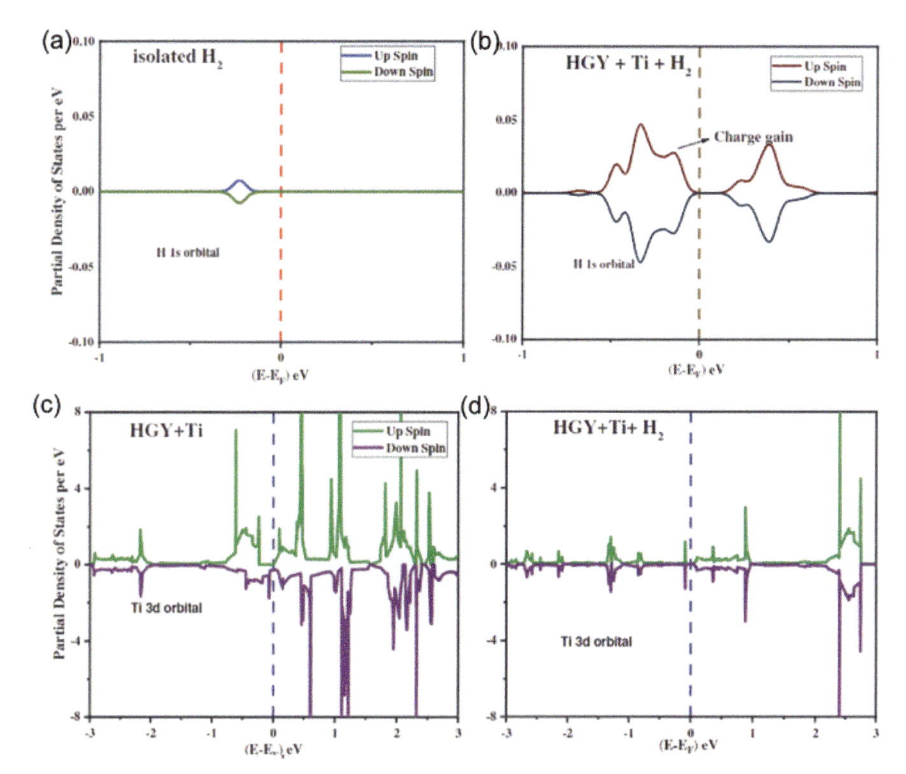

FIGURE 9.35 PDOS analysis of (a) H 1s orbital of isolated hydrogen, (b) H 1s orbital in HGY+Ti+H_2, (c) 3d orbital of HGY+Ti, (d) Sc 3d orbital of is HGY+Ti+H_2.

Source: Reproduced with permission from Ref. [89]

The Van't Hoff relation is used to calculate the average desorption temperature, which is 486 K, and the diffusion energy barrier was found to be 2.3 eV—the ideal values for practical applications.

9.4.6.4 YTTRIUM-DOPED HOLEY GRAPHYNE

Singh et al.[48] analyzed the hydrogen storage capacity of ultra-porous holey graphyne using first-principles DFT calculations and reported that 3 Y atoms can be loaded on each side of HGY and each Y can adsorb 7 H2 molecules, resulting in a total of 6 Y atoms and 42 H2 adsorption. 9.34 wt% gravimetric storage capacity was obtained on using both sides of HGY+Y system, with an average binding energy and desorption temperature of −0.34 eV and 438 K, respectively. According to the study of the PDOS (Figure 9.36) and Bader charge, the interaction between Y and HGY is caused via charge transfer from Y-4d to C-2p.[48]

TABLE 9.10 The Adsorption Analysis in HGY+Ti+H_2 system.[89]

System HGY+Y+ n H_2 n =	E_{ads}(eV)	H-Ti (Å)	H-H (Å)	Δq (e)
1	−0.59	1.94	0.82	0.15
2	−0.48	2.02	0.80	0.08
3	−0.34	2.22	0.82	0.03
4	−0.42	1.96	0.84	0.07
5	−0.32	4.30	0.82	0.003
6	−0.28	3.96	0.82	0.02
7	−0.20	4.27	0.82	0.004

Additionally, it has been shown that the interaction between the H2 molecule and the HGY + Y system is a Kubas type interaction via the charge transfer of Y-4d to the σ* of H-1s and a back donation takes place as well simultaneously from σ of H-1s to Y-4d.[48] Bader charge calculation shows that the amount of charge transfer is 0.007 e. Also, the diffusion barrier for checking metal clustering was calculated as 2.95 and 13.96 eV for two different paths, and both of them can be considered too high to allow metal clustering.

9.4.7 MXENE

Transition metal carbides, nitrides, or mixtures make up 2D MXene. The long-standing issue of transition metal aggregation in solid-state hydrogen storage media can be resolved in this system by the strong bonding between the transition metal and carbon/nitrogen in MXenes.[40,90,91] A great possibility of adsorbing molecular hydrogen with a desired binding energy of 10–20 kJ/mol for practical applications is presented by the partly metallic character of 2D MXenes. There have been several documented theoretical investigations into the storage of hydrogen in 2D MXenes. Only a small number of experimental investigations have, however, been published on the use of MXenes for hydrogen storage.[92]

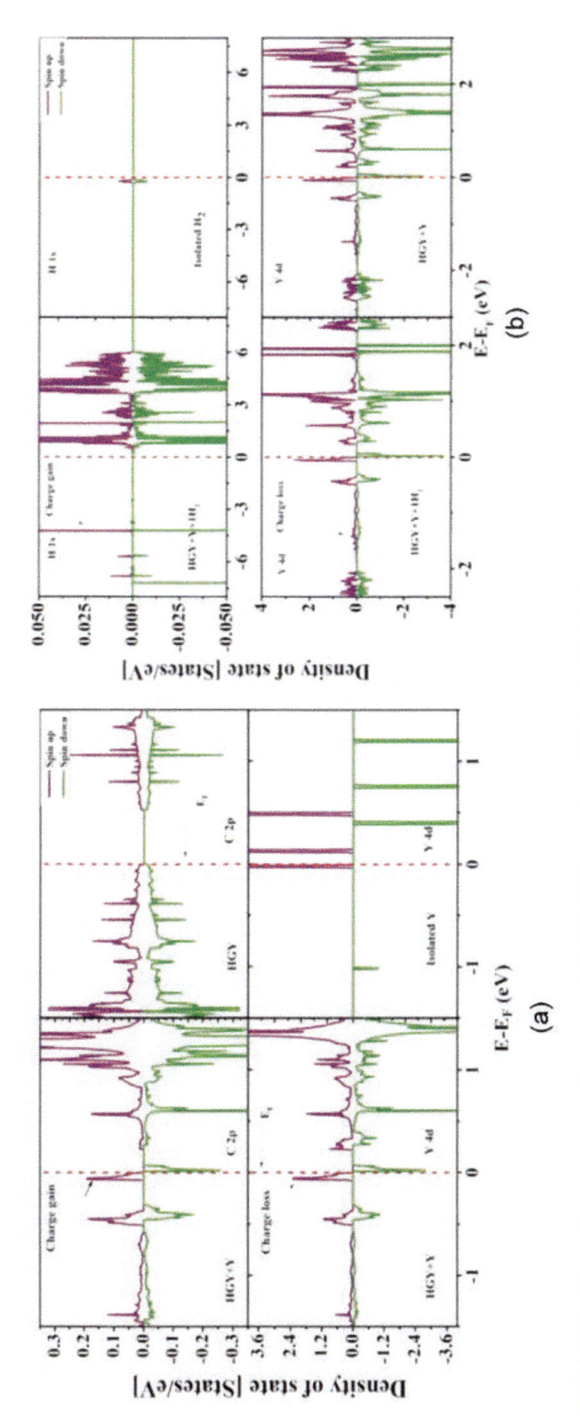

FIGURE 9.36 a) Interaction of Y atom with HGY, b) interaction of H2 with HGY+Y.

Source: Reproduced with permission from Ref. [48]

9.4.7.1 $TI_2C/SC_2C/V_2C$

First-principles calculations served as the foundation for the initial Ti_2C MXene hydrogen storage research.[93] Due to its large surface area and robust Ti-C bond, the Ti_2C MXene was anticipated to have a high gravimetric storage capacity. Calculations of the first-principles total-energy pseudopotential for Ti_2C were performed using the density functional theory (DFT). A Ti_2C 3×3 periodic supercell with six ideal hydrogen adsorption sites was the subject of calculations on hydrogen adsorption, as illustrated in Figure 9.37. According to calculations, the respective hydrogen adsorption values (and binding energies) are 1.7 wt% (5.027 eV), 3.4 wt% (0.272 eV), and 3.4 wt% (0.109 eV). Accordingly, the high binding energy of chemisorption under ambient circumstances prevents the chemically bound hydrogen from being released. In contrast, it is highly challenging to physically attach hydrogen to the host material. Therefore, the Kubas-type interaction, which is applicable in real-world applications, was used to get the 3.4 wt% reversible hydrogen storage capacity. The gravimetric capacity of 6.5 wt% defined by U.S. DOE is lower than the overall hydrogen adsorption capacity of 8.6 wt%.[93]

Similar calculation was also performed on Sc_2C and V_2C-type MXenes and the results were promising and analogous.

9.4.7.2 CR_2C

Density functional theory was used to examine the hydrogen storage capabilities of 2D Cr_2C MXene. For hydrogen adsorption, various calculation models for binding energies parallel and perpendicular to the Cr_2C surface were taken into account. High H-adsorption was shown by a partial density of states analysis. The density of states and binding energy supported the idea that weak electrostatic, physisorption, and chemisorption interactions between atomic and molecular hydrogen and a Cr_2C surface are conceivable.[94] The Cr_2C has a calculated 7.6 weight percent of H2 gravimetric storage capacity, of which 3.2 weight percent is attached with Kubas-type and 3.2 weight percent is H_2 adsorbed with weak electrostatic interaction (binding energy −0.26 eV/H2 with significant charge transfer between Cr and H2), which is useful for the reversible H2 storage near ambient condition. The result is that the reversible hydrogen storage capacity in ambient conditions (controlled by hydrogen bonded with energies ranging from 0.1 to 0.4 eV/H2, in the present case through Kubas and weak electrostatic interactions) is 6.4 wt.%, which is better than the previously studied Sc, Ti, and V-based MXene materials.[94]

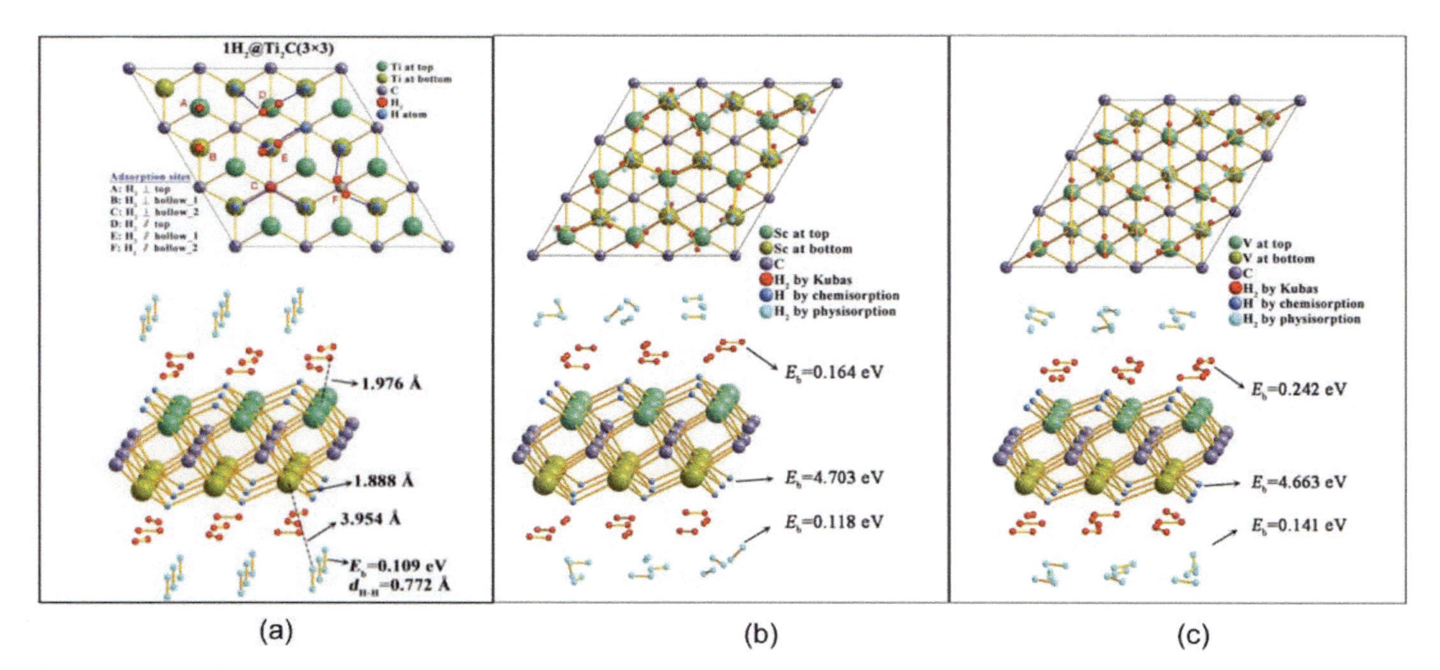

FIGURE 9.37 MXenes for hydrogen storage application: (a) Ti2C, (b) Sc_2C, (c) V_2C.

Source: Reproduced with permission from Ref. [93]

9.4.7.3 TI_2N

Hydrogen storage in 2D Ti2N was calculated by first-principles calculations. Simulation results revealed 8.55 wt% maximum hydrogen storage capacity in Ti_2N MXenes.[95] The interaction of the H-atoms or H2-molecules with the Ti2N monolayer is shown to be chemisorption in the first layer of H-atoms, Kubas interaction in the second layer of H_2-molecules, and physical interaction in the third layer of H_2-molecules, according to the adsorption energy, the charge transfer, and the projected electronic density of states. However, the reversible hydrogen storage capacity was 3.42 wt%, which is obtained from Kubas-type interactions. Li et al.[95] also investigated the hydrogen adsorption probability in Ti_2NO_2, Ti_2NF_2, and $Ti_2N(OH)_2$. The calculated comparable factors are listed in Table 9.11.[95]

TABLE 9.11 Hydrogen Storage in MXenes.[93–95]

Materials	Configurations	Adsorption energy (eV/H or eV/H2)	H_2 storage capacity	
			Reversible (wt%)	Total (wt%)
Ti_2C	Ti_2C18H	−5.027	1.7	8.6
	$Ti_2C\_18H\_18H_2$	0.272	3.4	
	$Ti_2C\_18H\_36\ H_2$	−0.109	3.4	
Sc_2C	Sc_2C_18H	−4.703	1.8	9.0
	$Sc_2C\_18H\_18\ H_2$	0.164	3.6	
	$Sc_2C\_18H\_36\ H_2$	−0.087	3.6	
Cr_2C	First layer	−0.96	1.2	7.6
	Second layer	−0.14	3.2	
	The last layer	−0.26	3.2	
Ti_2N	$Ti_2N\ \_18H$	−3.630	1.711	8.555
	$Ti_2N\ \_18H\_18\ H_2$	0.140	3.422	
	$Ti_2N\ \_18H\_36\ H_2$	−0.108	3.422	
Ti_2NO_2	$Ti_2NO_{2}\_18\ H_2$	−0.142	2.691	5.382
	$Ti_2NO_{2}\_36\ H_2$	−0.111	2.691	
Ti_2NF_2	$Ti_2NF_{2}\_18\ H_2$	−0.137	2.588	5.176
	$Ti_2NF_{2}\_36\ H_2$	−0.111	2.588	
$Ti_2N(OH)_2$	$Ti_2N(OH)_{2}\_18H;$	−0.153	2.656	5.312
	$Ti_2N(OH)_{2}\_36\ H_2$	−0.112	2.656	

9.4.8 ZEOLITE TEMPLATED CARBON

Zeolite templated carbon or ZTC is a newly investigated carbon 2D allotrope that consists of hexagons and pentagons in a periodical order.[37,96] The high surface area of 4000 m^2/g and uniform porosity are the two key properties that interest the researcher to investigate the system for hydrogen storage.[37] It can be synthesized by carbonizing alkanes or alkenes in the pores of zeolite and then the deposited carbon can be retracted by dissolution of the zeolite framework, using HCl+HF.[97]

9.4.8.1 ZR DECORATED ZTC

Mondal et al.[98] reported Zr decorated ZTC as a promising hydrogen storage system using first-principles DFT calculation and MD simulation. According to their calculation the ZTC structures were able to adsorb 8 Zr atoms, 4 on top and 4 on bottom. Each Zr atom is able to adsorb 7 H2 with an average binding energy of −0.433, which is in good agreement with DoE criteria. The overall wt% was calculated to be 9.24, which is way above the DoE limit. The Zr atom is adsorbed onto ZTC due to charge transfer of Zr 4d to C 2p.[98] But H2 is adsorbed on Zr+ZTC via charge transfer of Zr 4d to H 1s. Figure 9.38 shows the PDOS analysis where p-d hybridization for Zr addition and Kubas interaction of H2 with Zr+ZTC system is observed. Bader charge analysis shows that Zr loses 2.17 e charge and each H gains 0.12 e charge.[98] Also, the diffusion energy barrier for metal–metal clustering has been calculated to be 2.36 eV, which is high enough to neglect the clustering possibility even at high temperatures. MD simulation was done for the same system at 500 K, and the result shows the integrity of the structure remains the same even at that high temperature. Hence, this system can be considered for high-temperature fuel cell applications.

9.4.8.2 Y DECORATED ZTC

Yttrium-doped ZTC was studied as a hydrogen storage system by Kundu et al.[99] The ZTC +Y system may absorb up to $7H_2$ molecules per Y atom, according to DFT modeling. Adsorption energy is -0.35 eV/H_2 on average. With 8 Y atom loading (4 on top and 4 bottom), the system's hydrogen storage capability increases to 8.61 wt%, which is significantly more than the DoE-specified margin of 6.5 wt%. The calculated desorption temperature of 437 K is ideal for use in fuel cells. Metal–metal clustering is unlikely due

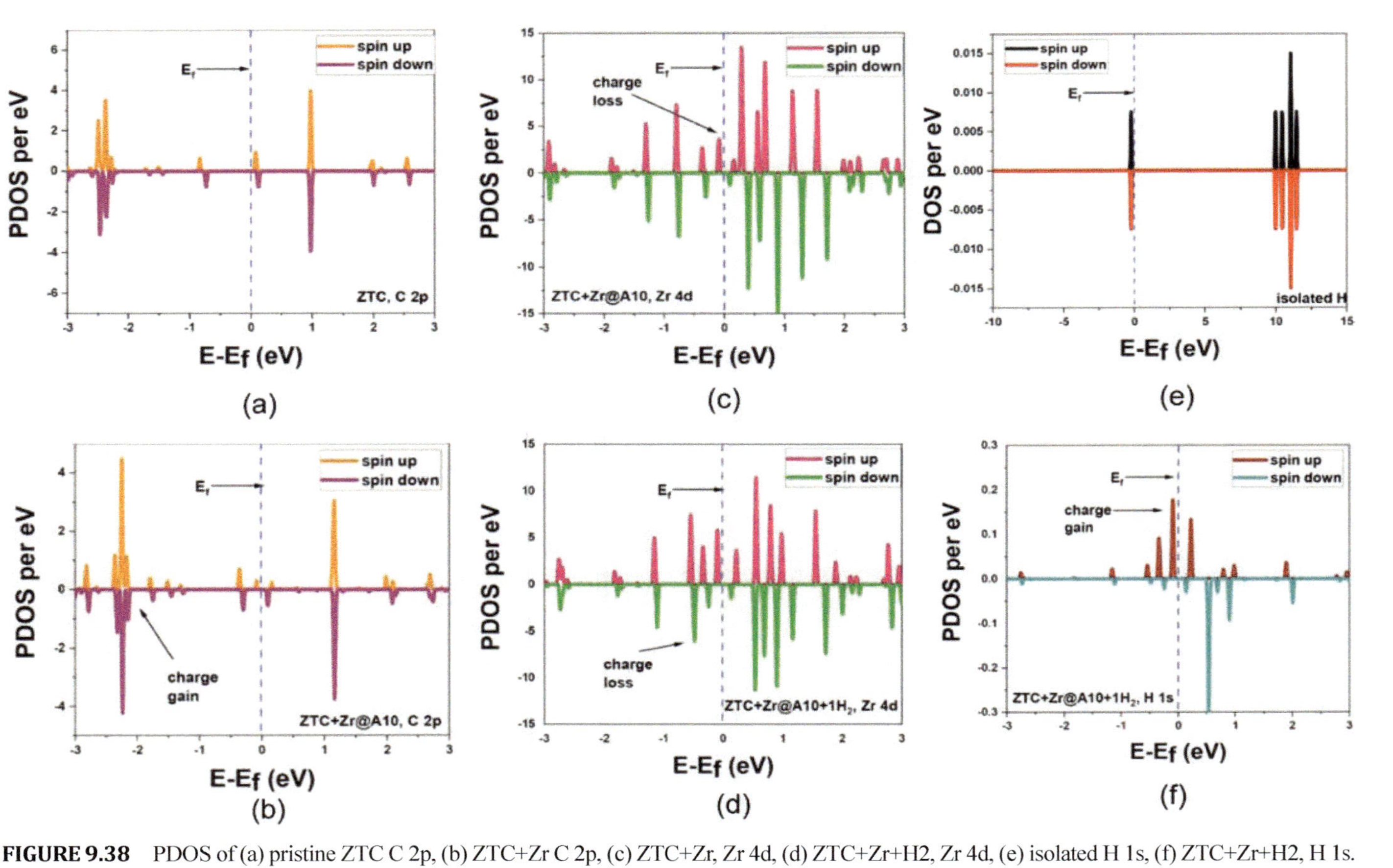

FIGURE 9.38 PDOS of (a) pristine ZTC C 2p, (b) ZTC+Zr C 2p, (c) ZTC+Zr, Zr 4d, (d) ZTC+Zr+H2, Zr 4d, (e) isolated H 1s, (f) ZTC+Zr+H2, H 1s.

Source: Reproduced with permission from Ref. [98]

to the presence of a substantial energy barrier for Y atom migration of 3.28 eV. Ab-initio MD simulations validated the structural stability of the ZTC + Y system at an increased temperature of 300K.[99] Similar to ZTC+Zr system Y 4d interacts with ZTC 2p and a p-d hybridization takes place. Also, for H2 adsorption charge is transferred from Y 4d to H 1s. It can be observed more clearly in suborbital PDOS analysis, as shown in Figure 9.39. It can be observed that after the addition of Y on ZTC, the DOS at d_{xz} and d_{yz} are significantly reduced (Figure 9.39.b). Bader charge analysis quantifies that Y atom loses 1.60 e.[99] Also, Figure 9.39.b shows that upon continuous addition of hydrogen, Y 4d losses charge near Fermi level. According to the calculations of Bader charge Y atom loses 0.01 e charge.

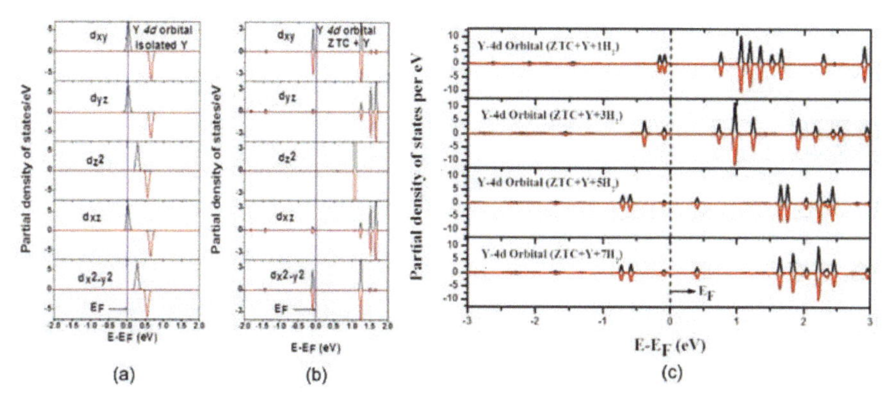

FIGURE 9.39 a) Suborbital analysis for Y 4d orbital in isolated Y, b) ZTC+Y, c) PDOS of ZTC+Y+H_2.

Source: Reproduced with permission from Ref. [99]

9.4.8.3 Li DECORATED ZTC

Ortega et al.[96] investigated Li decorated ZTC structures. According to their DFT results, the adsorption energy of Li on ZTC is −4.16 eV, which indicates strong adsorption of Li atoms on ZTC. Also, each Li can accommodate 6 hydrogen maximum and each ZTC was able to accommodate 3 Li atoms, resulting in the highest possible gravimetric wt% of hydrogen to be 6.78. Also, the average binding energy for this system is in the range of −0.1250 eV to −0.1320 eV.

We have summarized the 2D nanomaterials that have been investigated in recent years using DFT, for hydrogen storage applications in Table 9.12.

TABLE 9.12 Summary of Hydrogen Storage in 2D Nanomaterials Using DFT.

2D material	Doping	Wt %	BE/H_2 (eV)	Desorption T (K)	Ref
Graphene	Ti	7.8			54
Graphene	Y	6.17	−0.415	530	56
Graphene	Zr	11	−0.34	433	18
BN NS		6.7	−0.122		70
Graphyne	Y	10	−0.3	400	35
Graphyne	Ti	9.5	−0.54	895	73
Graphyne	Li	18.6	−0.26		73
Graphyne	Ca	10.5	−0.34		73
Graphyne	Sc	9.88	−0.60		73
ψ-graphene	Zr	11.3	−0.38	484.21	77
ψ-graphene	Ti	13.14	−0.30	387	78
ψ-graphene	Y	8.31	−0.39	496.55	79
g-C3N4	-	5.45			80
g-C3N4	Y	8.55	−0.331	384.24	81
g-C3N4	Sc	8.55	−0.39	458.28	84
Holey graphyne	Li	12.8	−0.22	282	86
Holey graphyne	Sc	9.8	−0.36	464	100
Holey graphyne	Ti	10.52	−0.38	486	89
Holey graphyne	Y	9.34	−0.34	438	48
Ti_2C	-	3.4 (Kubas type) 6.8 (Reversible) 8.6 (Total)	−0.272(Kubas)		93
V_2C	-		−0.242 (Kubas)		93

TABLE 9.12 *(Continued)*

2D material	Doping	Wt %	BE/H_2 (eV)	Desorption T (K)	Ref
Sc_2C	-	9.0 (Total)	−0.164 (Kubas)		93, 95
Cr_2C	-	7.6 (Total) 6.4 (Reversible) 3.2 (Kubas type)	0.26 (Kubas)		94
Ti_2N	-	8.55 (Total) 3.42 (Kubas type)	−0.140		95
Ti_2NO_2		5.382	−0.1265		95
Ti_2NF_2		5.176	−0.124		95
$Ti_2N(OH)_2$		5.312	−0.1325		95
ZTC	Zr	9.24	−0.433	-	98
ZTC	Y	8.61	−0.35		99
ZTC	Li	6.78	−0.132		96

9.5 CONCLUSIONS AND FUTURE DIRECTIONS

In this chapter, we have discussed about the state-of-the-art technologies for solid-state hydrogen storage methods, explored with density functional theory. But as we have already mentioned, we all know that technologies are kind of "merit–demerit" package. It is next to impossible to find something perfect and that has no drawback or bottlenecks. Similarly, for DFT studies hydrogen storage has the issue of measuring oxygen interference and often meta clustering for practical applicability analysis. Oxygen interference is a significant practical issue, and accounting for it inside the model is a challenging task.[101] As a result, it is difficult to extrapolate theoretical models' predictions of H2 storage capacity in terms of weight percentage. As we can see that the nanomaterials mainly show good performance when doped with metal, the incorporation of metal often increases the system cost. So, we need to find options to avoid metal doping and enhance the hydrogen storage still. Also, the replication techniques and setup parameters have an impact on the simulation outcomes. As an illustration, LDA exchange correlation, as was previously discussed, overbinds the H2 molecules, whereas GGA exchange correlation underbinds them. So, while choosing simulation methodologies and making predictions based on simulation details, one must be very cautious. This discussion gives the basis of future in-depth analysis in the field of computation studies for hydrogen storage.

KEYWORDS

- **hydrogen storage**
- **nanomaterials**
- **DFT**
- **metal doping**
- **Kubas interaction**

REFERENCES

1. Shafiee, S.; Topal, E. When will Fossil Fuel Reserves be Diminished? *Energy Policy* **2009,** *37* (1), 181–189.
2. Kober, T.; Schiffer, H. W.; Densing, M.; Panos, E. Global Energy Perspectives to 2060—WEC's World Energy Scenarios 2019. *Energy Strateg. Rev.* **2020,** *31,* 100523.

3. Energy, T. Alternative Energy 101: What It Is and Why It Is Important [Online]? https://taraenergy.com/blog/alternative-energy-101-what-why-important/ (accessed Sep 29, 2022).
4. Mazloomi, K.; Gomes, C. Hydrogen as an Energy Carrier: Prospects and Challenges. *Renew. Sustain. Energy Rev.* **2012,** *16* (5), 3024–3033.
5. Hensher, D. A. Climate Change, Enhanced Greenhouse Gas Emissions and Passenger Transport—What Can We Do to Make a Difference?. *Transp. Res. Part D Transp. Environ.* **2008,** *13* (2), 95–111.
6. Ahluwalia, R. K.; Wang, X.; Rousseau, A.; Kumar, R. Fuel Economy of Hydrogen Fuel Cell Vehicles. *J. Power Sources* **2004,** 130 (1–2), 192–201.
7. Williams, Q.; Hemley, R. J. Hydrogen in the Deep Earth. [Online] 2001. www.annual-reviews.org
8. Turner, J. A. Sustainable Hydrogen Production. *Science* 2004**,** *305* (5686), 972–974.
9. Kumar, R.; Kumar, A.; Pal, A. Overview of Hydrogen Production from Biogas Reforming: Technological Advancement. *Int. J. Hydrogen Energy* **2022,** *47* (82), 34831–34855.
10. Crabtree, G. W.; Dresselhaus, M. S. The Hydrogen Fuel Alternative. *MRS Bull.* **2008,** *33* (4), 421–428.
11. Acar, C.; Dincer, I. Review and Evaluation of Hydrogen Production Options for Better Environment. *J. Clean. Prod.* **2019,** *218,* 835–849
12. Andreas, R. Hydrogen Storage Methods. **2004,** 157–172.
13. Jena, P. Materials for Hydrogen Storage: Past, Present, and Future. *J. Phys. Chem. Lett.* **2011,** *2,* 23.
14. Kubas, G. Hydrogen Activation on Organometallic Complexes and H 2 Production, Utilization, and Storage for Future Energy. *J. Organomet. Chem.* **2009,** *694,* 2648–2653.
15. Niu, J.; Rao, B. K.; Jena, P. Binding of Hydrogen Molecules by a Transition-Metal Ion. *Phys. Rev. Lett.* **1992,** *68* (15), 2277–2280.
16. Mohan, M.; Sharma, V. K.; Kumar, E. A.; Gayathri, V. Hydrogen Storage in Carbon Materials—A Review. *Energy Storage* **2018,** pp. 1–26.
17. Kumar, S.; Kumar, T. J. D. Fundamental Study of Reversible Hydrogen Storage in Titanium- and Lithium-Functionalized Calix4]arene. *J. Phys. Chem. C* **2017,** *121* (16), 8703–8710.
18. Yadav, A.; Chakraborty, B.; Gangan, A.; Patel, N.; Press, M. R.; Ramaniah, L. M. Magnetic Moment Controlling Desorption Temperature in Hydrogen Storage: A Case of Zirconium-Doped Graphene as a High Capacity Hydrogen Storage Medium. *J. Phys. Chem. C* **2017,** *121* (31), 16721–16730.
19. Kumar, S.; Sathe, R. Y.; T. J. Dhilip Kumar, First Principle Study of Reversible Hydrogen Storage in Sc grafted Calix[4]arene and Octamethylcalix[4]arene. *Int. J. Hydrogen Energy 44* (10), 4889–4896, 2019.
20. Li, M. *et al.* Review on the Research of Hydrogen Storage System Fast Refueling in Fuel Cell Vehicle. *Int. J. Hydrogen Energy* **2019,** *44*. DOI: 10.1016/j.ijhydene.2019.02.208
21. Léon, A. *Hydrogen Technology: Mobile and Portable Applications*. Springer: Berlin, Germany, 2008 DOI: 10.1007/978-3-540-69925-5
22. Pirajan, M.; Dornheim, M.; Carlos, J. *Thermodynamics of Metal Hydrides: Tailoring Reaction Enthalpies of Hydrogen Storage Materials*. IntechOpen: Rijeka, 2011, p Ch. 33. DOI: 10.5772/21662.
23. Principi, G.; Agresti, F.; Maddalena, A.; Lo Russo, S. The Problem of Solid State Hydrogen Storage. *Energy* **2009,** *34* (12), 2087–2091.

24. Rusman, N. A. A.; Dahari, M. A Review on the Current Progress of Metal Hydrides Material for Solid-State Hydrogen Storage Applications. *Int. J. Hydrogen Energy* **2016,** *41* (28), 12108–12126.
25. Dematteis, E. M.; Barale, J.; Corno, M.; Sciullo, A.; Baricco, M.; Rizzi, P. Solid-State Hydrogen Storage Systems and the Relevance of a Gender Perspective. **2021,** DOI: 10.3390/en14196158
26. Reardon, H.; Hanlon, J. M.; Hughes, R. W.; Godula-Jopek, A.; Mandal, T. K.; Gregory, D. H. Emerging Concepts in Solid-State Hydrogen Storage: The Role of Nanomaterials Design. *Energy Environ. Sci.* **2012,** *5* (3), 5951–5979.
27. Ahmed, A.; *et al.* Balancing Gravimetric and Volumetric Hydrogen Density in MOFs †. *Energy Environ. Sci.* **2017,** *10,* p. 2459.
28. Go, D. A.; *et al.* Understanding Volumetric and Gravimetric Hydrogen Adsorption Trade-off in Metal−Organic Frameworks. **2017**. DOI: 10.1021/acsami.7b01190
29. Ahmed, A.; *et al.* Exceptional Hydrogen Storage Achieved by Screening Nearly Half a Million Metal-organic Frameworks. *Nat. Commun.* **2019,** 48109. DOI: 10.1038/s41467-019-09365-w
30. Zhang, X.; *et al.* Optimization of the Pore Structures of MOFs for Record High Hydrogen Volumetric Working Capacity. *Adv. Mater.* **2020,** *32* (1907995), 1–6.
31. Chen, Z.; *et al.* Balancing Volumetric and Gravimetric Uptake in Highly Porous Materials for Clean Energy. *Science* **2020,** *368* (6488), 297–303.
32. Chen, Z.; *et al.* Fine-Tuning a Robust Metal−Organic Framework toward Enhanced Clean Energy Gas Storage. *J. Am. Chem. Soc.* **2021,** *143,* 44.
33. Niemann, M. U.; Srinivasan, S. S.; Phani, A. R.; Kumar, A.; Goswami, D. Y.; Stefanakos, E. K. Nanomaterials for Hydrogen Storage Applications: A Review. *J. Nanomater.* **2008,** *2008,* 950967.
34. Chakraborty, B.; Modak, P.; Banerjee, S. Hydrogen Storage in Yttrium-Decorated Single Walled Carbon Nanotube. *J. Phys. Chem. C* **2012,** *116* (42), 22502–22508.
35. Gangan, A.; Chakraborty, B.; Ramaniah, L. M.; Banerjee, S. First Principles Study on Hydrogen Storage in Yttrium Doped Graphyne: Role of Acetylene Linkage in Enhancing Hydrogen Storage. *Int. J. Hydrogen Energy* **2019,** *44* (31), 16735–16744.
36. Cheng, H. M.; Yang, Q. H.; Liu, C. Hydrogen Storage in Carbon Nanotubes. *Carbon N. Y.* **2001,** *39* (10), 1447–1454.
37. Isidro-Ortega, F. J.; Pacheco-Sánchez, J. H.; Alejo, R.; Desales-Guzmán, L. A.; Arellano, J. S. Theoretical Studies in the Stability of Vacancies in Zeolite Templated Carbon for Hydrogen Storage. *Int. J. Hydrogen Energy* **2019,** *44* (13), 6437–6447.
38. Ren, H. J.; Cui, C. X.; Li, X. J.; Liu, Y. J. A DFT Study of the Hydrogen Storage Potentials and Properties of Na- and Li-doped Fullerenes. *Int. J. Hydrogen Energy* **2017,** *42* (1), 312–321.
39. Zhang, Y.; Cheng, X. Hydrogen Storage Property of Alkali and Alkaline-Earth Metal Atoms Decorated C24 Fullerene: A DFT Study. *Chem. Phys.* **2018,** *505,* 26–33.
40. Sun, Q.; Wang, Q.; Jena, P.; Kawazoe, Y. Clustering of Ti on a C60 Surface and Its Effect on Hydrogen Storage. *J. Am. Chem. Soc.* **2005,** *127* (42), 14582–14583.
41. Zacharia, R.; Kim, K. Y.; Fazle Kibria, A. K. M.; Nahm, K. S. Enhancement of Hydrogen Storage Capacity of Carbon Nanotubes via Spill-Over from Vanadium and Palladium Nanoparticles. *Chem. Phys. Lett.* **2005,** *412* (4–6), 369–375.
42. Ghosh, S.; Padmanabhan, V. Hydrogen Storage in Titanium-Doped Single-Walled Carbon Nanotubes with Stone-Wales Defects. *Diam. Relat. Mater.* **2017,** *77,* 46–52.

43. Gu, J.; Zhang, X.; Fu, L.; Pang, A. Study on the Hydrogen Storage Properties of the Dual Active Metals Ni and Al Doped Graphene Composites. *Int. J. Hydrogen Energy* **2019,** *44* (12), 6036–6044, doi: 10.1016/j.ijhydene.2019.01.057.
44. Tarasov, B. P.; *et al.* Hydrogen Storage Behavior of Magnesium Catalyzed by Nickel-Graphene Nanocomposites. *Int. J. Hydrogen Energy* **2019,** *44* (55), 29212–29223.
45. Hoang, T. K. A.; Antonelli, D. M. Exploiting the Kubas Interaction in the Design of Hydrogen Storage Materials. *Adv. Mater.* **2009,** *21* (18), 1787–1800.
46. Ziesche, P.; Kurth, S.; Perdew, J. P. Density Functionals from LDA to GGA. *Comput. Mater. Sci.* **1998,** *11* (2), 122–127.
47. DoE. DOE Technical Targets for Onboard Hydrogen Storage for Light-Duty Vehicles. *Office of Energy Efficiency & Renewable Energy* [Online]. https://www.energy.gov/eere/fuelcells/doe-technical-targets-onboard-hydrogen-storage-light-duty-vehicles
48. Singh, M.; Shukla, A.; Chakraborty, B. An Ab- Initio Study of the y Decorated 2D Holey Graphyne for Hydrogen Storage Application. *Nanotechnology 33* (40), 2022. DOI: 10.1088/1361-6528/ac7cf6
49. Henkelman, G.; Arnaldsson, A.; H. Jónsson, A Fast and Robust Algorithm for Bader Decomposition of Charge Density. *Comput. Mater. Sci.* **2006,** *36* (3), 354–360.
50. Mulliken, R. S. Electronic Population Analysis on LCAO–MO Molecular Wave Functions. I. *J. Chem. Phys.* **1955,** *23* (10), 1833–1840.
51. Kokalj, A. XCrySDen—A New Program for Displaying Crystalline Structures and Electron Densities. *J. Mol. Graph. Model.* **1999,** *17* (3–4), 176–179.
52. Momma, K.; Izumi, F. VESTA 3 for Three-dimensional Visualization of Crystal, Volumetric and Morphology Data. *J. Appl. Crystallogr.* **2011,** *44* (6), 1272–1276.
53. Zhou, D.; Cui, Y.; Han, B.-H. Graphene-based Hybrid Materials and their Applications in Energy Storage and Conversion. *Chinese Sci. Bull.* 2012, *57,* 2983–2994.
54. Liu, Y.; Ren, L.; He, Y.; Cheng, H. P. Titanium-Decorated Graphene for High-Capacity Hydrogen Storage Studied by Density Functional Simulations. *J. Phys. Condens. Matter.* **2010,** *22* (44). DOI: 10.1088/0953-8984/22/44/445301
55. Liu, W.; Liu, Y.; Wang, R. Prediction of Hydrogen Storage on Y-Decorated Graphene: A Density Functional Theory Study. *Appl. Surf. Sci.* **2014,** *296,* 204–208.
56. Desnavi, S.; Chakraborty, B.; Ramaniah, L. M. First Principles DFT Investigation of Yttrium-Doped Graphene: Electronic Structure and Hydrogen Storage. Articles You May Be Interested In First Principles DFT Investigation of Yttrium-Decorated Boron-Nitride Nanotube: Electronic Structure and Hydrogen Storage AIP Conference *First Principles DFT Investigation of Yttrium-Doped Graphene: Electronic Structure and Hydrogen Storage* **2014,** *1591,* 50115, doi: 10.1063/1.4873109.
57. Lale, A.; Bernard, S.; Demirci, U. B. Boron Nitride for Hydrogen Storage. *Chempluschem* **2018,** *83,* 893–903.
58. Weng, Q.; Wang, X.; Zhi, C.; Bando, Y.; Golberg, D. Boron Nitride Porous Microbelts for Hydrogen Storage. **2022,** *16,* 16.
59. Weng, Q.; *et al.* One-Step Template-Free Synthesis of Highly Porous Boron Nitride Microsponges for Hydrogen Storage 1. Introduction. *Adv. Energy Mater* **2014,** *4,* 1301525. DOI: 10.1002/aenm.201301525
60. Kurdyumov, A. V.; Britun, V. F.; Petrusha, I. A. Structural Mechanisms of Rhombohedral BN Transformations Into Diamond-like Phases. *Diam. Relat. Mater.* **1996,** *5* (11), 1229–1235.

61. Ma, R.; Bando, Y.; Zhu, H.; Sato, T.; Xu, C.; Wu, D. Hydrogen Uptake in Boron Nitride Nanotubes at Room Temperature. 2002. DOI: 10.1021/ja026030e.
62. Leela Mohana Reddy, A.; Tanur, A. E.; Walker, G. C. Synthesis and Hydrogen Storage Properties of Different Types of Boron Nitride Nanostructures. *Int. J. Hydrogen Energy* **2010,** *35* (9), 4138–4143.
63. Tang, C.; Bando, Y.; Ding, X.; Qi, S.; Golberg, D. Catalyzed Collapse and Enhanced Hydrogen Storage of BN Nanotubes. *J. Am. Chem. SOC*, **2002,** ***124,*** 14550–14551.
64. Lim, S. H.; Luo, J.; Ji, W.; Lin, J. Synthesis of Boron Nitride Nanotubes and Its Hydrogen Uptake. *Catal. Today* **2007,** *120* (3–4), 346–350.
65. Okan, B. S.; Özlem Kocabaş, Z.; Ergü, A. N.; Baysal, M.; Letofsky-Papst, I.; Yürüm. Y. Effect of Reaction Temperature and Catalyst Type on the Formation of Boron Nitride Nanotubes by Chemical Vapor Deposition and Measurement of Their Hydrogen Storage Capacity. 2012. DOI: 10.1021/ie301605z.
66. Ma, R.; Bando, Y.; Sato, T. Synthesis of Boron Nitride Nanofibers and Measurement of their Hydrogen Uptake Capacity Articles You May Be Interested In. *Appl. Phys. Lett.* **2002,** *81*, 5225.
67. Lian, G.; Zhang, X.; Zhang, S.; Liu, D.; Cui, D.; Wang, Q. Controlled Fabrication of Ultrathin-Shell BN Hollow Spheres with Excellent Performance in Hydrogen Storage and Wastewater Treatment †. DOI: 10.1039/c2ee03240f
68. Lei, W.; *et al.* Oxygen-Doped Boron Nitride Nanosheets with Excellent Performance in Hydrogen Storage. *Nano Energy* **2014,** *6,* 219–224.
69. Muthu, R. N.; Rajashabala, S.; Kannan, R. Hydrogen Storage Performance of Lithium Borohydride Decorated Activated Hexagonal Boron Nitride Nanocomposite for Fuel Cell Applications. *Int. J. Hydrogen Energy* **2017,** *42* (23), 15586–15596.
70. Chettri, B.; Patra, P. K.; Hieu, N. N.; Rai, D. P. Hexagonal Boron Nitride (h-BN) Nanosheet as a Potential Hydrogen Adsorption Material: A Density Functional Theory (DFT) Study. *Surf. Interf.* **2021,** *24*, 101043.
71. Ren, J.; Zhang, N.; Zhang, H.; Peng, X. First-Principles Study of Hydrogen Storage on Pt (Pd)-Doped Boron Nitride Sheet. DOI: 10.1007/s11224-014-0531-2
72. Baughman, R. H.; Eckhardt, H.; Kertesz, M. Structure-Property Predictions for New Planar Forms of Carbon: Layered Phases Containing sp 2 and sp Atoms Articles you may be Interested in. *J. Chem. Phys*, **1987,** *87,* 6687.
73. Guo, Y.; *et al.* A Comparative Study of the Reversible Hydrogen Storage Behavior in Several Metal Decorated Graphyne. *Int. J. Hydrogen Energy* **2013,** *38* (10), 3987–3993.
74. Desyatkin, V. G.; *et al.* Scalable Synthesis and Characterization of Multilayer γ-Graphyne, New Carbon Crystals with a Small Direct Band Gap. DOI: 10.1021/jacs.2c06583.
75. Li, X.; Wang, Q.; Jena, P. ψ-Graphene: A New Metallic Allotrope of Planar Carbon with Potential Applications as Anode Materials for Lithium-Ion Batteries. *J. Phys. Chem. Lett.* **2017,** *8,* 43. DOI: 10.1021/acs.jpclett.7b01364
76. Xie, L.; Sun, T.; He, C.; An, H.; Qin, Q.; Peng, Q. Effect of Angle, Temperature and Vacancy Defects on Mechanical Properties of PSI-Graphene. *Crystals* **2019,** *9* (5)..
77. Nair, H. T.; Jha, P. K.; Chakraborty, B. High-Capacity Hydrogen Storage in Zirconium Decorated Psi-Graphene: Acumen from Density Functional Theory and Molecular Dynamics Simulations. *Int. J. Hydrogen Energy* 2022. DOI: 10.1016/J.IJHYDENE. 2022.08.084

78. Chakraborty, B.; Ray, P.; Garg, N.; Banerjee, S. High Capacity Reversible Hydrogen Storage in Titanium Doped 2D Carbon Allotrope Ψ-Graphene: Density Functional Theory Investigations. *Int. J. Hydrogen Energy* **2021,** *46* (5), 4154–4167.
79. Chakraborty, B.; Vaidyanathan, A.; Kandasamy, M.; Wagh, V.; Sahu, S. High-Capacity Hydrogen Storage in Yttrium-Decorated Ψ-Graphene: Acumen from Density Functional Theory. *J. Appl. Phys.* **2022,** 132 (6), 65002.
80. Koh, G.; Zhang, Y. W.; Pan, H. First-Principles Study on Hydrogen Storage by Graphitic Carbon Nitride Nanotubes. *Int. J. Hydrogen Energy* **2012,** *37* (5), 4170–4178.
81. Mane, P.; Vaidyanathan, A.; Chakraborty, B. Graphitic Carbon Nitride (g-C3N4) Decorated with Yttrium as Potential Hydrogen Storage Material: Acumen from Quantum Simulations. *Int. J. Hydrogen Energy*, **2022**. DOI: 10.1016/J.IJHYDENE.2022.04.184
82. Habibi-Yangjeh, A.; Basharnavaz, H. Remarkable Improvement in Hydrogen Storage Capabilities of Graphitic Carbon Nitride Nanosheets under Selected Transition Metal Embedding: A DFT Study. *Int. J. Hydrogen Energy* **2021,** *46* (68), 33864–33876.
83. Panigrahi, P.; Kumar, A.; Karton, A.; Ahuja, R.; Hussain, T. Remarkable Improvement in Hydrogen Storage Capacities of Two-Dimensional Carbon Nitride (g-C3N4) Nanosheets Under Selected Transition Metal Doping. *Int. J. Hydrogen Energy* **2020,** *45* (4), 3035–3045.
84. Chakraborty, B.; Mane, P.; Vaidyanathan, A. Hydrogen Storage in Scandium Decorated Triazine Based g-C3N4: Insights from DFT Simulations. *Int. J. Hydrogen Energy* **2022,** no. xxxx. DOI: 10.1016/j.ijhydene.2022.02.185
85. Liu, X. *et al.* Direct Band Gap Semiconducting Holey Graphyne: Structure, Synthesis and Potential Applications. *arXiv Prepr. arXiv1907.03534*, 2019.
86. Gao, Y.; Zhang, H.; Pan, H.; Li, Q.; Zhao, J. Ultrahigh Hydrogen Storage Capacity of Holey Graphyne. *Nanotechnology* **2021,** *32* (21), DOI: 10.1088/1361-6528/abe48d
87. Mahamiya, V.; Shukla, A.; Chakraborty, B. Scandium Decorated C24 Fullerene as High Capacity Reversible Hydrogen Storage Material: Insights from Density Functional Theory Simulations. *Appl. Surf. Sci.* **2022,** *573* (2021), 151389, DOI: 10.1016/j.apsusc.2021.151389
88. Modak, P.; Chakraborty, B.; Banerjee, S. Study on the Electronic Structure and Hydrogen Adsorption by Transition Metal Decorated Single Wall Carbon Nanotubes. *J. Phys. Condens. Matter* **2012,** *24* (18). DOI: 10.1088/0953-8984/24/18/185505
89. Dewangan, J.; *et al.* Reversible Hydrogen Adsorption in Ti-Functionalized Porous Holey Graphyne: Insights from First-Principles Calculation. **2022**. DOI: 10.1002/est2.391.
90. Park, N.; Choi, K.; Hwang, J.; Kim, D. W.; Kim, D. O.; Ihm, J. Progress on First-Principles-Based Materials Design for Hydrogen Storage. *Proc. Natl. Acad. Sci.* **2012,** *109* (49), 19893–19899.
91. Krasnov, P. O.; Ding, F.; Singh, A. K.; Yakobson, B. I. Clustering of Sc on SWNT and Reduction of Hydrogen Uptake: Ab-Initio All-Electron Calculations. DOI: 10.1021/jp077264t.
92. Kumar, P.; Singh, S.; Hashmi, S. A. R.; Kim, K. H. MXenes: Emerging 2D Materials for Hydrogen Storage. *Nano Energy* **2021,** *85,* 105989.
93. Hu, Q.; et al. MXene: A New Family of Promising Hydrogen Storage Medium. *J. Phys. Chem. A* **2013,** *117,* 37.
94. Yadav, A.; Dashora, A.; Patel, N.; Miotello, A.; Press, M.; Kothari, D. C. Study of 2D MXene Cr2C Material for Hydrogen Storage Using Density Functional Theory. *Appl. Surf. Sci.* vol. **2016,** *389,* 88–95.

95. Li, Y.; Guo, Y.; Chen, W.; Jiao, Z.; Ma, S. Reversible Hydrogen Storage Behaviors of Ti 2 N MXenes Predicted by First-Principles Calculations. DOI: 10.1007/s10853-018-2854-7
96. Isidro-Ortega, F. J.; Pacheco-Sánchez, J. H.; Desales-Guzmán, L. A. Hydrogen Storage on Lithium Decorated Zeolite Templated Carbon, DFT Study. *Int. J. Hydrogen Energy* **2017,** *42* (52), 30704–30717.
97. Park, H.; Bang, J.; Han, S. W.; Bera, R. K.; Kim, K.; Ryoo, R. Synthesis of Zeolite-Templated Carbons Using Oxygen-Containing Organic Solvents. *Microporous Mesoporous Mater.* **2021,** *318,* 111038.
98. Mondal, B.; Kundu, A.; Chakraborty, B. High-Capacity Hydrogen Storage in Zirconium Decorated Zeolite Templated Carbon: Predictions from DFT Simulations. *Int. J. Hydrogen Energy* 2022. DOI: 10.1016/J.IJHYDENE.2022.09.056
99. Kundu, A.; Trivedi, R.; Garg, N.; Chakraborty, B. Novel Permeable Material 'Yttrium Decorated Zeolite Templated Carbon' for Hydrogen Storage : Perspectives from Density Functional Theory. *Int. J. Hydrogen Energy* **2022.** DOI: 10.1016/j.ijhydene.2022.06.159
100. Mahamiya, V.; Shukla, A.; Garg, N. High-Capacity Reversible Hydrogen Storage in Scandium Decorated Holey Graphyne : Theoretical Perspectives. *Int. J. Hydrogen Energy* **2022,** *47* (12), 7870–7883.
101. Sigal, A.; Villarreal, M.; Rojas, M. I.; Leiva, E. P. M. A New Model for the Prediction of Oxygen Interference in Hydrogen Storage Systems. *Int. J. Hydrogen Energy* **2014,** *39* (11), 5899–5905.
102. Sönnichsen, N. Hydrogen Consumption Worldwide in 2020, by Country. *Statista* [Online] 2022. https://www.statista.com/statistics/1292403/global-hydrogen-consumption-by-country/
103. Yun, T. Z. Is Malaysia Ready for the Hydrogen Economy?. *The edge Malaysia* [Online], 2021. https://www.theedgemarkets.com/article/cover-story-malaysia-ready-hydrogen-economy
104. Hydrogen Production Methods [Online]. https://medium.com/coinmonks/hydrogen-production-methods-9b4504d26269
105. Züttel, A. Hydrogen Storage Methods. *Naturwissenschaften* **2004,** *91,* 157–172.
106. Department of energy, Hydrogen Storage. *Hydrogen and Fuel Cell Technologies Office* [Online]. https://www.energy.gov/eere/fuelcells/hydrogen-storage

Index

W

Y

Z